Peter Kershaw, Bruno David, Nigel Tapper, Dan Penny
and Jonathan Brown (Editors)

Bridging Wallace's Line:

The Environmental and Cultural History and Dynamics of the SE-Asian-Australian Region

ADVANCES IN GEOECOLOGY 34

**A Cooperating Series of the
International Union of Soil Science (IUSS)**

IUSS – UISS – IBU
ISBN 3-923381-47-6

Cover: 'Tracing the Wallace Line: wing leaf and land', water colour painting
by John Wolseley

Die Deutsche Bibliothek - CIP Einheitsaufnahme

Bridging Wallace´s line : the environmental and cultural history and dynamics of
the SE-Asian-Australian region / Peter Kershaw ...(eds.). -
Reiskirchen : Catena Verl., 2002
(Advances in geoecology ; 34)
ISBN 3-923381-47-6

Managing Editor "Advances in GeoEcology":
Margot Rohdenburg

ISBN 3-923381-47-6

Contents

PART III   THE PEOPLING OF SUNDA AND SAHUL

Preface

Peter Kershaw, Bruno David, Nigel Tapper, Dan Penny and
Jonathan Brown

In the early 1990s, there was general interest in Australia in improving political
and economic links with the Asian region, particularly the countries of SE Asia,
Australia's nearest neighbours. Monash University was keen on extending these
linkages with social, cultural and environmental interests in mind. Consequently, support in the form of a Monash Development Fund grant was received to
examine historical relationships between the two regions, from an integrated
human and physical geographical perspective. Towards the end of the three-
year period of support, a conference was held in December 1996 to discuss
research findings, to place them within a more national and international
research context, and to establish priorities for future investigation. Attracting
some 50 contributors from various parts of the world additional to Australia and
SE Asia, it was clear from the meeting that exciting new discoveries were being
made which would substantially change perceptions about the history of SE Asia
and its relationships with Australia. It was decided that a book should be published to encapsulate these new data and ideas and share them with the broader
research community and general public. However, time was needed to complete
some research projects and clarify some emerging concepts. Time has also resulted in new information. This volume incorporates some of the original contributions that have matured considerably since the mid-1990s, as well as new
contributions that have emerged in the intervening years.

Volume structure

The introduction 'Bridging Wallace's Line: bringing home the Antipodes'
provides a context for examination of the SE Asian-Australian region, traditionally seen in Western thought as the 'other' side of the world, a realm where
things are supposed to be 'different'. With global processes and local histories
in mind, we emphasise linkages between Australia and SE Asia rather than the
differences that have long been formalised in Wallace's, Huxley's and
Lydekker's Lines. Through the exclusion of faunal distributions and history in
this volume – apart from a consideration of modern humans and their immediate
ancestors – contributions from the physical, biological and social sciences tend
to reflect important inter-regional ties and continuities.

The main body of the book is divided into three Parts. Part 1 includes contributions that explore present-day processes or long-term geological frameworks,
providing contextual information for the later contributions that concentrate on
the dynamics of the environment and people during the Quaternary. Climatically, the region, particularly the humid core from peninsular Malaya to northern
Australasia (frequently referred to as the Maritime Continent), has long been

recognised as critical to an understanding of the dynamics of the global atmosphere. This region has been coined the 'boiler box' of the world. It also holds a central position in the operation of monsoon systems and the El Niño-Southern Oscillation (ENSO) phenomenon that together have a major influence over the distribution and well-being of a large proportion of the world's population. Some details on the contemporary operation of these climate systems and their impacts on drought, fire, and haze are provided by Tapper. However, it is only recently that the importance of monsoons and ENSO in the production and impact of Quaternary-scale climatic fluctuations has begun to be realised. Events in other parts of the world, including the closing of the isthmus between North and South America and small global radiation changes leading to the expansion and contraction of ice sheets in the northern parts of Europe and America, have been seen as the primary controls over the onset and pattern of glacial-interglacial cyclicity. As details emerge on the nature and ages of individual cycles, other explanations have to be sought and attention is turning to the SE Asian-Australian region.

One good reason for the dynamism of the Maritime Continent region is its geological instability. The concept of continental drift, whereby Australia was shown to have broken away from the southern supercontinent of Gondwana and drifted northwards towards the Asian continent, was originally seen as providing the explanation for most of the biogeographic differences between the two regions. However, as outlined by Metcalfe, the movement of continents and associated terranes has been shown to have been much more complicated. Much of SE Asia is actually composed of fragments of Gondwana, derived from the Australian continent, while volcanic activity in association with tectonic movements have resulted in ever changing positions and numbers of emergent islands within the Maritime Continent. These movements have been instrumental in changing atmospheric circulation patterns directly and indirectly through their influence on the oceanic thermo-haline belt that influences the direction and intensity of currents over the world's oceans. Morley considers the effects of these changes on the vegetation and climate of the critical Tertiary period from evidence that has only recently become available, most of it generated from his palynological research within the region. The present-day relationships of the flora, derived from available evidence on both plate tectonics and palaeoclimate change, are presented by Whiffin.

Part 2 focuses on physical and biological changes in SE Asia-Australia during the Quaternary period, the time when modern humans and their ancestors have been present in the region. It provides a partial framework for understanding human occupation, but also reveals something about the nature, timing and degree of human impact in the latter parts of the period. Polhaupessy provides the only evidence of early Quaternary environments in her review of pollen data from sites on the pivotal island of Java. This island is situated close to both Wallace's Line and to the major vegetation divide between rainforest and savannah vegetation. It also has the most substantial evidence for early Quaternary vertebrate faunas, including *Homo erectus*, in the whole SE Asian-Australian region. The pollen data provide both environmental contexts and ages for the early Quaternary vertebrate assemblages. The very different environments from those of the latter part of the Quaternary, demonstrated by comparisons of pollen records from the two periods, help explain the diverse faunas of this time and the conditions conducive to human colonization and settlement.

Kershaw *et al.* pick up the Quaternary story towards the end of the Middle Pleistocene through analyses of pollen and charcoal preserved in long continuous records from lake, swamp and marine cores extending from the humid parts of Indonesia and northeastern Australia to the drier northwest Australian-Indonesian region. They demonstrate that, unlike the early Quaternary when generally drier conditions than those of today prevailed with presumably limited rainforest extent, the latter part of the Quaternary was dominated by vegetation changes as substantial as those of any other part of the world, predominantly controlled by glacial/interglacial oscillations. During interglacial periods, rainforest expanded to at least its present day extent while during glacial periods, rainforest was very much reduced and grassland and sclerophyll vegetation spread over the Sahul continental shelves. Although precipitation was the major variable, increased relative representation of montane vegetation within the rainforest pollen component suggests significant temperature lowering during glacial periods. There is also evidence of a general trend towards more open vegetation through the late Quaternary superimposed on glacial cyclicity, suggesting increased drying or climatic variability over this period, and possibly resulting from positional changes within the Maritime Continent as a result of Australia's continued northward movement. These changes may have involved the development of the West Pacific Warm Pool (WPWP) and consequent initiation or amplification of ENSO activity. The possibility that human activity was responsible for this trend towards more open vegetation cannot be entirely ruled out, and it is likely that anthropogenic burning at least exacerbated the trend within the last 40,000 years or so.

The same marine palynological records used by Kershaw *et al.* to outline dry land vegetation changes are used by Grindrod *et al.*, in association with more detailed Holocene records from northeastern Australia, to examine patterns and causes of change in mangrove communities within the late Quaternary. In contrast to previous explanations that emphasise the relationship between mangrove extent and precipitation in tropical regions, they conclude that mangrove representation is best explained as a result of changing coastal physiognomy with the spread of mangroves corresponding with marine transgressions. This conclusion has application to both sequence stratigraphic studies of past environments and to the understanding of present and predicted future mangrove status.

Two contributions examining environmental conditions during the Last Glacial Maximum (LGM), centred on 18,000 years BP, within the West Pacific Warm Pool area primarily address the vexed question of tropical ice age temperatures. As the warmest marine water on earth, the WPWP is critical to resolving the debate as to whether tropical temperatures were reduced only marginally (i.e. 1-2°C) as suggested by a number of studies on sea surface temperatures, or whether temperatures were lowered 6°C or more as suggested by terrestrial evidence. The answer is important to an understanding of changing global latitudinal temperature gradients, and consequently on the stability of climatic and oceanic circulation patterns. Martinez *et al.* focus on foraminiferal signals from marine cores within the region. They demonstrate that the WPWP was reduced in size, but are equivocal about the general degree of climate cooling during the LGM. However, results demonstrate that oceanic and atmospheric systems could have been very different to those of today even without a substantial reduction in WPWP temperatures. Peterson *et al.* concentrate on evidence for glaciation and lowering of vegetation belts during the LGM from New Guinea, adjacent to the WPWP. Both new data and reviewed existing information support the belief that high mountain temperatures were some 6-8°C lower than today and at least 3-5°C lower in tropical lowlands, and consequently re-enforce the dilemma. However, both chapters make important suggestions as to how future research may lead to some reconciliation.

The final two chapters of Part 2 deal specifically with those components of climate that have a major influence on human populations: monsoon and ENSO variability. Maxwell and Liu review available proxy, particularly pollen, data for monsoon activity since the LGM in the whole of the East Asian region affected by monsoon influences, and compare these with results of climate modelling. They find a general pattern of monsoon onset around the Pleistocene-Holocene boundary, an early Holocene peak in activity and monsoon waning between about 5000 and 3500 years ago. There is some regional variation but less than that existing between this Asian and the related African monsoon. Although major human migration and settlement patterns as well as cultural changes have been attributed to past monsoon activity, these are difficult to discern in the palaeoecological record. By contrast, Godley is able to relate changing settlement patterns and cultural developments in Thailand to past drought and flood frequencies over the last several hundred years from indices constructed from past ENSO records or events within the Pacific rim region. These records, although climatological in nature, come from documented historical sources as well as high-resolution proxy sequences derived from tree rings and tropical ice caps.

Two very different chapters on the peopling of the region, united by popularised theoretical controversy and complexity, open Part 3. Walters focuses on controversies that have been and are still involved in interpretation of early hominoid evidence, while van Dijk and Thorne model the sequential Asian settlement of Australia-New Guinea-Island Melanesia and Polynesia. Walters briefly provides a history of fossil discovery and interpretation of pre-modern hominoids in Java, and an assessment of the current, problematic situation concerning human evolutionary models over the tremendous time range of occupation in SE Asia. The major conclusion reached is that there was likely to have been a number of species of hominoids represented over some 2 million years. Van Dijk and Thorne review the evidence for, and ideas about, the migration of people into Australia, New Guinea, Island Melanesia and Polynesia. They avoid ongoing debates about the timing of first arrival, focusing instead on sources of colonisation, concluding that many migrations ultimately from Asia contributed to the make up of the peoples of these lands.

A regional account of colonisation is presented by O'Connor *et al.* in their reconstruction of human and environmental history from cave deposits on the Aru Islands, between New Guinea and Australia. The location of the study is important in that the Aru Islands are within, or near, suggested preferred routes of human entry into Australia when the major continental shelves were exposed during the last glacial period. The archaeological record shows discontinuous human occupation relatable to changing environmental conditions over the last 27,000 years.

The human history of the whole of the Australian continent is explored at a general spatial and temporal scale by Lourandos and David, who analyse the frequency of some 700 radiocarbon dates from nearly 200 occupied rockshelters through time and space. Temporal patterns are remarkably similar between regions, despite the range of environments sampled, from the arid interior to humid continental fringes and from tropical savannas to cool temperate vegetation adjacent to glaciers during the LGM. Populations appear to have been very low everywhere during the last glacial period, increasing with climatic amelioration after this time until the mid Holocene when populations increased substantially but climate 'deteriorated'. Lourandos and David consider that the mid Holocene decoupling of human population and environment apparent at this analytical scale, was due to the development of increasingly dynamic socio-demographic systems and processes, likely associated with increasing populations, rather than simply to 'adaptation' to altered climatic conditions.

The final paper, by Potter, exemplifies the acceleration of impact that people are having on the environment and the potential dangers of unmanaged activity to biodiversity and ecosystem integrity. These threats to the heart of the region have major implications for the future climatic, economic and social stability of the whole region in the future.

Acknowledgments

We thank Donna Aitken for her organisational skills in bringing some semblance of order and structure to the original meeting, and Paul Bishop who was very instrumental in our maritime continent research and the promotion of the meeting before his appointment to a chair at the University of Glasgow. The meeting benefited greatly from the input of many contributors who for various reasons, including exhaustion of the then topical debate over the significance of the Jinmium archaeological site for the timing of human arrival in Australia, have not contributed to this volume. Many of us who have contributed derived a great deal of knowledge from our resident SE Asian expert, Gale Dixon, who has subsequently retired. We dedicate this volume to his passion for the region. We are very grateful to the vision of the then Vice-Chancellor of Monash University, Mal Logan, for seeing the potential of this research area and for facilitating financial support for it. We hope that the acquisition of large grants from the Australian Research Council for continuation of environmental research in the SE Asian-Australian region since initial Monash support, and the production of this book, have appropriately demonstrated the faith he placed in us.

A number of people, in addition to the editors, assisted in the production of this volume. Bryce Barker, Paul Bishop, G. Cresswell, Gale Dixon, Patrick De Deckker, John Flenley, Mike Green, Simon Haberle, Ernst Loffler, Bob Morley,

Ian McNiven, Bob Wasson, Peter Whetten, Trevor Whiffin, Martin Williams and Wyss Yim kindly acted as external referees for submitted papers, Gary Swinton came to the rescue of a number of papers with his drafting skills, while Margot Rohdenburg of Catena Verlag sustained her interest over the years in publishing this volume.

Bridging Wallace's Line:
Bringing Home the Antipodes

Bruno David, Peter Kershaw and Nigel Tapper

Two and a half thousand years ago, the philosopher and mathematician Pythagoras of Samos (*c.* 569-475 BC) believed that the earth was a sphere at the centre of the Universe. He held that the earth was geometrically ordered, its dynamics following laws based on the interaction of opposites, and that the world's climates could be organised into a latitudinal sequence of geographic zones (polar, temperate and equatorial) (O'Connor and Robertson 1999). With balanced geometry in mind, Pythagoras coined the term 'Antipodes' to describe an imagined inhabitable land on the opposite side of the globe to the Northern Hemisphere, as the then-known world. Literally meaning 'having the feet opposite', the Antipodes accommodated people on a sphere, standing in the Southern Hemisphere with the soles of their feet set in opposition to those in the north.

From this comes the Oxford English Dictionary's definition of Antipodes as 'places on the surfaces of the earth directly opposite to each other, or the place which is directly opposite to another; *esp.* the region directly opposite our own'. It notes that to be at antipodes is to be 'in direct opposition', and that the antipodean world is 'of or pertaining to the opposite side of the world; *esp.* Australasian'. As the Great Southern Land long predicted to exist as a necessary counter-balance to the northern land masses of Europe, Africa and Asia, Sahul – with Australia its largest extant land mass – represents the antipodean lands of the ancient Greeks *par excellence*.

Sunda and Sahul

From a Pythagorean world of counter-balanced continents, it is appropriate that in due course two land masses should come to be recognised and named at the Western world's end, each lying on one side of the equatorial divide: Sunda and Sahul. At the edge of the land mass that begins with Europe and ends with Asia is an expansive submerged plain that spans the region from the Gulf of Thailand to the Java Sea. Known as the Sunda Platform, its shallow waters are broken by a series of islands that constitute the western Malay Archipelago, and known as the Greater Sunda Islands to the west – Java, Sumatra, Borneo, Sulawesi – and the Lesser Sunda Islands to the east of Java, from Bali to Timor. Lying in the Northern Hemisphere at the eastern end of the Europe-Asia land mass, Sunda came to represent the end of the familiar world.

ISBN 3-923381-47-6

Beyond Sunda lies Sahul. For more than two thousand years before its European discovery, it was already a land imagined and imaginary, diametrically opposed to the known world. But it was not until the voyages of discovery of the late Enlightenment that its lands and peoples began to be known through first-hand reports.

Where the lands of Sunda were the world's end, Sahul was the land beyond, a land of unbridged opposites. While Sunda was populated by a fauna which, while not entirely familiar, could well be situated in a tried and proven system of knowledge, with its black swans and ducked-billed egg-laying mammals that of Sahul from the onset threatened the very ordering principles. This, then, was the setting – Sunda and Sahul's opposition represented not a seamless world but one with margins. By identifying the nature of those margins, the world could be reunited into a single unified, meaningful and ordered whole.

Wallace's Line

In 1856, two years after arriving in Singapore to describe and collect faunal specimens from the Greater and Lesser Sunda Islands, the English surveyor and pioneer biogeographer Alfred Russel Wallace (1823-1913) observed with interest a curious geographic fact. On one side, to the west of the island of Lombok and contiguous with the Asian mainland, could be found a suite of fauna – tigers, monkeys, tree shrews and, among the birds, the rosy barbet and the Javanese three-toed woodpecker, to name but two. But to the east,

> On crossing over to Lombok, separated from Bali by a Strait less than twenty miles wide, I naturally expected to meet with some of these birds again; but during a stay there of three months I never saw one of them, but found a totally different set of species, most of which were utterly unknown not only in Java, but also in Borneo, Sumatra, and Malacca. (Wallace 1869: 155).

Wallace recognised a biogeographic disjuncture that severed the familiar European-African-Asian realm of placentals from a New Guinea-Australian realm of marsupials, friarbirds and cockatoos. He drew an imaginary line to distinguish these two worlds (see Whiffin, this volume, figure 5.1), dividing Bali from Lombok, Borneo from the Celebes (Sulawesi). In the islands of the Malay archipelago he identified

> two distinct faunas rigidly circumscribed which differ as much as do those of Africa and South America and more than those of Europe and North America; yet there is nothing on the map or on the face of the islands to mark their limits. The boundary line passes between islands closer together than others belonging to the same group. I believe the western part to be a separated portion of continental Asia while the eastern part is a fragmentary prolongation of a former west Pacific continent. In mammalia and birds, the distinction is marked by genera, families, and even orders confined to one region; insects by a number of genera and little groups of peculiar species, the families of insects having generally a very wide or universal distribution. (Wallace 1869: 358-9).

We have come to know this line of faunal discontinuity as Wallace's Line. But it soon came to notice that the division was more complex, for in reality there is much geographical variation across the divide, embracing an area we now conveniently know as Wallacea. To the west is Huxley's Line, which delimits the western extent of the Australian fauna; to the east is Lydekker's Line, which marks the eastern extent of the Oriental fauna (including most of the placentals). Within Wallacea, a number of additional lines have been identified based on the distribution of various faunal groups, such as Weber's Line, which delineates equal proportions of Oriental and Australian faunas. Stretching more than 2000 km east to west and 3000 km north to south, Wallacea marks a zone of mixture of Oriental and Australian, as well as indigenous faunas. As Wallace's Line symbolises a divided world, the zone of transition signals instead a varied but contiguous realm, rich in difference and change, but not adequately understood as a simple juxtaposition of discrete geographies. The biosphere may be geographically structured, a patchwork of biomes presently distinctive, but it is linked in a single world fabric, opportunistic and with the landscape ever contingent and shifting, with a structure whose origin can be found in the depths of the earth's history.

Sunda and Sahul, the Oriental and Australian biomes of Alfred Russel Wallace, may be considered more or less distinctive biogeographic zones. Yet like Wallacea they are bridged by a global history, and by common and ongoing processes. The 'uttermost ends of the earth' (Gamble 1992) may have once been beyond Europe's reach, a realm of imagination populated by imaginary beings, but in time Sunda and Sahul, both sides of Wallace's Line, have come to be understood as part of a united world rich in variety. Wallace's Line was not merely a division, but formalised a conceptual *crossing* from one world to another. In that crossing a link was made, enabling the West to incorporate both worlds into a single understanding. The centre was slowly moving away from Europe, the world as a diverse and structured unit becoming more integrated into a single but ordered biogeographical whole. Through Wallace's Line, Sunda and Sahul symbolically represents this intellectual shift. Wallace's Line symbolises not a biogeographic crossing but a conceptual shift, an accommodation of variability and difference in an inter-connected world. In climate, geology, biogeography, in the face of climate processes, continental drift, eustatic and isostatic fluctuations and human exploration, this book explores the links and the differences by which we have come to know both sides of Wallace's Line, bridging the divide in its unity as in its difference.

The concept of Wallace's Line emerged from a Eurocentric understanding of the world, where Australia represented a land upside-down. As long as the Antipodes stood, it followed that its peoples would suffer a fate always in relation to a Western norm. Wallace's Line is a useful symbol of a first crossing beyond this conceptual division, for a seam can be approached from either side. If the Great Southern Land represented the European Antipodes, the other side of the world, the reverse must also be true. In this sense this volume explores geographical processes and patterns from a *southern* perspective, but with global implications. Pythagorus believed in the interaction of opposites, a structured but unified world. Wallace's Line symbolises well this order, bringing home an Antipodes that has always stood upright.

References

Gamble, C. 1992. Archaeology, history and the uttermost ends of the earth – Tasmania, Tierra del Fuego and the Cape. *Antiquity,* 66: 712-20.

George, W. 1964. *Biologist Philosopher: A Study of the Life and Writings of Alfred Russel Wallace.* Abelard-Schuman, London: 320 pp.

O'Connor, J.J. and E.F. Robertson. 1999. *Pythagoras of Samos.* http://www-groups.dcs.st-and.ac.uk/~history/Mathematicians/Pythagoras.html

Wallace, A.R. 1869 (Reprinted 1962). *The Malay Archipelago.* Dover, New York: 515 pp.

PART I

ENVIRONMENTAL BACKGROUND

Climate, Climatic Variability and Atmospheric Circulation Patterns in the Maritime Continent Region

Nigel Tapper

Introduction and background meteorology

The term 'Maritime Continent' was proposed by Ramage (1968) to define the coherent climatic region stretching from tropical Australia, northward and eastward to include the Indonesian archipelago, peninsular Malaysia, the large islands of Borneo and New Guinea, along with adjacent waters extending into the western Pacific Ocean. This is one of three regions of enhanced tropical convective activity located around the world. Figure 2.1, showing mean outgoing long-wave radiation (deep tropical clouds have low outgoing long-wave radiation) in the austral summer identifies Central Africa and the Amazon Basin as the other areas of intense tropical cloudiness and precipitation. However both of these areas are continental by comparison with the largely maritime region to the north of Australia, hence the name of this region.

The enhanced cloudiness in these regions represents massive exchanges of energy that are fundamentally important in the general circulation of the global atmosphere. Of these regions, none is more important in global climate dynamics than the Maritime Continent region, because of its role in providing energy for the operation of the north-south tropical Hadley cell and the east-west Walker circulation, both important components of global circulation (Sturman and Tapper 1996). McBride (1999) further points out the significance of the region as a major source of latent heat release for driving the planetary-scale monsoon circulation and as a location for convection deep enough to allow the exchange of moisture and trace gases between the troposphere and stratosphere (Figure 2.2). The role of regional ocean currents and in particular Indonesian through-flow, which redistributes warm water from the western Pacific Ocean into the Indian Ocean, is also widely acknowledged as playing a major role in global climate (Godfrey *et al.* 1993; Meyers 1996). Despite its modest dimensions, this flow passing through the Lombok Strait between Bali and Lombok

ISBN 3-923381-47-6
© 2002 by CATENA VERLAG, 35447 Reiskirchen

and through the Timor and Sawu Seas on either side of Timor, transports a substantial portion of the heat absorbed by the equatorial Pacific Ocean into the Indian Ocean where it is transported poleward.

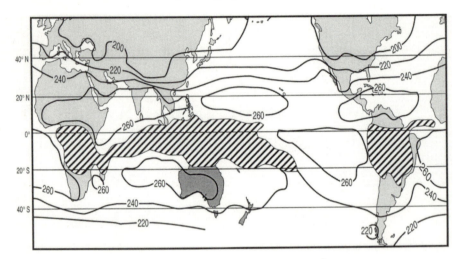

Fig. 2.1 Mean annual outgoing long-wave radiation (W/m²) for December-February. Tropical areas with outgoing long-wave radiation <240 W/m² are shaded (from Sturman and Tapper 1996).

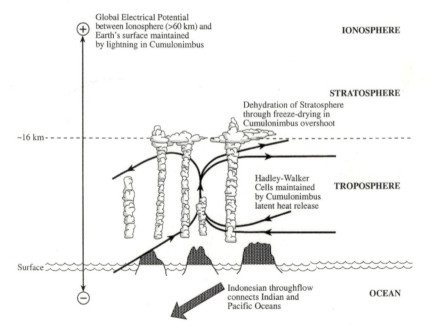

Fig. 2.2 Schematic representation of contributions of various phenomena in the Maritime Continent region to global scale circulations (from McBride 1999).

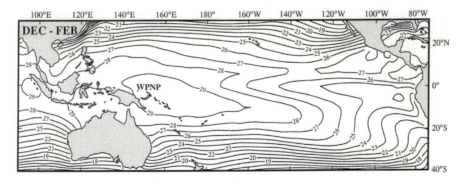

Fig. 2.3 Sea surface temperatures in °C for December-February, averaged for the years 1985-90 (from Vincent 1999).

The intense cloudiness of the region is related to its co-location with the vast area of very warm ocean surface water known as the Western Pacific Warm Pool (WPWP) (Figure 2.3). There is a clear positive relationship between sea surface temperature (SST) and amounts of tropical cloud convection (Waliser and Graham 1993; Webster *et al.* 1998). Deep oceanic convection is at a maximum with SSTs between 26.5 and 29°C, but a negative feedback between cloud and SST appears to regulate WPWP temperatures to a maximum of ~29.5°C. Surface airflow convergence associated with the mean position of the Intertropical Convergence Zone (ITCZ) and South Pacific Convergence Zone (SPCZ), which intersect in the Maritime Continent region (Figure 2.4), also contributes to the overall cloudiness of the region.

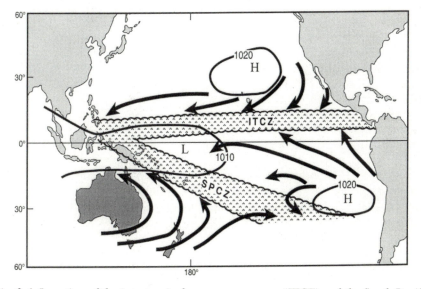

Fig. 2.4 Location of the intertropical convergence zone (ITCZ) and the South Pacific convergence zone (SPCZ) in the South Pacific region (from Trenberth 1991).

Peter Kershaw, Bruno David, Nigel Tapper, Dan Penny and Jonathan Brown (Eds): Bridging Wallace´s Line

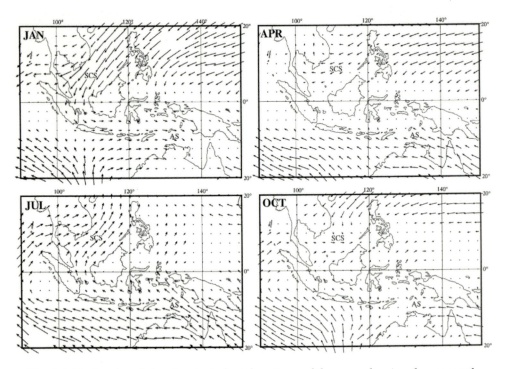

Fig. 2.5 Regional scale low-level winds at four times of the year, showing the seasonal reversal of the monsoon flow. Wind strength is indicated by the length of the arrow (from McBride 1992).

The other major climatic characteristic of the Maritime Continent region is the seasonal reversal of wind flow across the region, a pivotal part of the Asian-Australian monsoon. Figure 2.5 shows winds at 500 m, clearly illustrating the seasonal reversal in low-level monsoon flow. The reversal is particularly marked in the South China Sea (northeast Trade Winds in January and southwest monsoon winds in July) and the Arafura Sea (moist northwest monsoon flow in January and dry southeast Trade Wind flow in July). Winds across the region are relatively light (Figure 2.6), particularly in Indonesia and in the monsoon transition periods of March-April and October-November as the ITCZ moves north and south across the region. The ITCZ, which represents the broad zone of convergence of tropical easterly flow from each hemisphere, achieves its greatest seasonal movement in the Asian-Australian region (Figure 2.7) mainly because of the vast differences in seasonal heating between the respective regions. As easterly flow from one hemisphere crosses the equator, earth rotation (Coriolis deflection) causes the flow to recurve to become westerly monsoon flow in the other hemisphere.

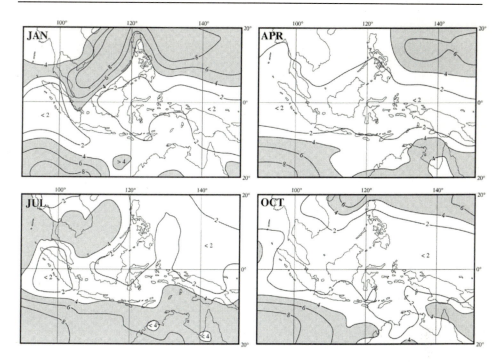

Fig. 2.6 *Regional scale wind speed at 500 m. at four times of the year. Isotachs are at 2 m/s intervals and winds >4 m/s are shaded (from McBride 1992).*

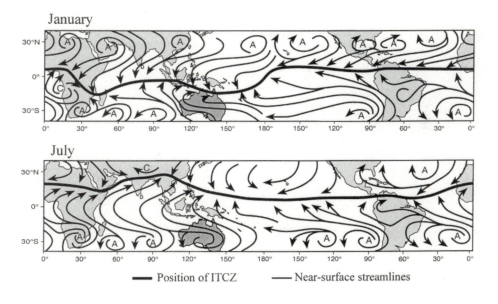

Fig. 2.7 *Mean position of the ITCZ during January and July and streamlines of near-surface windflow in the tropics (from Sturman and Tapper 1996).*

Peter Kershaw, Bruno David, Nigel Tapper, Dan Penny and Jonathan Brown (Eds): Bridging Wallace´s Line

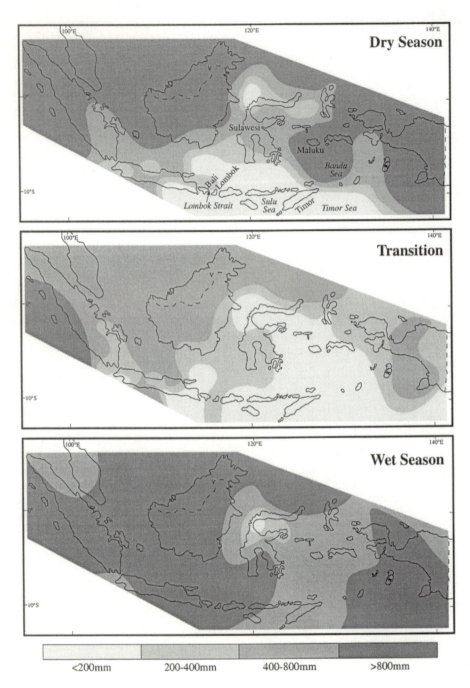

Fig. 2.8 Spatial distribution of seasonal rainfall for Indonesia, averaged for the years 1947-98. Dry season = May to September, Transition season = October to November, Wet season = December to March (from Kirono 2000).

The climatological characteristics of regional airflow described above are present more or less every year, and are important both in the seasonal distribution of rainfall and in the transport of terrestrial materials such as aerosols from vegetation fires and pollen. As long ago as the 1920s, a Dutch colonial meteorologist raised the possibility of Australian dust and aerosols from biomass burning contributing to dry season haze in the Dutch East Indies (Braak 1929).

Southern and eastern parts of Indonesia and northern Australia generally have low rainfall May-October under the influence of the dry Trade Winds blowing out of central Australia, with rainfall increasing into the transition seasons and during the monsoon period. Figure 2.8 shows the spatial distribution of mean seasonal rainfall for the central Maritime Continent area. Dry season rainfall shows a strong north to south gradient since the ITCZ lies just to the north of the region at this time. High rainfall over the Maluku region at this time of year occurs because of warming and moistening of the dry southeast Trade Winds as they flow over the Banda Sea to the northeast of Timor. During the transition season there is a strong east to west rainfall gradient because the monsoon starts in the north and west around August-September, migrating to eastern areas by December (see discussion on seasonal influences below). During the wet season there is no strong regional north to south or east to west gradient of rainfall. The strongest local gradients occur because of exposure to the prevailing westerly monsoon flows. For example there is anomalous low rainfall over central Sulawesi because of the rain-shadow effect. Rainfall regions demarcated for the central Maritime Continent region are shown in Figure 2.9. This distribution, recently defined using rotated principal component analysis by Kirono (2000), is similar to the rainfall typology maps prepared fifty years ago by Schmidt and Ferguson (1951).

Weak low-level winds across much of the Maritime Continent, especially during the transition season, means that there is considerable air stagnation that allows haze to reach unacceptable levels when fires are burning in the region. This effect is compounded by persistent upper air inversions that are characteristic of most Trade Wind flow regimes. These are seen at heights of ~2000 m over northern Australia during the dry season, but are less distinct over Indonesia (Tapper 2000). Nevertheless characteristic maximum atmospheric mixing depths that help define dispersion heights of burning products away from the surface during the dry season would appear to be about 1500 m over Indonesia (Nurhayati 1994).

Major influences on the weather and climate of the maritime continent region

The climate of the Maritime Continent region exhibits a general degree of predictability. It is a region of equatorial or near-equatorial warmth, is at least seasonally wet, and it has a seasonal reversal of prevailing low-level winds. However, there is a range of influences operating over varying time scales that affect the region's local and regional climate and in particular the temporal and spatial distribution of rainfall. These are discussed in more detail in the following sections, beginning with those factors having influence over the shorter term.

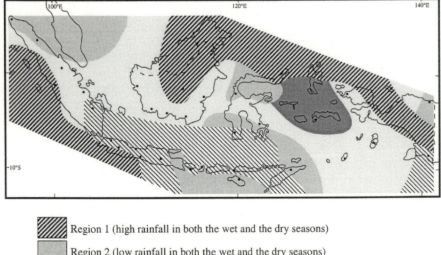

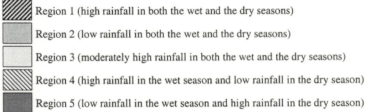

Region 1 (high rainfall in both the wet and the dry seasons)

Region 2 (low rainfall in both the wet and the dry seasons)

Region 3 (moderately high rainfall in both the wet and the dry seasons)

Region 4 (high rainfall in the wet season and low rainfall in the dry season)

Region 5 (low rainfall in the wet season and high rainfall in the dry season)

Fig. 2.9   Regional seasonality of rainfall for Indonesia, determined by principal component analysis of rainfall data for the years 1947-98 (after Kirono 2000).

Diurnal and mesoscale influences

The atmosphere of the Maritime Continent region is typical of near-equatorial oceanic areas. Air masses are relatively moist and horizontally homogeneous, with weak Coriolis deflection of horizontal air motion because of the low latitudes involved. These conditions, along with a regional gradient wind flow that is often quite weak, mean that local circulation systems such as land and sea breezes operating over daily time scales become relatively more important in local climate than at higher latitudes (Sturman and Tapper 1996).

There are two distinct modes of cloud convection in the Maritime Continent region (McBride 1999). During the transition season to monsoon outbreak and during breaks in the monsoon, deep convective thunderstorm activity occurs mainly over land, and preferentially over islands and peninsulas in the region (Figure 2.10). These isolated thunderstorms develop in a highly unstable atmosphere with high values of convective available potential energy (CAPE) (Beringer *et al.* 2001), and can penetrate the tropopause to heights of more than 20 km. In contrast, convection during the active monsoon period occurs in a low CAPE regime, is broader and shallower and associated with monsoon rain over large areas (Figure 2.10). Island thunderstorms have long been used as navigation aids by traditional seafarers of the region (Simpson *et al.* 1993) since they are strongly diurnally modulated by sea breeze convergence and reach their maximum vertical extent in the early afternoon (Beringer *et al.* in press; Oliphant *et al.* 2001). Vertical motion generated by convergence of sea breezes across islands and peninsulas or by interacting gust fronts from existing thunderstorms provides the trigger for the release of instability and the initial growth of the thunderstorms (Keenan *et al.* 2000). These island thunderstorms produce immense quantities of precipitation in a very short time period.

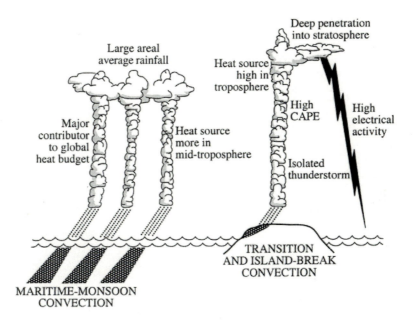

Fig. 2.10   Schematic representation of the properties of monsoon and monsoon-break convection over the maritime continent region (from McBride 1999).

The importance of terrestrial convection is confirmed by Wyrtki (1956) who found the average value of rainfall over land in the Maritime Continent region to be 1.5 times that over water. The dominance of water in the region reduces the ratio of average rainfall over the whole region to that over the sea to about 1.16 (McBride 1999).

Intra-seasonal influences

A range of intra-seasonal meteorological phenomena play an important role in rainfall processes in the region, particularly as they influence monsoon onset and the timing of major rainfall periods within the monsoon season. At the synoptic scale, northerly surges of low-level wind from the South China Sea, associated with fluctuations in the intensity of the Siberian high-pressure system are known to be important in regulating convective activity and rainfall over northern parts of the Maritime Continent. This is also thought to be a mechanism for monsoon onset over southern Indonesia and northern Australia (Manton and McBride 1992). Australian west coast surges and frontal activity over the Australian continent during the transition season have also been shown to be important (Sturman and Tapper 1996). Another influence on monsoon onset over the region and on intra-seasonal rainfall is the 40-50 day oscillation in equatorial wind and rainfall first identified by Madden and Julian (1971, 1972). This large-scale wave of enhanced convection is first seen in the Indian Ocean and moves from west to east at equatorial latitudes with a propagation speed of ~20 metres per second. The Madden-Julian Oscillation (MJO) has a horizontal structure similar to the Walker Circulation (discussed below), and the intensity of the zonally oriented circulation varies markedly with longitude, with deepest convection being achieved over the Indian Ocean and Western Pacific (Figure 2.11).

Seasonal influences

Clearly the most important seasonal influence on the meteorology of the Maritime Continent region is the Asian-Australian monsoon, part of the planetary monsoon circulation. Broad characteristics of the circulation relevant to the region were discussed above, as were some of the synoptic scale triggers of onset. The monsoon is responsible for a large component of annual rainfall over much of the region. For example Australia's 'Top End' receives more than 80% of its rainfall in the December to March monsoon period. Much of this rainfall falls over the mainland and islands that provide additional uplift of air masses (e.g. through topographic effects and/or heating).

The Asian-Australian monsoon is a massive thermal circulation driven by the seasonal surface temperature difference between (broadly) central Asia and Australia. During the austral summer, the very warm temperatures of continental Australia and adjoining oceans, coupled with cold temperatures over Asia, draw surface airflow from Asia toward Australia. Above ~3000 m this very moist, warm monsoon northwest airflow is overlain by easterly airflow, effectively the return flow of the thermal circulation. The circulation broadly reverses during the austral winter.

One marker of monsoon onset in the Maritime Continent region is the simultaneous observation of westerly wind at 850 hPa (~1500 m) and easterly wind at 200 hPa (~7500 m). Such a definition has been used to produce Figure 2.12 that shows mean monsoon onset and retreat dates across the region.

One of the most significant seasonal influences is the impact of characteristic wind flow regimes on the transport of biomass burning products throughout the region. The Maritime Continent region has a long history of haze problems associated with vegetation fires, most recently during the drought years of 1994 and 1997-98. Recent developments in numerical weather prediction (NWP) and atmospheric transport modelling (ATM) allow long distance transport patterns (atmospheric trajectories) to be calculated. An example of the seasonal variability of atmospheric transport in the Maritime Continent region is given in

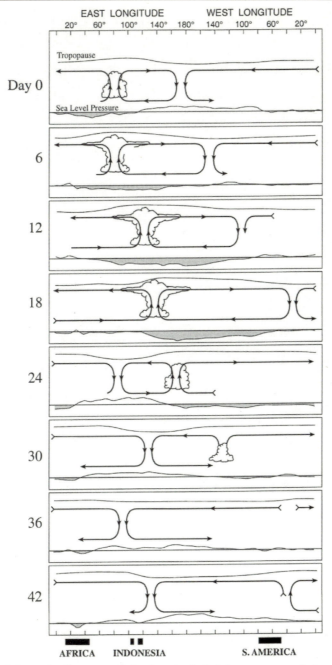

Fig. 2.11 Schematic diagram of the longitude-height structure of the atmospheric circulation for different phases of the 40-50 day oscillation. Tropospheric height is shown at the top and sea level pressure at the bottom of each frame. Clouds indicate zones of enhanced convection and the approximate day number from 0 is shown on the left margin (from Madden and Julian 1972).

Figure 2.13 which shows monthly trajectory composites (~30 daily forward trajectories originating at ~500 m, each extending out for five days) for February 1995 (north Australian wet season) and July 1994 (north Australian dry season) for three locations in the Maritime Continent region: eastern Indonesia on the equator (0.0°N, 125.00°E), Jabiru in the Top End (12.86°S, 132.35°E), and east-central Northern Territory close to Tennant Creek (20.00°S, 135.00°E). It should be noted that there is considerable flow variability from year to year, mainly associated with El Nino-Southern Oscillation (ENSO) influences (Wain 2000), so these patterns for individual months should not necessarily be regarded as a reliable transport climatology.

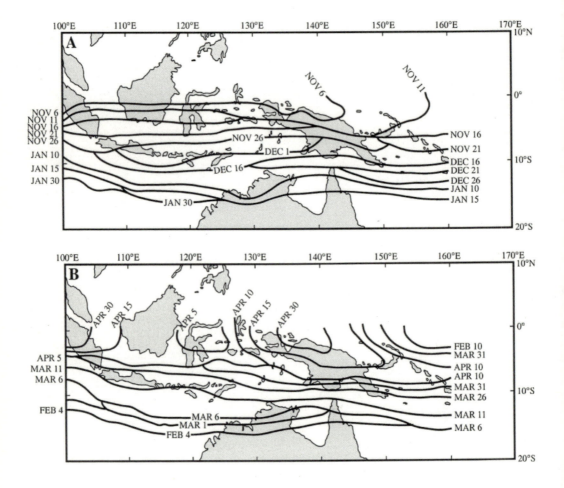

Fig. 2.12 (a) Monsoon onset and (b) monsoon retreat dates defined by observations of wind at 850 and 200 hPa (from Tanaka 1994).

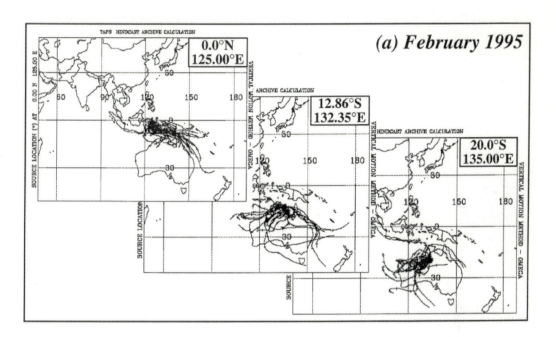

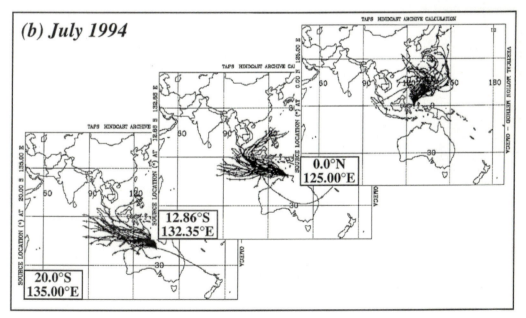

Fig. 2.13   Daily 950 hPa forward trajectories (5 day runs) from three locations in the maritime continent region.   Upper boxes are for February 1995, lower boxes are for July 1994 (from Tapper 2000).

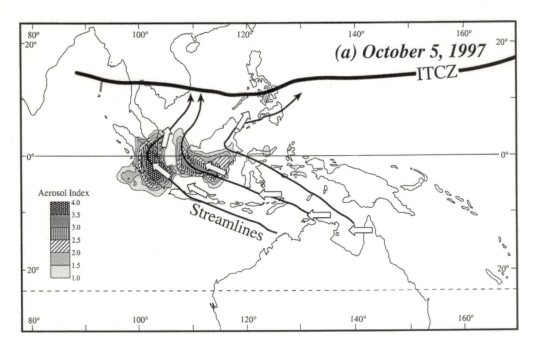

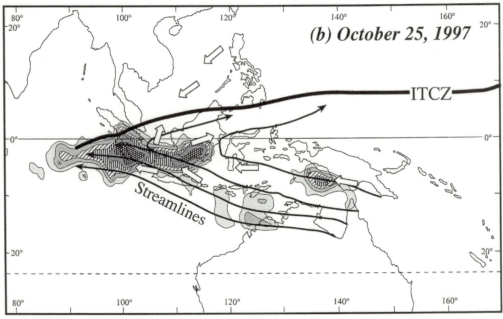

Fig. 2.14   *The locations of the intertropical convergence zone (ITCZ), major haze concentrations, airflow streamlines (black lines) and actual surface wind directions (white arrows) on (a) October 5, 1997, and (b) October 25, 1997 (from BAPPENAS 1999).*

During February 1995 transport from the northern point (located in the centre of the Maritime Continent region) is dominantly toward the southeast in the north-west monsoon flow. Many of the five-day trajectories impinge on mainland Australia, but a lack of burning and rainfall washout at this time of year means that the air is likely to be relatively clean. Transport from the mainland Australian locations is more variable, possibly reflecting monsoon and monsoon-break conditions. However a large amount of re-circulation occurs, consistent with persistent continental low pressure at this time of the year (Sturman and Tapper 1996).

Dry season airflow patterns (Figure 2.13) are less complex. The Australian locations exhibit consistent transport to the northwest in the southeasterly Trade Wind flow, but some re-curvature toward the northeast is seen as trajectories cross the equator. Clearly there is some potential for transport of products of biomass burning from Australia into the Indonesian region at this time of year, as there is also for transport within SE Asia. Transport from the northern location is dominantly toward the northeast as the Trade Winds re-curve over the equator to become the southwest Asian monsoon flow.

Clearly the position of the ITCZ and its monsoonal movement has a major influence on the transport patterns, as was seen during the major SE Asian haze episode of September-October 1997. Through much of this period the ITCZ lay well north over Indo-China and the Philippines (Figure 2.14a), allowing smoke from fires in Kalimantan and Sumatra to be transported northward over Singapore and Peninsula Malaysia. Smoke haze cleared from these areas and was pushed well into the Indian Ocean as the ITCZ moved southward during October (Figure 2.14b).

Inter-seasonal and inter-annual influences

The Walker circulation is a major circulation cell moving air between the eastern and western sides of the South Pacific Ocean. It is oriented northwest to south-east across the equatorial Pacific with a rising limb normally centred over the warm waters of the Indonesian region and descending motion over the cool eastern Pacific (Figure 2.15a). Fluctuations in the intensity and even the direction of this circulation occur during ENSO events (Figure 2.15b), which have a quasi-periodic time scale of 2-3 years. El Niño refers to the rapid rise in SST in the tropical central and eastern Pacific that occurs during the breakdown of the normal Walker circulation, while the Southern Oscillation refers to the associated global scale pressure fluctuations. Rising air motion and lower surface pressures are always associated with the regions of warmer water.

The strength of the Walker circulation is proportional to the difference in pressure between Indonesia and the eastern South Pacific. The Southern Oscillation Index (SOI), derived from the pressure difference between Tahiti and Darwin, is a measure of the strength of the Walker circulation and its direction. ENSO has been shown to be the major control on rainfall variability across large parts of the globe (Nicholls and Wong 1990), with the strongest amplification of variability occurring at lower latitudes and in relatively lower rainfall regions. This is therefore particularly relevant to the more strongly monsoonal parts of the Maritime Continent region. It has been long documented that extremes in rainfall for much of Indonesia and northern and eastern Australia are strongly related to variations in the Southern Oscillation (Braak 1919; Berlage 1927; Nicholls 1981, 1983; McBride and Nicholls 1983; Hastenrath 1987; Drosdowsky and Williams 1991; Yamanaka 1998; Kirono *et al.* 1999). Kirono and Tapper (1999) have shown statistically significant relationships between rice production and ENSO for Indonesia and report agricultural losses of US$2.75 billion associated with the 1997-98 ENSO drought out of total losses to the Indonesian economy of US$9 billion.

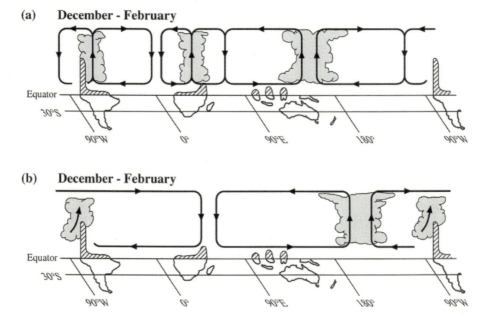

Fig. 2.15 (a) Normal Walker circulation and (b) its breakdown during El Niño-Southern Oscillation (ENSO) events (from Sturman and Tapper 1996).

For the Indonesian region, correlations between rainfall and SOI are strongest in the south and east, and during the dry and monsoon transition seasons (Figure 2.16). The major reason for rainfall suppression over eastern Indonesia during ENSO years is a delayed monsoon arrival and early dry season onset (Kirono and Tapper 1999), which produces a much longer dry season. For northern Australia, correlations between SOI and rainfall are also strongest during the transition (monsoon build-up) season (McBride and Nicholls 1983). It is not surprising that there are close links between the strength of the monsoon and ENSO. Webster (1994) outlines the three components to monsoon forcing: differential heating between the land and ocean during the annual cycle of solar heating, moist convection processes, and planetary rotation. The first two of these influences are also fundamental in the operation of ENSO. During an El

Niño event, the warmest SSTs move east out of the Maritime Continent region, along with the focus of cloud convection that provides energy to the atmosphere through latent heat processes. This weakens both the monsoon and the normal Walker circulation.

Table 2.1. Documented occurrence of El Niño-Southern Oscillation (ENSO), drought and fire in Indonesia since 1877 (sources: Denis 1998; Holmes 1998).

Year(s) and ENSO?	Drought Severity and Broad Location	Fires and Broad Location
1877 Strong ENSO	Java (no data elsewhere)	Central Kalimantan
1888 Strong ENSO	South – severe	
1896 Strong ENSO	South – not severe	
1902 Weak ENSO	South and east – severe	
1911 Weak ENSO	Localized – not severe	
1914 Strong ENSO	East – severe	South and East Kalimantan
1918 Weak ENSO	South and east – moderate	
1930 Weak ENSO	East – moderate	
1940-41 Strong ENSO	East – severe	
1940s-50s	Lack of records	
1961 No ENSO	South and east – severe	
1963 Weak ENSO	South and east – severe	
1965 Strong ENSO	South and east – severe	
1967 No ENSO	South and east – moderate	
1969 Weak ENSO	South – moderate	
1972 Strong ENSO	South and east – severe	
1976-77 Strong ENSO	South and east – severe	
1982-83 Strong ENSO	Widespread – severe	East Kalimantan
1986-87 Strong ENSO	South and east – moderate	All Kalimantan
1991-94 Strong ENSO	Widespread – severe	Sumatra, Kalimantan, Java
1997-98 Strong ENSO	Widespread – severe	Sumatra, Borneo generally

Strong ENSO = dry season SOI below -10
Weak ENSO = dry season SOI between 0 and -10
No ENSO = dry season SOI above 0

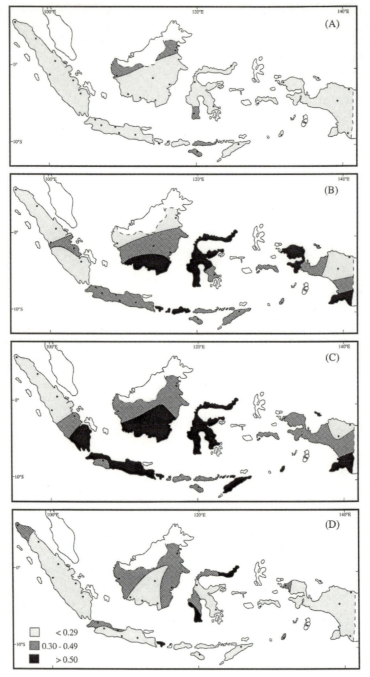

Fig. 2.16   The magnitude of the simultaneous correlation between seasonal rainfall and the Southern Oscillation Index (SOI) for various parts of Indonesia (correlations >0.29 are significant at the 0.05 level). Data are for the period 1947-1998 (from Kirono et al. 1999).

There is strong documentary evidence linking major forest fire episodes in Indonesia to ENSO. Since systematic record keeping began in Indonesia in the 1860s there have been 20 droughts of national significance recorded (Holmes 1998, Denis 1998) (Table 2.1). During the same period five major fire events have been recorded in the Indonesian region. Large fires occurred in Central Kalimantan in 1877, in South and East Kalimantan in 1914, in East Kalimantan in 1982-83, throughout Kalimantan in 1986-87, in Sumatra, Java and Kaliman-tan in 1991-94, and in Sumatra and Borneo generally in 1997-98 (Denis 1998). Table 2.1 reveals that all of these fire events were associated with strong ENSO events (defined as having dry season SOI below -10). Interestingly, none of these large fire events, with the possible exception of the 1997-98 event, saw much large scale burning in eastern Indonesia. This suggests that it is the drying out of the more heavily forested areas further north and west, particularly in Kalimantan and Sumatra, that is most problematic from a forest fire perspective. The link between ENSO and fire activity in Indonesia is further documented in Figure 2.17. This Figure indicates a curvilinear relationship between forest area burnt in Indonesia and SOI, with relatively little burning occurring in seasons with SOI above -5.

Apart from impacts on drought and fire occurrence, the two extreme modes of the Walker circulation also have implications for haze dispersion. During the 'normal' phase, enhanced vertical motion (convection) over Indonesia and stronger Trade Winds provide better conditions for moving haze out of the region both horizontally and vertically, as well as for providing rainout of particulate matter. During ENSO years the Trade Winds slacken and convection is reduced, particularly over Indonesia, resulting in greater stagnation of haze such as was seen during September-October 1997 (Figure 2.18).

Some modelling simulations coupling atmosphere and ocean have raised con-cerns about possible changes in ENSO under enhanced greenhouse conditions. Meehl *et al.* (1993), Knutson and Manabe (1994, 1995) and Smith *et al.* (1996) all produced enhanced ENSO-like events in their greenhouse simulations. This view has been supported by recent analysis of global warming trends, SST and ENSO severity (Kerr 1999). On the contrary, the work of Gordon and O'Farrell (1996) indicated weakened ENSO-like phases. Clearly, this highly sophisticated modelling and observational work must continue for us to be better informed about the likely future impacts on ENSO of any global warming.

Finally, it should be noted that a quasi-biennial oscillation (QBO) signal has also been seen in regional rainfall (Yasunari 1981). However, in turn the QBO also appears to be at least partly related to fluctuations in the Walker circulation (Sturman and Tapper 1996). ENSO therefore remains the greatest source of year to year variability in rainfall in the Maritime Continent region.

Peter Kershaw, Bruno David, Nigel Tapper, Dan Penny and Jonathan Brown (Eds): Bridging Wallace's Line

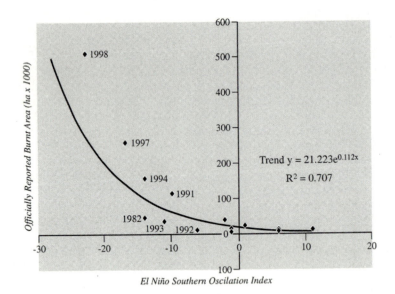

Fig. 2.17   *The relationship between the Southern Oscillatoion Index (SOI) and forest area burnt in Indonesia (from Badawi et al. 1998).*

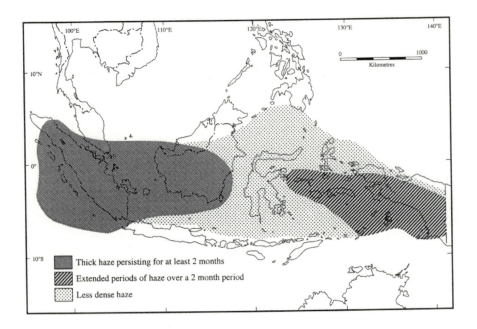

Fig. 2.18   *The distribution of haze across the maritime continent region during the haze event of September-October 1997 (from BAPPENAS 1999)*

Concluding Comments

This chapter has provided a broad overview of the climate of the Maritime Continent region, concentrating in particular on its climate variability and characteristic atmospheric circulation patterns. The region is shown to be a focus for two of the most significant global atmospheric circulations; the Asian-Australian monsoon and the Walker circulation. Together these two circulations and variations in their strength have a profound influence on the social and economic well-being of a large proportion of the global population, both those living within the Maritime Continent region, and through atmospheric tele-connections, large populations living elsewhere in Asia, Africa and South America. While this chapter has attempted to synthesise much of what is known about the climatology and meteorology of this important region, it is clear that the atmospheric processes affecting this region are fundamentally very complex and that there are many gaps in our knowledge of their operation. This has been recognised by the meteorological community and the drive to increase fundamental knowledge of the meteorology of the region is reflected in a raft of field campaigns during the last 15 years, conducted largely under the auspices of the World Climate Research Program (WCRP). These include the Australian Monsoon Experiment (AMEX), the Island Thunderstorm Experiment (ITEX), the Tropical Ocean Global Atmosphere-Coupled Ocean Atmosphere Response Experiment (TOGA-COARE), the Maritime Continent Thunderstorm Experiment (MCTEX), and the GEWEX (Global Energy and Water Cycle Experiment) Asian Monsoon Experiment (GAME). Detailed reference to many of these field campaigns can be found in Keenan *et al.* (1989, 2000), Webster and Lukas (1992), Yasunari (1994) and Webster *et al.* (1998).

References

Badawi, W., Walsh, T. and Jhamtani, H. 1998. *Forest and Land Fires in Indonesia, Impacts, Factors and Evaluation, Volume 1.* State Ministry for Environment, Republic of Indonesia; United Nations Development Programme: 185 pp.

BAPPENAS. 1999. *Planning for Fire Prevention and Drought Management.* Final Report of TA 2999-INO. Asian Development Bank and BAPPENAS (Indonesian National Development Planning Agency), Jakarta (April 1999).

Berlage, H.P. 1927. *East Monsoon Forecasting in Java.* Verhandelingen No. 20, Koninklijk Magnetisch en Meteorologisch Observatorium te Batavia, Indonesia.

Beringer, J., Tapper, N. and Keenan, T. 2001. Evolution of maritime continent thunderstorms under varying meteorological conditions over the Tiwi Islands. *International Journal of Climatology,* 21: 1021-1036.

Braak, C. 1919. *Atmospheric Variations of Short and Long Duration in the Malay Archipelago and Neighboring Regions, and the Possibility to Forecast Them.* Verhandelingen No. 5, Koninklijk Magnetisch en Meteorologisch Observatorium te Batavia, Indonesia.

Braak, C. 1929. *The Climate of the Netherland Indies; Volume I and II.* Verhandelingen No.8, Koninklijk Magnetisch en Meteorologisch Observatorium te Batavia, Indonesia.

Denis, R. 1998. *A Review of Fire Projects in Indonesia 1982-1998.* Centre for International Forestry Research, Bogor: 110 pp.

Drosdowsky, W. and Williams, M. 1991. The Southern Oscillation in the Australian region. Part I: anomalies at extremes of the oscillation. *Journal of Climate*, 4: 619-638.

Godfrey, J., Hirst, A. and Wilkin, J. 1993. Why does the Indonesian throughflow appear to originate from the North Pacific. *Journal of Physical Oceanography*, 23: 1087-1098.

Gordon, C. and O'Farrell, S. 1996. Transient climate change in the CSIRO coupled model with a coupled sea ice model. *Monthly Weather Review*, 125(5): 875-901.

Hastenrath, S. 1987. Predictability of Java monsoon anomalies: a case study. *Journal of Climate and Applied Meteorology*, 26: 133-141.

Holmes, D. 1998. *An Analysis of the Drought of 1997-98 in Riau and East Kalimantan, Comparison with Previous Droughts, and Comparison with Other Regions of Equatorial Indonesia.* Working Paper of ADBTA2999-INO, Asian Development Bank and BAPPENAS (Indonesian National Development Planning Agency), Jakarta, (April 1999).

Keenan, T., Morton, B., Manton, M., and Holland, G. 1989. The Island Thunderstorm Experiment (ITEX) – A study of tropical thunderstorms in the maritime continent. *Bulletin of the American Meteorological Society*, 70: 152-159.

Keenan, T., Rutledge, S., Carbone, R., Wilson, J., Takahashi, T., May, P., Tapper, N., Platt, M., Hacker, J., Sekelsky, S., Moncrieff, M., Saito, K., Holland, G., Crook, A. and Gage, K. 2000. The Maritime Continent Thunderstorm Experiment (MCTEX): Overview and some results. *Bulletin of the American Meteorological Society*, 81: 2433-2455.

Kerr, R. 1999. Big El Niños ride the back of slower climate change. *Science*, 283: 1108-1109.

Kirono, D. 2000. *Indonesian Seasonal Rainfall Variability, Links to ENSO and Agricultural Impac*t. Unpublished Ph.D. Thesis, School of Geography and Environmental Science, Monash University, Australia.

Kirono, D. and Tapper, N. 1999. ENSO rainfall variability and impacts on crop production in Indonesia. *Physical Geography*, 20(6): 508-519.

Kirono, D., Tapper, N. and McBride, J. 1999. Documenting Indonesian rainfall in the 1997-1998 El Niño. *Physical Geography*, 20(5): 422-435.

Knutson, T. and Manabe, S. 1994. Impact of increasing CO_2 on simulated ENSO-like phenomena. *Geophysical Research Letters*, 21: 2295-2298.

Knutson, T. and Manabe, S. 1995. Time-mean response over the tropical Pacific to increased CO_2 in a coupled ocean-atmosphere model. *Journal of Climate*, 8: 2181-2199.

McBride, J. 1992. The meteorology of Indonesia and the Maritime Continent. *Extended Abstracts, Fourth International Symposium on Equatorial Observations Over Indonesia (4th ICEAR Symposium)*, Jakarta, Indonesia (November 10-11, 1992).

McBride, J. 1999. Indonesia, Papua New Guinea and Tropical Australia: the Southern Hemisphere Monsoon. In: D. Vincent and D. Karoly, D. (Editors) *Meteorology of the Southern Hemisphere.* American Meteorological Society, Boston: 89-99.

McBride, J. and Nicholls, N. 1983. Seasonal relationships between Australian rainfall and the Southern Oscillation. *Monthly Weather Review*, 111: 1998-2004.

Madden, R. and Julian, P. 1971. Detection of a 40-50 day oscillation in the zonal wind in the tropical Pacific. *Journal of the Atmospheric Sciences*, 28: 702-708.

Madden, R. and Julian, P. 1972. Description of global-scale circulation cells in the tropics with a 40-50 day period. *Journal of the Atmospheric Sciences*, 29: 1109-1123.

Manton, M. and McBride, J. 1992. Recent research on the Australian monsoon. *Journal of the Meteorological Society of Japan*, 70: 275-285.

Meehl, G., Branstator, G. and Washington, W. 1993. Tropical Pacific inter-annual variability and CO_2 climate change. *Journal of Climate*, 6: 42-63,

Meyers, G. 1996. Variation of Indonesian throughflow and the El Niño-Southern Oscillation. *Journal of Geophysical Research*, 101(65): 12,255-12 ,263.

Nicholls, N. 1981. Air-sea interaction and the possibility of long-range weather prediction in the Indonesian Archipelago. *Monthly Weather Review*, 109: 2435-2443.

Nicholls, N. 1983. The Southern Oscillation and Indonesian sea surface temperatures. *Monthly Weather Review*, 112: 424-432.

Nicholls, N. and Wong, P. 1990. Dependence of rainfall variability on mean rainfall, latitude, and the Southern Oscillation. *Journal of Climate*, 3: 163-170.

Nurhayati. 1994. *Visual Air Quality in Metropolitan Jakarta.* Unpublished M.Sc. thesis, School of Geography and Environmental Science, Monash University, Australia.

Oliphant, A., Sturman, A. and Tapper, N. 2001. The evolution and structure of a tropical island sea/land breeze system, Northern Australia. *Meteorology and Atmospheric Physics*, 78: 45-59.

Ramage, C. 1968. Role of a tropical 'maritime continent' in the atmospheric circulation. *Monthly Weather Review*, 96: 365-369.

Schmidt, F. and Ferguson, J. 1951. *Rainfall types based on wet and dry period ratios for Indonesia with Western New Guinea.* Verhandelingen, No. 42, Kementerian Perhubungan Djawatan Meteorologi dan Geofysik, Jakarta.

Simpson, J., Keenan, T. Ferrier, B., Simpson, R. and Holland, G. 1993. Cumulus mergers in the maritime continent. *Meteorology and Atmospheric Physics*, 51: 73-99.

Smith, T, Reynolds, R. and Ropelewski, C. 1996. Reconstruction of historical sea surface temperatures using empirical orthogonal functions. *Journal of Climate,* 9: 1403-1420.

Sturman, A. and Tapper, N. 1996. *The Weather and Climate of Australia and New Zealand.* Oxford University Press, Melbourne, 476 pp.

Tapper, N. 2000. Atmospheric issues for fire management in eastern Indonesia and northern Australia. In: J. Russell-Smith, G. Hill, S. Djoeroemana and B. Myers (Editors) *Fire and Sustainable Agricultural and Forestry Development in Eastern Indonesia and Northern Australia Conference*, Australian Centre for International Agricultural Research, Canberra: 21-30.

Tanaka, M. 1994. The onset and retreat dates of the Austral summer monsoon over Indonesia, Australia and New Guinea. *Journal of the Meteorological Society of Japan*, 72: 255-267.

Trenberth, K. 1991. General characteristics of El Niño-Southern Oscillation. In: M. Glantz, R. Katz and N. Nicholls (Editors) *Teleconnections Linking Worldwide Climate Anomalies*. Cambridge University Press, Cambridge: 13-41.

Vincent, D. 1999. Pacific Ocean. In: D. Vincent and D. Karoly (Editors) *Meteorology of the Southern Hemisphere.* American Meteorological Society, Boston: 101-117.

Wain, A. 2000. A Trajectory Climatology for the Maritime Continent Region. Unpublished Ph.D. thesis, School of Geography and Environmental Science, Monash University.

Waliser, D. and Graham, N. 1993. Convective cloud systems and Warm-Pool sea surface temperatures: coupled interactions and self regulation. *Journal of Geophysical Research*, 98: 12,881-12,893.

Webster, P. 1994. The role of hydrological processes in ocean-atmosphere interaction. *Reviews of Geophysics*, 32: 427-476.

Webster, P. and Lukas, R. 1992. The Coupled Ocean-Atmosphere Response Experiment. *Bulletin of the American Meteorological Society*, 73: 1377-1416.

Webster, P., Magana, V., Palmer, T., Shukla, J., Tomas, R., Yanai, M. and Yasunari, T. 1998. Monsoons: processes, predictability and the prospects for prediction. *Journal of Geophysical Research,* 103: 14,451-14,510.

Wyrtki, K. 1956. *The Rainfall Over The Indonesian Waters.* Verhandelingen, No. 49, Lembaga Meteorologi dan Geofysik, Jakarta.

Yamanaka, M. (Editor) 1998. *Climatology of Indonesian Maritime Continent.* Kyoto University, Kyoto: 90 pp.

Yasunari, T. 1981. Temporal and spatial variations of monthly rainfall in Java, Indonesia. *Southeast Asian Studies,* 19: 170-186.

Yasunari, T. 1994. Scientific strategy of GEWEX Asian Monsoon Experiment (GAME). *Eos,* 75: 31.

Tectonic History of the
SE Asian-Australian Region

Ian Metcalfe

Introduction

Pre-Quaternary connections between SE Asia and Australia go back more than 1000 million years. Since plate-tectonics began operating back in the early Pre-Cambrian, continents created in the Archaean have collided, split, wandered around the earth and re-collided several times over. The continents are carried by constantly moving lithospheric plates that are created at mid-ocean ridges and destroyed by subduction at convergent plate boundaries. The present-day config-uration of East and SE Asia is the result of these long drift histories, and of lost oceans now preserved as narrow remnant suture zones where 'worlds' have collided. Changing continent-ocean and land-sea configurations caused by plate tectonic movements, by continental collisions, volcanism and global eustatic sea-level changes have influenced the evolution and dispersal of faunas and floras, the environment and climate and ultimately the cultural history of the Southeast Asian-Australian region. The modern distribution of fauna and flora in SE Asia and Australia has its roots in the geological and tectonic history of the region. In this paper, I will provide an overview of the tectonic history of the region, beginning 1000 million years ago and ending just prior to the arrival of *Homo erectus* to the region, and will present palaeogeographic maps depicting the waxing and waning SE Asia-Australia connections and changing continent-ocean and land-sea configurations that have affected past climates.

Present day tectonic setting and plate interactions

East and SE Asia lie at the zone of convergence of three major tectonic plates, the Eurasian Plate, the Indian-Australian Plate and the Pacific Plate (Figure 3.1). The broad kinematic framework of East and SE Asia is determined by the relative motions of these major plates but the large scale tectonics of the region are determined also by the interactions of these major plates with the smaller Philippine Sea and Caroline Plates. The Eurasian plate is relatively stable and the northwards movement of the Indian-Australian plate causes continued northwards collision of continental India with Eurasia and subduction of the oceanic plate beneath the SE Asian part of Eurasia. Oblique subduction in the Sumatra region gives way to a major transform boundary in the Burma region

ISBN 3-923381-47-6
© 2002 by CATENA VERLAG, 35447 Reiskirchen

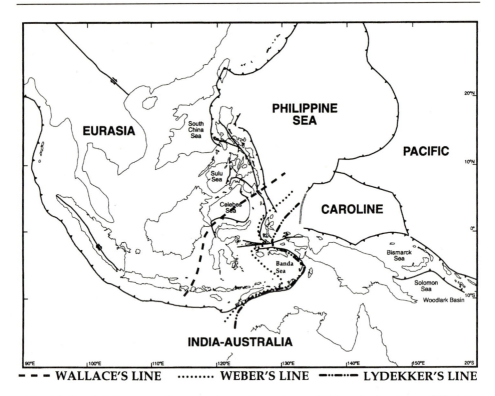

- - - WALLACE'S LINE   ········· WEBER'S LINE   ━··━··━ LYDEKKER'S LINE

Fig. 3.1 Simplified present-day tectonic configuration and lithospheric plates of SE Asia. Also shown are Wallace's Line (1863) separating Australian and Asian faunas, now taken as the eastern boundary of strictly Asian fauna; Weber's Line, the boundary of faunal balance between Asian:Australian animals (50:50) for mammals and molluscs; Lydekker's Line, western boundary of strictly Australian faunas. After Hall (1998) and Moss and Wilson (1998).

and the Andaman Sea basin has been interpreted as a 'leaky transform'. The Philippine Sea plate is actively subducting westwards beneath the Philippines, Taiwan and Japanese Eurasia. Other active subduction zones in the region are found to the west of the Philippines, to the north of Sulawesi and forming the western and eastern margins of the Molucca Sea basin (Figure 3.2). Australia, still moving north, continues its collision with the region which began 25 million years ago. An important regional problem is whether SE Asia moves independently of Eurasia and recent geodetic data have suggested that it moves very slowly (less than 1cm/year northwards relative to Eurasia (McCaffrey 1996). Present day motion of the Philippine Sea plate is still poorly constrained but a convergence rate between the Philippine Sea plate and Eurasia of 86±2 mm/annum is indicated (McCaffrey 1996). Recent slip partitioning studies and slip vector analysis in the SE Asian region (McCaffrey 1996) indicate that the Caroline plate is presently kinematically part of the Pacific plate. For further details of the present-day tectonics of the region see Part 1 of Hall and Blundell (1996).

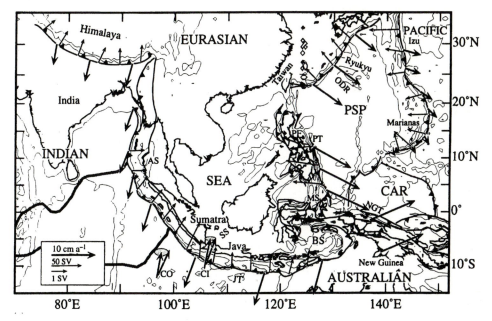

Fig. 3.2 Tectonic map of SE Asia. Thrust faults have barbs on the hanging wall. Heavy black arrows show relative plate motions at boundaries (vectors show upper plate moving relative to lower). Grey arrows with error ellipses show GPS vectors of Cocos Island (CO) and Christmas Island (Cl) relative to Java (Tregoning et al. 1994) (black arrows show expected Australia-Eurasia vector). Small arrows pointing landward of trenches show average slip vector (SV) azimuths of interplate thrust earthquakes along 400 km long segments of the forearcs (lengths of arrows are scaled to number of SV in average). Thick grey lines show outline of broad deforming region between the Indian and Australian plates. SEA, SE Asian plate; PSP, Philippine Sea plate; ODR, Oki Daito ridge; CAR, Caroline plate; JT, Java trench; BS, Banda Sea; MS, Molucca Sea; PF, Philippine fault; PT, Philippine trench; AS, Andaman Sea; SS, Sunda Strait; NGT, New Guinea trench. Bathymetry contours at 2000 m intervals. From McCaffrey (1996).

The Giant East and SE Asian Continental 'Jigsaw Puzzle'

East and SE Asia comprises a complex aggregation of pieces of continental lithosphere or terranes (Figures 3.3 and 3.4) that have progressively collided and amalgamated to one another over the last 400 million years. The boundaries of these terranes are marked by major geological discontinuities such as major strike-slip faults or by suture zones that represent the actual remains or the former sites of ancient, long-lost oceans. During the last decade multidisciplinary data, including palaeobiogeographic, palaeomagnetic, palaeoclimatic, stratigraphic, and geochemical data show that the various terranes of East and SE Asia had their origins on the Indian-Australian margin of Gondwanaland (Burrett 1973; Burrett and Stait 1985; Metcalfe 1988, 1996, 1998; Burrett *et al.* 1990). This data suggests that North China, South China, Tarim (here taken to include the Qaidam, Kunlun and Ala Shan blocks), Sibumasu (with the contiguous Lhasa and Qiangtang blocks), and Indochina formed the outer margin of northern Gondwanaland in the Early Palaeozoic and probably formed part of the India-Australian element of the ancient supercontinent Rodinia as far back as 1000 million years ago (Figure 3.5).

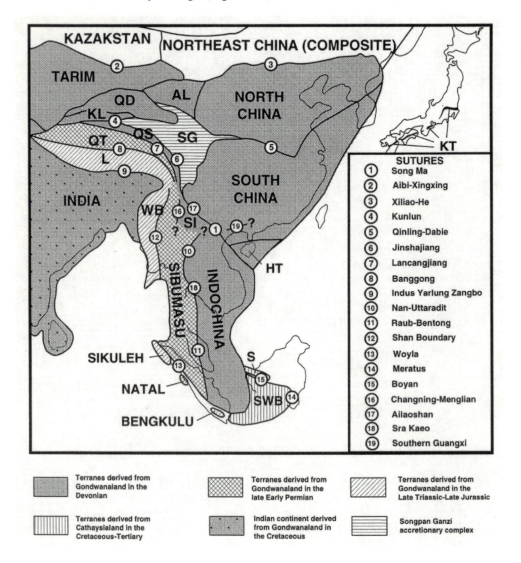

Fig. 3.3 Distribution of principal continental terranes and sutures of East and SE Asia. WB = West Burma, SWB = South West Borneo, S = Semitau Terrane, HT = Hainan Island terranes, L = Lhasa Terrane, QT = Qiangtang Terrane, QS = Qamdo-Simao Terrane, SI= Simao Terrane, SG = Songpan Ganzi accretionary complex, KL = Kunlun Terrane, QD = Qaidam Terrane, AL = Ala Shan Terrane, KT = Kurosegawa Terrane.

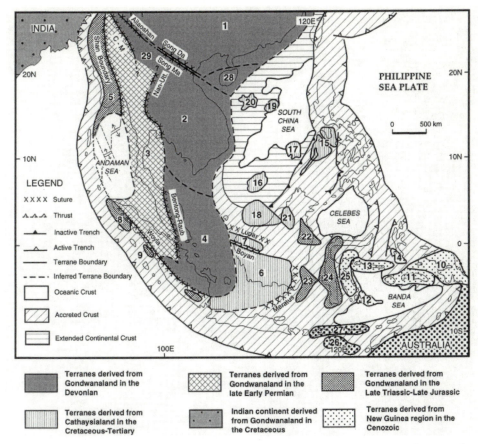

Fig. 3.4 Distribution of continental blocks and fragments (terranes) and principle sutures of SE Asia (modified after Metcalfe 1990). 1) South China, 2) Indochina, 3) Sibumasu, 4) East Malaya, 5) West Burma, 6) SW Borneo, 7) Semitau, 8) Sikuleh, 9) Natal, 10) West Irian Jaya, 11) Buru-Seram, 12) Buton, 13) Bangai-Sula, 14) Obi-Bacan, 15) North Palawan, 16) Spratley Islands-Dangerous Ground, 17) Reed Bank, 18) Luconia, 19) Macclesfield Bank, 20) Paracel Islands, 21) Kelabit-Longbowan, 22) Mangkalihat, 23) Paternoster, 24) West Sulawesi, 25) East Sulawesi, 26) Sumba, 27) Banda Allochthon, 28) Hainan Island terranes, 29) Simao terrane.

Tectonic evolution

Proterozoic (2500-545 Ma)

India, Australia, and the East Asian terranes formed an integral part of the ancient supercontinent Rodinia that was formed by collision and amalgamation of the world's continents around 1000 million years ago. Fragmentation of this ancient supercontinent occurred about 700 million years ago and former elements of that supercontinent, including Australia, India, Antarctica and elements that now constitute South Africa and South America, collided and coalesced to form Gondwanaland about 500 million years ago (Figure 3.5).

Peter Kershaw, Bruno David, Nigel Tapper, Dan Penny and Jonathan Brown (Eds): Bridging Wallace´s Line

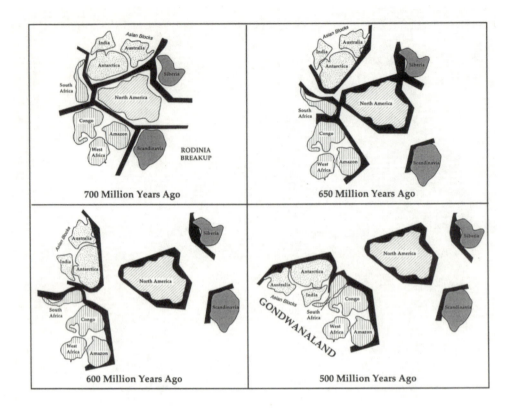

Fig. 3.5 The supercontinent Rodinia 700 million years ago and its subsequent breakup showing the formation of Gondwanaland about 500 million years ago.

Phanerozoic (545-0 Ma)

The East and SE Asian terranes represent three collages that successively rifted and separated from the India-Australian margin of Gondwanaland as three elongate continental slivers. As these three slivers separated from Gondwanaland in the Devonian, late Early Permian and Late Triassic-Late Jurassic, three ocean basins, the Palaeo-Tethys, Meso-Tethys and Ceno-Tethys were opened and subsequently closed (Metcalfe 1990, 1996, 1998; Figure 3.6). Australia remained part of eastern Gondwanaland (attached to Antarctica and India) during the Palaeozoic and early Mesozoic and finally separated from Antarctica about 45 million years ago after which it drifted northwards eventually to begin its currently ongoing collision with SE Asia. The principal events that have occurred in the SE Asian-Australasian region during the Phanerozoic are outlined below.

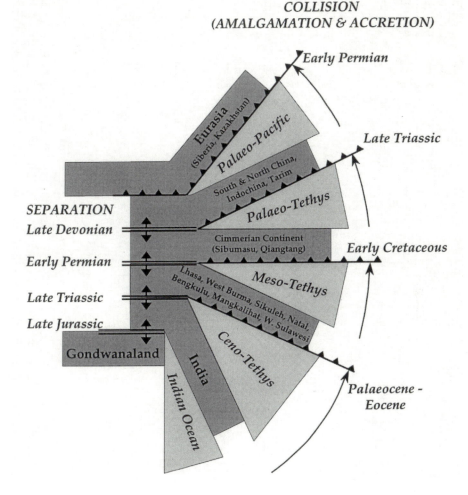

Fig. 3.6 Schematic diagram showing times of separation and subsequent collision of the three continental slivers/collages of terranes that rifted from Gondwanaland and translated northwards by the opening and closing of three successive oceans, the Palaeo-Tethys, Meso-Tethys and Ceno-Tethys.

Camrian-Ordovician-Silurian (545-410 Ma)

Australia was located in low northern/equatorial latitudes during the early Palaeozoic. The East and SE Asian terranes formed part of Himalayan-Australian 'Greater Gondwanaland' (Figures 3.7 and 3.8). Faunas of this age on the Asian blocks and Australasia define Asian-Australian 'provinces', for example the Sino-Australian brachiopod province in the Silurian of Rong *et al.* (1995) – see Figures 3.7 and 3.8.

Peter Kershaw, Bruno David, Nigel Tapper, Dan Penny and Jonathan Brown (Eds): Bridging Wallace´s Line

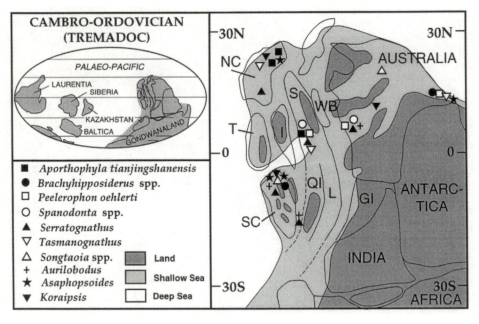

Figure 3.7

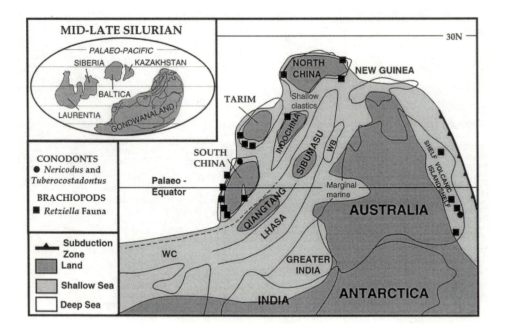

Figure 3.8

← *Fig. 3.7 Reconstruction of eastern Gondwanaland for the Cambro-Ordovician (Tremadoc) showing the postulated positions of the East and SE Asian terranes, distribution of land and sea, and shallow-marine fossils that illustrate Asia-Australia connections at this time. NC = North China SC = South China T = Tarim I = Indochina QI = Qiangtang L = Lhasa S = Sibumasu WB = West Burma GI = Greater India. Present day outlines are for reference only. Distribution of land and sea for Chinese blocks principally from Wang (1985). Land and sea distribution for Pangea/ Gondwanaland compiled from Golongka et al. (1994), Smith et al. (1994); and for Australia from Struckmeyer and Totterdell (1990).*

← *Fig. 3.8 Reconstruction of eastern Gondwanaland for the Mid-Late Silurian showing the postulated positions of the East and SE Asian terranes, distribution of land and sea, and shallow-marine fossils that appear to define an Australasian province at this time. WC = Western Cimmerian Continent WB = West Burma. Present day outlines are for reference only. Distribution of land and sea for Chinese blocks principally from Wang (1985). Land and sea distribution for Pangea/Gondwanaland compiled from Golongka et al. (1994), Smith et al. (1994); and for Australia from Struckmeyer and Totterdell (1990).*

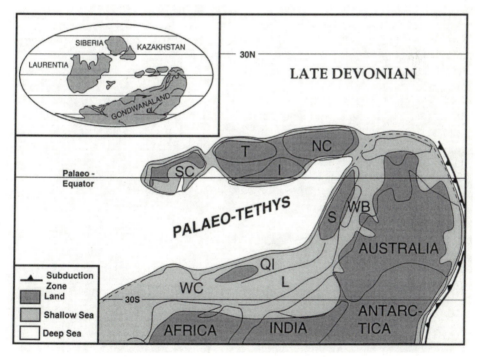

Fig. 3.9 Reconstruction of eastern Gondwanaland for the Late Devonian showing the postulated positions of the East and SE Asian terranes, distribution of land and sea, and opening of the Palaeo-Tethys ocean at this time. Present day outlines are for reference only. Distribution of land and sea for Chinese blocks principally from Wang (1985). Land and sea distribution for Pangea/Gondwanaland compiled from Golongka et al. (1994), Smith et al. (1994); and for Australia from Struckmeyer and Totterdell (1990). Symbols as for Figures 3.4 and 3.5.

Peter Kershaw, Bruno David, Nigel Tapper, Dan Penny and Jonathan Brown (Eds): Bridging Wallace's Line

Devonian (410-354 Ma)

Australia continued to reside in low southern latitudes during the Devonian but rotates counter clockwise. This counter clockwise rotation is concurrent with the clockwise rotation of the Chinese terranes as they separate from Gondwanaland as an elongate continental sliver. As this sliver searated from Gondwanaland, the Palaeo-Tethys ocean opened up (Figure 3.9). Devonian faunas on the Chinese terranes still have some Australian connections. Early Devonian endemic faunas of South China are interpreted (Metcalfe 1998) as a result of the rifting process and do not necessarily imply continental separation of South China from the other Chinese blocks and Australia at this time (Metcalfe 1998).

Carboniferous (354-298 Ma)

During the Carboniferous, Gondwanaland rotates clockwise and collides with Laurentia in the west to form Pangea. Australia, still attached to NE Gondwanaland, drifts from low southern latitudes (0 to 40°S) in Tournaisian-Visean times to high southern latitudes (50° to 75°S) in middle-late Namurian times. The major Gondwanan glaciation commenced in the Namurian and extended through into the Early Permian. There were major global shifts in both plate configurations and climate during this time and a change from warm to cold conditions in Australasia. This is reflected in the change from high to low diversity of faunas in Australasia (Jones *et al.* 2000) and especially eastern Australia where low diversity, endemic faunas developed in the Upper Carboniferous (Figure 3.10). North and South China, Indochina and Tarim faunas and floras are tropical/subtropical Cathaysian/Tethyan types during the Carboniferous and show no Gondwanaland affinities. These terranes had already separated from Gondwanaland and were located in low lattitude/equatorial positions during the Carboniferous (Figure 3.10). Indochina and South China collided and amalgamated within the Tethys during the early Carboniferous along the Song Ma suture zone now located in Laos and Vietnam. Ice sheets and glaciers extended across much of eastern Gondwanaland during the Late Carboniferous and ice reached the marine environment of the Indian-Australian continental shelf of Gondwanaland and glacial-marine sediments (diamictites; pebbly mudstones interbedded with normal marine shales and sands) were deposited on the continental shelf of eastern Gondwanaland.

Early Permian (298-270 Ma)

During the Permian, Australia remained in high southern latitudes. Glacial ice continued to reach the marine environment of the NE Gondwanaland margin and glacial-marine sediments continued to be deposited on the Sibumasu, Qiangtang and Lhasa terranes. Gonwanaland cold-climate faunas and floras characterised the Sibumasu, Qiangtang and Lhasa terranes at this time. In addition, the distinctive cool-water tolerant conodont genus *Vjalovognathus* defines an eastern peri-Gondwanaland cold-water province at this time (Figure 3.10).

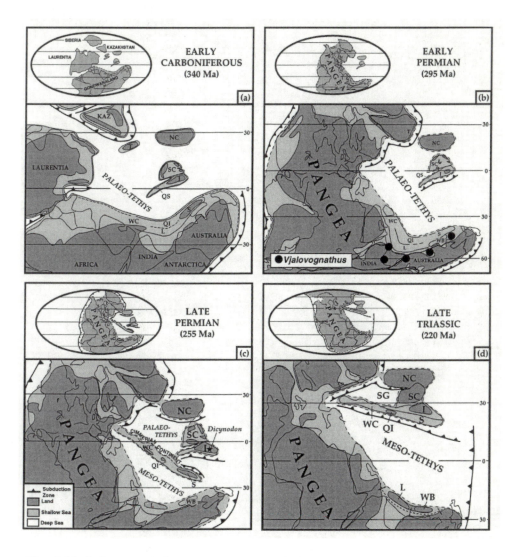

Fig. 3.10 Palaeogeographic reconstructions of the Tethyan region for (a) Early Carboniferous, (b) Early Permian, (c) Late Permian and (d) Late Triassic showing relative positions of the East and SE Asian terranes and distribution of land and sea. The distribution of the Lower Permian cold-water tolerant conodont genus Vjalovognathus, and the location of the Late Permian Dicynodon from Laos are also shown. Present day outlines are for reference only. Distribution of land and sea for Chinese blocks principally from Wang (1985). Land and sea distribution for Pangea/Gondwanaland compiled from Golongka et al. (1994), Smith et al. (1994); and for Australia from Struckmeyer & Totterdell (1990). Symbols as for Figures 3.4 and 3.5.

Late Permian (270-253 Ma)

During the late Early Permian, the Cimmerian continental sliver begins to rift away from the northeastern margin of Gondwanaland and by early Late Permian the Sibumasu and Qiangtang terranes had separated from Gondwanaland and the Meso-Tethys opened between this continental sliver and Gondwanaland. The Palaeo-Tethys continues to be destroyed by northwards subduction beneath Laurasia, North China and the amalgamated South China and Indochina terranes. Following separation, and during their northwards drift, the Sibumasu and Qiangtang terranes develop initially a Cimmerian Province fauna and then are absorbed into the Cathaysian Province. North and South China begin to collide during the Late Permian and a connection between mainland Pangea and Indochina, via South and North China or via the western Cimmerian continent, is indicated by the occurrence of the genus *Dicynodon* in the Upper Permian of Indochina (Figure 3.10).

Triassic (252-205 Ma)

Australia was in low to moderate southern latitudes during the Triassic. The Sibumasu and Qiangtang terranes collided and sutured to the Indochina/South China amalgamated terrane. The North and South China collision was nearly complete with exhumation of ultra-high pressure metamorphics along the Qinling-Dabie suture zone. Also, sediment derived from the North-South China collisional orogen poured into the Songpan Ganzi accretionary complex basin producing huge thicknesses of flysch turbidites.

Jurassic (205-141 Ma)

Australia remained in low to moderate southern latitudes in the Jurassic. Rifting and separation of the Lhasa, West Burma, Sikuleh, Natal, Mangkalihat and West Sulawesi terranes from NW Australia occurred progressively during the Late Triassic to Late Jurassic. The Ceno-Tethys ocean basin opened behind these terranes as they separated from Gondwanaland (Figure 3.11). Final welding of North China to Eurasia also took place in the Jurassic and the initial pre-breakup rifting of the main Gondwanaland supercontinent began at this time.

Cretaceous (141-65 Ma)

Lhasa block collides and sutures to Eurasia in earliest Cretaceous times. Gondwanaland breaks up and India drifts north, making initial contact with Eurasia at the end of the Cretaceous. This early contact between India and Eurasia is indicated by palaeomagnetic data from the Ninetyeast Ridge (Klootwijk *et al.* 1992) and also by Late Cretaceous frogs being able to hop from Eurasia on to India. The small West Burma, Sikuleh and Natal terranes also accrete to Sibumasu during the Cretaceous. Australia begins to separate from Antarctica and drift northwards but a connection with Antarctica via Tasmania still remained.

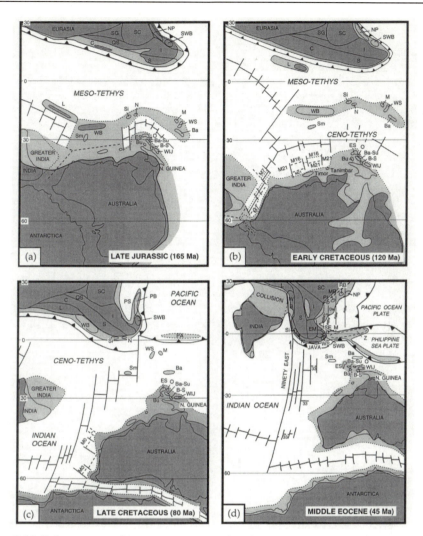

Fig. 3.11 Palaeogeographic reconstructions for the East Asia-Australia region in (a) Late Jurassic, (b) Early Cretaceous, (c) Late Cretaceous and (d) Middle Eocene times showing distribution of land and sea. SG = Songpan Ganzi accretionary complex SWB = South West Borneo (includes Semitau) NP = North Palawan and other small continental fragments now forming part of the Philippines basement Si = Sikuleh N = Natal M = Mangkalihat WS = West Sulawesi Ba = Banda Allochthon ES = East Sulawesi O = Obi-Bacan Ba-Su = Bangai-Sula Bu = Buton B-S = Buru-Seram WIJ = West Irian Jaya Sm = Sumba PA = Insipient Philippine Arc. M numbers represent Indian Ocean magnetic anomalies. Other terrane symbols as in Figures 1.4 and 1.5. Modified from Metcalfe (1990, 1996, 1998) and partly after Smith et al. (1981), Audley-Charles (1988) and Audley-Charles et al. (1988). Present day outlines are for reference only. Distribution of land and sea for Chinese blocks principally from Wang (1985). Land and sea distribution for Pangea/Gondwanaland compiled from Golongka et al. (1994), Smith et al. (1994); and for Australia from Struckmeyer and Totterdell (1990).

Cenozoic (65-0 Ma)

Cenozoic evolution of East and SE Asia involved substantial movements along, and rotations of strike-slip faults, rotations of continental blocks and oceanic plates, and the development and spreading of 'marginal' seas. This evolution was essentially due to the combined effect of the interaction of the Eurasian, Pacific and Indo-Australian plates and the collision of India with Eurasia.

Various tectonic models have been proposed for the Cenozoic evolution of the region which utilise different mechanisms for the India-Eurasia collision and also different interpretations on rotations of continental blocks and oceanic plates. Three basic mechanisms have been proposed for the northwards motion of India into Eurasia. The first model involves underthrusting of greater India beneath Eurasia (Argand 1924; Powell and Conaghan 1975; Zhao and Morgan 1987); the second involves crustal shortening and thickening (Dewey and Burke 1973; England and McKenzie 1982; Dewey *et al.* 1988); and the third involves eastwards lateral extrusion (Molnar and Tapponier 1975; Tapponier *et al.* 1986).

Cenozoic clockwise rotations of crustal blocks in mainland SE Asia are observed in Indochina and western Thailand and major progressive anticlockwise rotations are seen in Borneo and the Malay Peninsula (Schmidtke *et al.* 1985; Schmidtke *et al.* 1990; Metcalfe 1993; Richter and Fuller 1996; Richter *et al.* 1999). These anticlockwise rotations observed in Borneo and Malaya are at variance with the extrusion model. Other models for the region, including those of Daley *et al.* (1991) and Lee and Lawver (1994), neglect the clockwise rotation of the Philippine Sea plate. The model by Rangin *et al.* (1990) did include the clockwise rotation of the Philippine Sea plate but did not take into account the counter-clockwise rotation of Borneo. The reconstruction model proposed by Hall (1996) and illustrated in Figure 13, which takes into account both the clockwise rotation of the Philippine Sea plate and the anticlockwise rotation of Borneo and Peninsular Malaysia, is here preferred. The various processes that have affected the region and their ages are given in Table 3.1. For recent overviews on the Cenozoic evolution of the region, see Packham (1996) and Hall (1996, 1998).

India travelled rapidly (10 cm/annum) NE relative to Eurasia during the Late Cretaceous and Early Eocene (between 84 and 45 Ma) and began its collision with Eurasia between 60 and 50 million years ago (Klootwijk *et al.* 1992; Rowley 1996). From about 45 million years ago it pursued a northwards motion relative to Eurasia at a slower (5cm/annum) rate (Dewey *et al.* 1989). Australia also moved northwards at this time but at a much more sedate pace, being separated from India by major transform faults (Figure 3.12).

Table 3.1. Cenozoic evolution.

PROCESS	AGE
Collision of India with Eurasia.	Initial contact around 60 Ma, major indentation from around 45 Ma onwards.
Final separation of Australia from Antarctica and establishment of the circum-Antarctic ocean current.	45 Ma Late Eocene.
Northwards drift of the isolated Australian continent over 27 degrees of latitude. Gradual drying of the continent and evolution of the distinctive Australian fauna and flora from Gondwanaland ancestry.	45-0 Ma Late Eocene to present.
Clockwise rotation and extrusion of Indochina and parts of northern Sibumasu.	During the early Cenozoic but precise age not known.
Major strike-slip faulting.	30-15 Ma Middle Oligocene to Middle Miocene
Anti-clockwise rotation of Borneo and the Malay Peninsula.	Progressively during the Cenozoic but mostly during the Miocene.
Initial spreading of the South China Sea basin	30 Ma Middle Oligocene.
Proto-South China Sea destroyed by southwards subduction.	40-15 Ma Middle Eocene to Middle Miocene.
Collision of the Australian continent with the Philippine Sea plate (initiating clockwise rotation of Philippine Sea plate and causing structural inversion in many Cenozoic basins of the region).	25 Ma Late Oligocene.
Clockwise rotation of the Philippine Sea plate.	From about 20 Ma onwards.
Opening of the Sulu Sea.	20 Ma Early Miocene.
Molucca Sea double subduction established.	10 Ma Late Miocene.
Collision of the Philippine arc and Eurasian continental margin in Taiwan.	5 Ma End Miocene.

During the last 50 million years of SE Asian evolution, there have been two major episodes of change in configuration of the region and character of plate boundaries, caused principally by arc-continent collisions around 25 and 5 million years ago. About 25 million years ago (Figure 3.13), the Australian continent collided with the Philippine Sea plate volcanic arc and caused major effects that propagated far to the west. This collision resulted in the initiation of the clockwise rotation of the Philippine Sea plate, the counter-clockwise rotation of Borneo, and initiation of the 'bacon-slicer' Sorong fault system which led to the westwards movement of continental fragments including East Sulawesi, Buton, Bangai-Sula, and Obi-Bacan to the west where they now form part of

Peter Kershaw, Bruno David, Nigel Tapper, Dan Penny and Jonathan Brown (Eds): Bridging Wallace´s Line

eastern Indonesia (Hall 1996). About five million years ago, the Philippine arc collided with the Eurasian continental margin in Taiwan changing the motion of the Philippine Sea plate and initiating new subduction systems on the east and west sides of the Philippines.

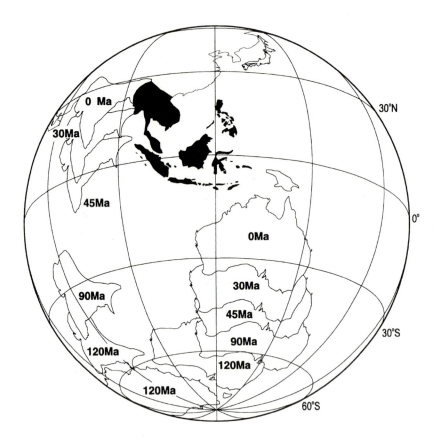

Fig. 3.12 Diagram showing the nothwards drift of India and Australia during the last 120 million years. SE Asia is shown in its present position for reference (from Hall 1998).

The complex plate movements, block rotations, orogenic activity and changes in sea level have resulted in major changes in the distribution of land, sea and changes in relief (montane versus lowland) in the region. This has had a marked effect on the distribution, dispersal and evolution of fauna and flora in the region and on the climate. The arrival of Australia and its collision with SE Asia has also had a profound effect on the faunas and floras of the SE Asian-Australian transition zone and to a large extent, the extant distribution of organisms in SE Asia-Australia directly reflect the geological and tectonic history of the region.

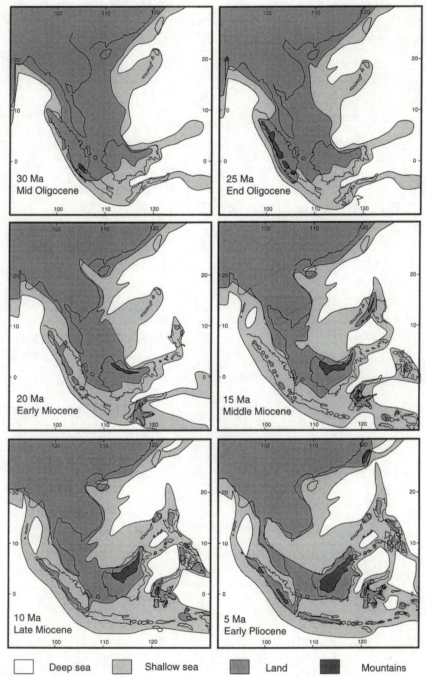

Fig. 3.13 Reconstructions of East and SE Asia for the last 30 million years showing the variation in distributions of deep sea, shallow sea, land and mountains. After Hall (1998).

Peter Kershaw, Bruno David, Nigel Tapper, Dan Penny and Jonathan Brown (Eds): Bridging Wallace´s Line

Observed endemism of recent and Quaternary faunas on specific islands in the region (e.g. Sulawesi, Flores), the demarkation of extant faunas and floras by the Wallace Line have their explanation in the continued changes in the geological configuration, climate, sea-level changes and land-sea distributions in the Quaternary. The colonisation of the Australian continent by humans was restricted by the lack of Asia-Australia land connections and by the width of marine barriers between mainland Asia, SE Asian islands and Australia. A major current debate centres on whether *Homo erectus* required the intellectual development for language and seamanship to cross from Asia to Flores by 800,000 years ago or was the dispersal of early humans affected by natural rafting or some other means? A greater understanding of the geological evolution, especially changes in land-sea distributions and changes in the widths of marine barriers, is fundamental to understanding the migrations of humans to Australia.

References

Argand, E. 1924. La tectonique de l'Asie. *Proceedings of the 13th International. Geological Congress, Brussels*: 171-372.

Audley-Charles, M.G. 1988. Evolution of the Southern Margin of Tethys (North Australian Region) from Early Permian To Late Cretaceous. In: M.G. Audley-Charles, and A. Hallam (Editors) *Gondwana and Tethys*, Geological Society Special Publication No. 37, Oxford University Press, Oxford: 79-100.

Audley-Charles, M.G., Ballantine, P.D. and Hall, R. 1988. Mesozoic-Cenozoic rift-drift sequence of Asian fragments from Gondwanaland. *Tectonophysics*, 155: 317-330.

Burrett, C. 1973. Ordovician biogeography and continental drift. *Palaeogeography, Palaeoclimatology, Palaeoecology*, 13: 161-201.

Burrett, C. and Stait, B. 1985. South-East Asia as part of an Ordovician Gondwanaland - a palaeobiogeographic test of a tectonic hypothesis. *Earth and Planetary Science Letters*, 75: 184-190.

Burrett, C., Long, J. and Stait, B. 1990. Early-Middle Palaeozoic Biogeography of Asian Terranes Derived from Gondwana. In: W.S. McKerrow and C.R. Scotese (Editors) *Palaeozoic Palaeogeography and Biogeography*. Geological Society Memoir No. 12: 163-174.

Daley, M.C., Cooper, M.A., Wilson, I, Smith, D.G. and Hooper, B.G.D. 1991. Cenozoic plate tectonics and basin evolution in Indonesia. *Marine and Petroleum Geology*, 8: 2-21.

Dewey, J.F. and Burke, K.C.A. 1973. Tibetan, Variscan and Precambrian basement reactivation: products of a continental collision. *Journal of Geology,* 81: 683-692.

Dewey, J.F., Cande, S. and Pitman, W.C. 1989. Tectonic evolution of the India/Eurasia Collision Zone. *Eclogae Geologicae Helvetiae.* 82: 717-734.

Dewey, J.F., Shackleton, R.M., Chang Chengfa, and Sun Yiyin. 1988. The tectonic evolution of the Tibetan Plateau. *Philosophical Transactions of the Royal Society of London*, A327: 379-413.

England, P.C. and McKenzie D.P. 1982. A thin viscous sheet model for continental de-formation. *The Geophysical Journal of the Royal Astronomical Society.* 70: 295-321.

Golongka, J., Ross, M.I. and Scotese, C.R. 1994. Phanerozoic Paleogeographic and Paleoclimatic Modeling Maps. In: A.F. Embry, B. Beauchamp and D.J. Glass (Editors) *Pangea: Global Environments and Resources*, Canadian Society of Petroleum Geologists Memoir No. 17: 1-47.

Hall, R. 1996. Reconstructing Cenozoic SE Asia. In: R. Hall and D. Blundell (Editors) *Tectonic Evolution of Southeast Asia*. Geological Society Special Publication No. 106: 153-184.

Hall, R. 1998. The Plate Tectonics of Cenozoic SE Asia and the Distribution of Land and Sea. In: R. Hall and J.D. Holloway (Editors) *Biogeography and Geological Evolution of Southeast Asia*, Backhuys Publishers, Amsterdam: 99-131.

Hall, R. and Blundell, D. (Editors) 1996. *Tectonic Evolution of Southeast Asia*. Geological Society Special Publication No. 106: 566 pp.

Jones, P.J., Metcalfe, I., Engel, B.A., Playford, G., Rigby, J., Roberts, J., Turner, S., and Webb, G.E. 2000. Carboniferous palaeobiogeography of Australasia. Memoir of the Association of Australasian Palaeontologists, 23: 259-286.

Klootwijk, C.T., Gee, J.S., Peirce, J.W., Smith, G.M. and McFadden, P.L. 1992. An early India-Asia contact: Paleomagnetic constraints from Ninetyeast Ridge, ODP Leg 121. *Geology*, 20: 395-398.

Lee, T.Y. and Lawver, L.A. 1994. Cenozoic plate reconstruction of Southeast Asia. *Tectonophysics*, 251: 85-138.

McCaffrey, R. 1996. Slip Partitioning at Convergent Plate Boundaries in SE Asia. In: R. Hall and D. Blundell (Editors) *Tectonic Evolution of Southeast Asia*. Geological Society Special Publication No. 106: 3-18.

Metcalfe, I. 1988. Origin and Assembly of Southeast Asian Continental Terranes. In: M.G. Audley-Charles and A. Hallam (Editors) *Gondwana and Tethys*. Geological Society of London Special Publication No. 37: 101-118.

Metcalfe, I. 1990. Allochthonous terrane processes in Southeast Asia. *Philosophical Transactions of the Royal Society of London,* A331: 625-640.

Metcalfe, I. 1993. Southeast Asian Terranes: Gondwanaland Origins and Evolution. In: R.H. Findlay, R. Unrug, M.R. Banks and J.J. Veevers (Editors) *Gondwana 8 - Assembly, Evolution, and Dispersal* (Proceedings of the 8th Gondwana Symposium, Hobart, 1991). A.A. Balkema, Rotterdam: 181-200.

Metcalfe, I. 1996. Pre-Cretaceous Evolution of SE Asian Terranes. In: R. Hall and D. Blundell (Editors) *Tectonic Evolution of Southeast Asia*. Geological Society Special Publication No. 106: 97-122.

Metcalfe, I. 1998. Palaeozoic and Mesozoic Geological Evolution of the SE Asian Region: Multidisciplinary Constraints and Implications for Biogeography. In: R. Hall and J.D. Holloway (Editors) *Biogeography and Geological Evolution of Southeast Asia*. Backhuys Publishers, Amsterdam: 25-41.

Molnar, P. and Tapponnier, P. 1975. Cenozoic tectonics of Asia: effects of a continental collision. *Science*, 189: 419-426.

Moss, S.J. and Wilson, M.E.J. 1998. Biogeographic Implications of the Tertiary Palaeogeographic evolution of Sulawesi and Borneo. In: R. Hall and J.D. Holloway (Editors) *Biogeography and Geological Evolution of Southeast Asia*. Backhuys Publishers, Amsterdam: 133-163.

Packham, G. 1996. Cenozoic SE Asia: Reconstructing its Aggregation and Reorganization. In: R. Hall and D. Blundell (Editors) *Tectonic Evolution of Southeast Asia*. Geological Society Special Publication No. 106: 123-152.

Powell, C.M. and Conaghan, P.J. 1975. Tectonic models of the Tibetan Plateau. *Geology*, 3: 727-731.

Rangin, C., Jolivet, L., Pubellier, M. and Group, T.P.W. 1990. A simple model for the tectonic evolution of Southeast Asia and Indonesian region for the past 43 m.y. *Bulletin de la Société géologique de France*, 8: 889-905.

Richter, B. and Fuller, M. 1996. Palaeomagnetism of the Sibumasu and Indochina blocks: implications for the extrusion tectonic model. In: R. Hall and D. Blundell (Editors) *Tectonic Evolution of Southeast Asia*. Geological Society Special

Publication No. 106: 203-224.

Richter, B., Schmidtke, E., Fuller, M., Harbury, N. and Samsudin, A.R. 1999. Paleo-
magnetism of Peninsular Malaysia. *Journal of Asian Earth Sciences*, 17: 477-519.

Rong J-Y., Boucot, A.J., Su Y-Z. and Strusz, D.L. 1995. Biogeographical analysis of
Late Silurian brachiopod faunas, chiefly from Asia and Australia. *Lethaia*, 28(1): 39-
60.

Rowley, B.B. 1996. Age of initiation of collision between India and Asia: A review of
stratigraphic data. *Earth and Planetary Science Letters*, 145: 1-13.

Schmidtke, E., Dunn, J.R., Fuller, M.D. and Haston, R. 1985. Preliminary paleomagnetic
results from Sabah, E.Malaysia. *EOS*, 66: 864.

Schmidtke, E.A., Fuller, M.D. and Haston, R.B. 1990. Paleomagnetic data from
Sarawak, Malaysia (Borneo) and the Late Mesozoic and Cenozoic tectonics of
Sundaland. *Tectonics*, 9: 123-140.

Smith, A.G., Hurley, A.M. and Briden, J.C. 1981. *Phanerozoic Palaeocontinental World
Maps*. Cambridge University Press, Cambridge: 102pp.

Smith, A.G, Smith, D.G. and Funnell, B.M. 1994. *Atlas of Mesozoic and Cenozoic
Coastlines*. Cambridge University Press, Cambridge: 99pp.

Struckmeyer, H.I.M. and Totterdell, J.M. (Coordinators) and BMR Palaeogeographic
Group. 1990. *Australia: Evolution of a Continent*. Bureau of Mineral Resources,
Australia: 97pp.

Tapponnier, P., Peltzer, G. and Armijo, R. 1986. On the mechanics of the collision
between India and Asia. In: M.P. Coward and A.C. Ries (Editors) *Collision
Tectonics*. Geological Society of London Special Publication No. 19: 185-201.

Tregoning, P., Brunner, F.K., Bock, Y., Puntodewo, S.S.O., McCaffrey, R. 1994. First
geodetic measurement of convergence across the Java Trench. *Geophysical Research
Letters*, 21: 2135-2138.

Wang, H. 1985. *Atlas of the Palaeogeography of China*. Cartographic Publishing House,
Beijing: 27pp.

Zhao, W. and Morgan, W.J. 1987. Injection of Indian crust into Tibetan lower crust: a
two-dimentional finite element model study. *Tectonics* 6: 489-504.

Tertiary Vegetational History of Southeast Asia, with Emphasis on the Biogeographical Relationships with Australia

Robert J. Morley

Introduction

According to the classical view of biogeography in the Australasian region, the megathermal taxa, which characterise the tropical rainforests of NE Australia, dispersed from tropical latitudes during the Tertiary. This hypothesis now requires re-examination, as extensive palaeobotanical and palynological studies from Australia conclusively demonstrate that Australian rainforests had a very long geological history as summarised in Hill (1994). Other misconceptions with respect to the biogeographical relationships of Australasia and SE Asia have come about following the widely accepted hypothesis of Takhtajan (1969), that angiosperms evolved in the tropical Far East, 'somewhere between Assam and Fiji based on the concentration of primitive angiosperm taxa in that area'. Current views however, based on the fossil record, suggest that Western Gondwanaland was a much more likely area of initial radiation (Hickey and Doyle 1977), and that the primitive taxa used by Takhtajan (1969) to infer that the tropical Far East was the 'cradle' of the angiosperms actually dispersed into their present areas of distribution well after the time of their initial radiation (Morley 2000, 2001). Takhtajan (1987) tried to resurrect his theory by suggesting that angiosperms originated on an isolated Gondwanan microcontinent, which subsequently became embedded in SE Asia, but this was dismissed by Truswell *et al.* (1987) on the grounds that the earliest angiosperm records for Australia post-date those from western Gondwana by 10 million years.

In order to more clearly understand the past interactions of the Australian and SE Asian floras, a detailed fossil database for the SE Asian region is required. Such a database is available in the form of unpublished palynological records produced for petroleum exploration companies to help resolve stratigraphic problems. This database has several drawbacks compared to carefully documented palaeoecological studies since only identifications necessary to resolve stratigraphic problems are usually recorded. Nevertheless, the database is sufficient to differentiate some of the main events which have determined the character of the present day rain forests in both regions, and this has recently been used by Morley (2000) to outline the history of the SE Asian flora over the last 100 million years. This paper presents a summary of the Tertiary history of the flora and vegetation of SE Asia based heavily on that review. It also

ISBN 3-923381-47-6
© 2002 by CATENA VERLAG, 35447 Reiskirchen

discusses the timing and nature of plant dispersals between SE Asia and Australasia during the Tertiary and Late Cretaceous, as far as can be deduced from the palynological record, and also the manner in which the East Malesian flora, which is rich in elements of Asian affinity, became established to the east of the biogeographical divide of Wallace's Line. The ages of the epochs of the Tertiary, and stages of the Cretaceous, follow those described in Harland *et al.* (1990).

The importance of the Indian Plate as a dispersal vector from Africa

The history of the Indian Plate has been particularly crucial in determining the character of the SE Asian flora, and also the nature of biogeographic distributions within the palaeotropical province. The Indian Plate separated from Gondwanaland during the Aptian (125-112 Ma), and drifted slowly northward, so that by the Cenomanian (97-91 Ma), it lay adjacent to Madagascar. At this time there must have been a clear dispersal path for plants from the African Plate to Madagascar and the Indian Plate, for several West Gondwanan pollen types, such as *Afropollis* (Winteraceae) and the *Constantisporis* group (family indeterminate), show similar distribution patterns in West Africa and India during the Late Cretaceous. The presence of the West Gondwanan elaterospore *Elateropollenites africaensis* (Gnetaceae) in the Turonian (91-89 Ma) of New Guinea raises the possibility of a dispersal path via the Indian Plate to Australasia at this time, possibly via an island arc along the leading edge of the Indian Plate (Morley 1998). This path might also have been followed by early angiosperms. Before the Indian Plate finally separated from Madagascar during the mid Maastrichtian (about 70 Ma), many typically African taxa were established there, including members of Sapindaceae and Myrtaceae, *Ctenolophon* (Ctenolophonaceae) and several Palmae. As the Indian Plate drifted toward the Asian Plate during the Late Maastrichtian (70-65 Ma) and Paleocene (65-60 Ma), it bore a flora with three distinct elements: an autochthonous element, with gymnosperms, carried from Antarctica, an allochthonous element, which dispersed from Africa via Madagascar, and also a significant endemic element, which became established as India drifted across the different climatic belts of the equatorial zone.

An important feature of Palaeocene (and late Cretaceous) tropical vegetation was the abundance of Palmae in all three tropical regions, indicated by the common occurance of palm pollen. The wide representation of palms at this time might suggest that some early palm dispersals may have occurred independently of continental drift, especially with respect to those taxa that grew along coasts, such as the mangrove palm *Nypa*, and its probable relatives that sourced pollen referred to the form-genus *Proxapertites*. However, other palms show regional geographical variation, which suggest that the Late Cretaceous dispersal paths are as yet poorly understood. For instance, whereas iratoid and mauritioid palms appear to have been restricted to African and South America, the presence of calamoid pollen from the Maastrichtian of the Horn of Africa and also in SE Asia suggests some dispersal between Africa and Asia prior to the Indian collision. The same path may also have been followed by some Myrtaceae.

The character of the SE Asian Tertiary flora prior to the collision of India with the Asian Plate is best illustrated by reference to palynomorph assemblages recorded from the Kayan (formerly Plateau Sandstone) Formation in Sarawak (Muller 1968), which is now considered to be essentially Paleocene (65-57 Ma) to Early Eocene (57-50 Ma) in age, rather than including a significant representation of Cretaceous sediments (Morley 1998, 2000). Palynomorph assemblages are dominated by palm pollen, referable to *Nypa, Proxapertites* and Calamoidae. Other important taxa were *Celtis* (Ulmaceae), Sapotaceae, *Ilex* (Aquifoliaceae). Myrtaceae and Apocynaceae, in addition to Laurasian gymnosperms and Ephedraceae, emphasising the essentially Asian affinity of this flora. There were also a significant number of pollen types which cannot be referred to extant taxa.

For comparison with areas to the east, a pollen record is also available for about the same time interval from West Papua, from a lateral equivalent of the Early Eocene Waripi Formation (Morley 1998), and this has yielded assemblages of essentially Australian aspect, with common *Casuarina*, as well as diverse Proteaceae, Myrtaceae, *Dissiliaria* or *Austrubuxus* (Euphorbiaceae) and rare *Nothofagus*, together with a tropical shoreline element with *Nypa*. Evidence therefore suggests that during the earliest Tertiary, the floras of SE Asia and Australia were widely separated, a view also supported by geological data (Hall 1995, 1996). The palynomorph assemblages from both the Paleocene to Early Eocene of Sarawak, and of West Papua differed markedly from those reported from the Indian Plate prior to its collision with Asia (Morley 2000).

Middle and Late Eocene, subsequent to the collision of the Indian and Asian plates

At the time of the collision of the Indian Plate with Asia, during the Middle Eocene (50-39 Ma), both India and SE Asia lay at similar latitudes and within the same moist climatic belt (Figure 4.1), as evidenced by the widespread occurrence of coals within the sedimentary record of both areas. At this time, SE Asian palynofloras show a sudden appearance of pollen of taxa which were characteristic of the Paleocene and Early Eocene of India, or were of Gondwanan affinity (Figure 4.2) such as *Durio* (Bombacaceae), *Gonystylus* (Gonystylaceae), Iguanuroid palms, aff *Beauprea* (Proteaceae), Restionaceae and *Mischocarpus* type (Sapindaceae). At the same time, most of those taxa characteristic of the Kayan Formation which cannot be referred to modern taxa disappear from the SE Asian record. This is thought to reflect the establishment of a moist corridor between India and SE Asia, with the dispersal of a somewhat aggressive Indian flora into SE Asia and the extinction of many local elements. Middle and Late Eocene (39-35 Ma) palynofloras from Java are very diverse, suggesting that at this time, under the influence of a warm and moist climate, the SE Asian flora had become quite species-rich.

Dipterocarpaceae are also likely to have dispersed to SE Asia from Africa via the Indian Plate, since they currently have representatives in both Africa and South America, and fossil woods referable to the SE Asian subfamily Diperocarpoidae are reported from the Tertiary of East Africa. In SE Asia, their earliest fossil record is from the Middle Eocene of Myanmar (Curiale *et al.* 1994) in the form of the geochemical biomarker bicadenane (thought to be derived from diperocarp resins; van Aarssen *et al.* 1990) and is consistent with dispersal via India.

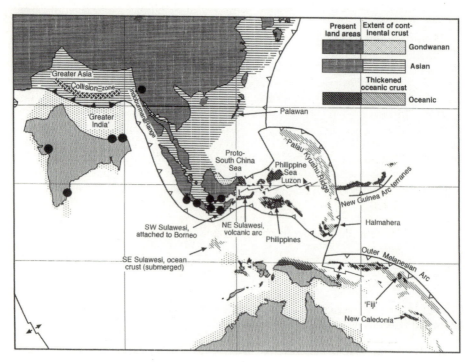

Fig. 4.1 Palaeogeography of the SE Asian and Australian region about 40 million years ago, (reconstruction based on Hall 1998), with geographical distribution of Iguanurinoid palm pollen (Palmaepollenites kutchensis) shown by filled circles.

Oligocene and earliest Miocene

The terminal Eocene cooling event had a major impact on the flora of the SE Asian region and resulted in many extinctions, for instance of iguanuroid palms, the parent plants of *Proxapertites*, and Restionaceae. Oligocene (35-25 Ma) floras were species-poor, and the climate much drier than in the Eocene, with most of the SE Asian region bearing open, seasonal vegetation, sometimes with grasses. Extensive moist rainforests were present only in Assam and Myanmar (Figure 4.3), the only region of South and SE Asia where Oligocene coal deposits are widespread. At this time, the SE Asian region was characterised by extensive freshwater lakes, bordered by herbaceous marshes and swamp forests. The latter were often dominated by *Barringtonia* (Lecythidaceae) and ancestral *Lagerstroemia* (Lythraceae). Parallels to such a setting occur in the seasonal swamp forests bordering the Tonle Sap today in Kampuchea.

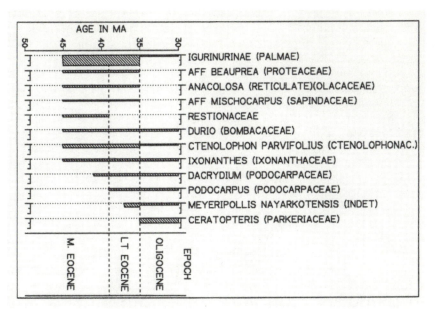

Fig. 4.2 Ranges of taxa appearing in the Eocene of Java either with Gondwanan affinities, or from the Early Eocene and Paleocene of India.

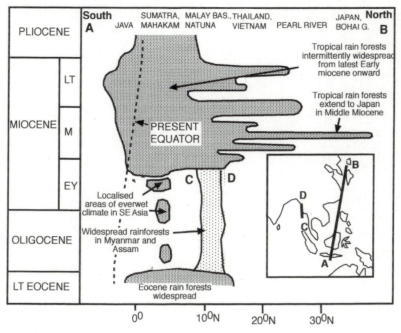

Fig. 4.3 Schematic, and simplified distribution of tropical rainforest climates in SE and East Asia during the Tertiary. Position of palaeoequator according to Smith et al. (1994).

Peter Kershaw, Bruno David, Nigel Tapper, Dan Penny and Jonathan Brown (Eds): Bridging Wallace´s Line

The Late Oligocene (30-25 Ma) and earliest Miocene (25-21 Ma) periods contain common dipterocarp pollen, comparable to that of *Shorea* or *Hopea*, often associated with common grasses, suggesting that at this time, dipterocarps were common in SE Asia, but were components of open, seasonal forests rather than closed rainforests.

Temperate angiosperm and Laurasian conifer pollen also forms a characteristic feature of the SE Asian Oligocene and earliest Miocene. This element includes common *Pinus*, together with pollen of *Picea*, *Abies*, *Tsuga* and *Alnus* (Betulaceae). Muller (1966, 1972) suggested that these were derived from montane forests, and that during the Oligocene and Early Miocene (25-16 Ma), high mountain ranges occurred, now denuded, which would have provided the source. This suggestion is, however, in conflict with geological data, which suggests that the main period of mountain building and uplift occurred subsequent to the decline of the Laurasian montane pollen element (Hall 1998). A solely climatic explanation is therefore much more appropriate. *Pinus* pollen is thought chiefly to be derived from moisture deficient, lowland sources (Morley 1991), with settings analogous to the *Pinus merkusii* forests in Thailand today. The remaining types, however, which display belts of increasing abundance and diversity to the north (Morley 1998, 2000), are best explained in terms of the southern expansion of frost-tolerant temperate conifer belt at the expense of lowland and lower montane tropical vegetation. Such an extension of frost-tolerant vegetation into tropical latitudes suggests that Oligocene and earliest Miocene climates were substantially cooler than either the Late Eocene, or later Early Miocene, and also cooler than the present day climate.

Later Early Miocene to Early Pliocene

After about 21 million years ago, the climate of SE Asia changed substantially. Peat swamps became widespread, after being virtually absent during the Oligocene. The climate also became much warmer, with tropical rainforests not only becoming widespread over most of the SE Asian region, and replacing the low diversity seasonal vegetation of the Oligocene, but also expanding northward into East Asia. The warmest climates were experienced during the earliest Middle Miocene, about 16 million years ago, at which time diverse mangroves with some other tropical rainforest elements occurred as far north as Hokkaido in Japan. A predominance of moist, warm climates were the rule for most of the remainder of the Miocene and Early Pliocene (5-3.5 Ma), although phases of drier, and cooler climates occurred intermittently (Figure 4.3). It was during this period that the flora of SE Asia began to take on its present form, gradually increasing in species-richness through time (Morley 2000), and with Dipterocarpaceae forming widespread components within moist rainforests.

Late Pliocene and Quaternary

The Late Pliocene (3.5-1.6 Ma) and Quaternary (1.6-0 Ma) period of global cooling, coinciding with periods of ice expansion in polar regions, again brought drier climates to the SE Asian region. This is clearly illustrated by the occurrence of mid-Pleistocene pine savanna in the area of Kuala Lumpur (Morley and Flenley 1987; Morley 1998), and of alternating phases of grassland and forest expansion in the region of Nusa Tenggara (van der Kaars 1991a). The development of seasonal climates in the region of Java and Nusa Tenggara, however, is also thought to be due to the northward drift of the Australian Plate into the southern hemisphere low pressure zone, since grass-dominated pollen assemblages become widespread at approximately the same time in both Java and Australia, at the beginning of the Pliocene, and today, the strong dry season in this area is controlled by the presence of high pressure over northern Australia. Current evidence (Polhaupessy this volume) suggests that seasonal climates were more pronounced during the Late Pliocene and Early Quaternary than today. Recent work based on grass cuticle studies (Pribatini and Morley 1999) suggests that during the Early Quaternary, widespread (natural) burning of grasslands took place within the region of Central Java.

Wallace's Line and the origin of Asian affinity taxa in eastern Indonesia and New Guinea

The dominance, in terms of species, of taxa of Asian origin in the lowland flora of New Guinea has attracted attention for several decades (e.g. Good 1962), especially since the Makasar Straits, followed by Wallace's Line, has been a significant barrier to plants as well as animals (e.g. van Balgooy 1987). A possible explanation to this anomaly is forthcoming from consideration of the geological history and palynological record of the southwestern arm of the island of Sulawesi (Morley 1998, 2000). During the Early and Middle Eocene, SW Sulawesi was firmly attached to SE Kalimantan and formed part of the old Sunda Craton. It was only during the Late Eocene that the Makassar Straits formed as a result of rifting and crustal extension along the eastern margin of Sundaland. Pollen assemblages of Middle Eocene age from SW Sulawesi, from before the time of the formation of the Makasar Straits, have much in common with penecontemporaneous assemblages from Java and SE Kalimantan, with pollen of the palm subtribe Iguanurinae, Calamoidae, *Durio* and *Gonystylus* as characteristic Asian elements (Morley 1998). As a result of the formation of the Makasar Straits, this Asian affinity flora became stranded to the east of Wallace's Line on emergent parts of SW Sulawesi (Figure 4.4). With the subsequent collision of the Australian and Sunda/Philippine Plates during the latest Oligocene/earliest Miocene (Hall 1995, 1996), it is most likely that elements of the SW Sulawesi flora were able to disperse to the east, and colonise the islands of Eastern Indonesia and New Guinea as they became raised above sea level, without the need for these taxa to cross the barrier of Wallace's Line (Figure 4.5). It is most likely through this mechanism that the eastern Indonesia, and New Guinea lowland flora became so rich in plant taxa of Asian origin, despite much of the region consisting of a substrate derived from Gondwanaland. The floras of the islands of eastern Indonesia and New Guinea were probably also sourced by elements originating in the Philippines and Halmahera, which have a much older geological history than other eastern Malesian islands, and would have borne floras of tropical affinity throughout the Tertiary. There is also a clear element best established in the uplands which was derived from mainland Australia.

It is thought unlikely that Dipterocarpaceae were a part of this early migration, but probably arrived much later following rare sweepstakes dispersal events. This is suggested by much lower dipterocarp diversities to the east of the Line, especially since some grouips of species, such as most members of the genus *Hopea* in New Guinea, are very closely related, and have probably all decende following a single dispersal event. It is also noteworthy that although the geochemical biomarker bicadinane (derived from dipterocarp resins) is common in most petroleum source rocks from the Sunda region, it has not been reported from the Late Miocene and Early Pliocene oil fields of the Bird's Neck area of Irian Jaya, a scenario which is also consistent with a very late Neogene or Pleistocene dispersal for members of this family.

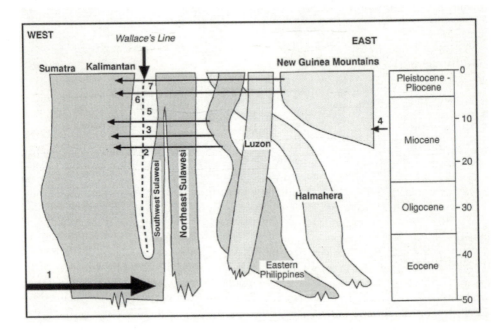

Fig. 4.4 Proximity of some Sundanian and East Malesian land masses through time (data source Hall 1995): South Sulawesi and Sunda were joined during the Middle Eocene and shared the same flora; whereas most of the islands of Eastern Indonesia were not formed until the Middle Miocene, Halmahera and the Philippines have a much older history, and would have borne a flora of tropical aspect throughout the Tertiary. Main migrations: 1) dispersals from Indian Plate, c.45 million years ago; 2) Myrtaceae, c.17 million years ago; 3) Camptostemon, c.14 million years ago; 4) Nothofagus, Middle Miocene; 5) Stenochlaena milnei, c.9 million years ago; 6) Dacrycarpus imbricatus, c.3.5 million years ago; 7) Phyllocladus, c.1.0 million years ago.

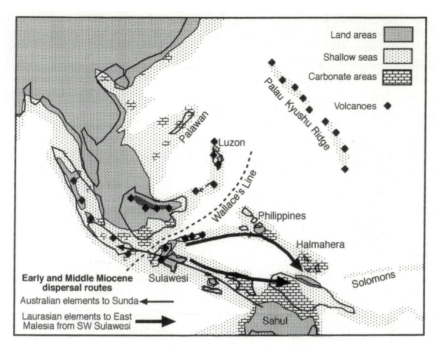

Fig. 4.5 Palaeogeography of the SE Asian and Australian region about 20 million years ago, (reconstruction based on Hall 1998), showing Wallace's Line and with suggested dispersal routes from SW Sulawesi to the islands of Eastern Indonesia and New Guinea after uplift above sea level during the Miocene. The likely route of dispersals from Australia to Sunda, following the collision of the Australian Plate with the Philippine and Sunda Plates, is also shown.

Plant dispersals between India/SE Asia and eastern Indonesia/ Australasia

The palynological record identifies relatively few instances of dispersal across Wallace's Line during the Neogene, suggesting that the Makasar Straits have formed a considerable barrier to plant dispersal within the Malesian region for a major part of the Tertiary. Since most palynological studies have been undertaken to the west of the Line, all dispersal events suggested by the fossil record are from east to west. However, it is equally likely that dispersals also occurred from west to east. The earliest dispersal event so far observed is within the earlier part of the Early Miocene, about 21 million years ago (Morley 2000), with the sudden appearance in Java of pollen of the Australasian taxa *Phormium* (Phormiaceae) and *Casuarina* (Casuarinaceae). The timing of this dispersal event suggests that it is linked with the latest Oligocene/earliest Miocene collision of the Australian Plate with the Philippine and Sunda plates (Figure 4.4). *Dacrydium* pollen shows an appearance at the same time, and may have dispersed from Australasia together with the other taxa, since it is absent from the Oligocene of Java.

Following the dispersal event of 21 million years ago, subsequent dispersals (Figure 4.4) are thought to have occurred about 17 million years ago, at which time Myrtaceae pollen shows a sudden increase in abundance, previously interpreted by Muller (1972) as due to dispersal from the east (and thus Myrtaceae dispersed into SE Asia both from the west and east); at 14 million years ago, with the westward dispersal of the mangrove tree *Camptostemon* (Bombacaceae), followed about 9 million years ago by the climbing fern *Stenochlaena milnei* (Blechnaceae). The distinctive spores of *S. milnei* are found west of Wallace's Line as far north as South China, and westward to India (inexplicably, this species suddenly became extinct during the earliest Quaternary).

Two subsequent dispersals are recorded 3.5 and about 1 million years ago, with the westward dispersal of the podocarp *Dacrycarpus imbricatus,* which subsequently reached Indochina, and *Phyllocladus,* which dispersed westward only to Borneo. These taxa dispersed from New Guinea at a time when there was considerable uplift in the southern Philippines (van der Kaars 1991b). The presence of islands in this area probably allowed *D. imbricatus* to 'island hop' to the Sunda region, with dispersal between islands being effected by birds. It is noteworthy that *Nothofagus* (Fagaceae) accompanied *D. Imbricatus* in New Guinea, where it first appears during the Middle Miocene, but since *Nothofagus* is more poorly dispersed, it was never able to disperse beyond New Guinea in the manner of *Dacrycarpus* (Figure 4.4).

Since lowland dispersals across Wallace's Line are of a small group of unrelated species, these are perhaps best explained in terms of occasional sweepstakes events, whereas the upland route, followed by gymnosperms, could be envisaged as dispersal by filter.

The Australian pollen record also provides relatively few instances of dispersal from the SE Asian area which might relate to Neogene plate collision (Truswell *et al.* 1987). Possible examples of dispersal from the north, based on mid-Tertiary appearances in Australia, are the climbing fern *Stenochlaena palustris* (Blechnaceae), *Acacia, Caesalpinia* and *Crudia* (Leguminosae) and *Merremia* (Convolvulaceae).

Conclusions

The SE Asian region has undergone five main phases of development during the course of the Tertiary as follows:
1. Prior to the collision of the Indian Plate with Asia, the flora was rich in palms, and many angiosperm taxa were present which have no modern analogue. The Sunda region and New Guinea were widely separated geographically, and bore very different floras.
2. Following the Indian Plate's collision with Asia, many Indian taxa were able to disperse into the SE Asian region along a moist corridor which became established between the two areas, and numerous local taxa became extinct. The Middle and Late Eocene flora of SE Asia was very diverse, and the climate warm and wet.
3. Many extinctions occurred as a result of the terminal Eocene event. The Oligocene to earliest Miocene climate was cool, and much drier than for the

Eocene, and the Sunda region was characterised by seasonal climate vegetation. Moist rainforests were restricted to Assam and Myanmar.

4. During most of the Miocene and Early Pliocene, the region experienced a much warmer and wetter climate, with highest temperatures occurring in the earliest Middle Miocene. It is within this period that the SE Asian flora as we know it today began to take form.

5. During the Late Pliocene and Quaternary, the climate was alternately wet/ warm, and dry/cool, coinciding with 'interglacial' and 'glacial' phases at higher latitudes; the Pliocene development of seasonal climates in Java and Nusa Tenggara is probably also linked to the northward drift of the Australian Plate.

6. SW Sulawesi was attached to SE Kalimantan prior to the Middle Eocene, and at that time bore a flora of Asian affinity which became marooned to the east of Wallace's Line following the Late Eocene opening of the Makasar Straits. It is thought that this flora largely sourced the islands of Eastern Indonesia and New Guinea as they were formed during the course of the Neogene.

7. Following the collision of the Australian Plate with the Philippine and Sunda Plates at the end of the Oligocene, a few Australian plant taxa were able to disperse into SE Asia. It is unlikely that there were earlier dispersal routes between SE Asia and Australia for angiosperms. However, it is possible that there was a dispersal route to Australasia from the Indian Plate (and Africa) via an island arc along the leading edge of the Indian Plate during the Turonian. This route was followed by Gnetalean gymnosperms, and could also have been followed by angiosperms.

Acknowledgements

The author gratefully acknowledges Professor Peter Kershaw for financial assistance, making it possible to attend this conference in Melbourne.

References

Curiale, J.A., Kyi, P. Collins, I.D., Din, A., Nyein, K., Nyunt, M. and Stuart, C.J. 1994. The central Myanmar (Burma) oil family – composition and implications for source. *Organic Geochemistry*, 22: 237-55.

Good, R. 1962. On the geographical relationships of the angiosperm flora of New Guinea. Bulletin of the British Museum (Natural History). *Botany*, 12: 205-26.

Hall, R. 1995. Plate tectonic reconstructions of the Indonesian region. *Proceedings of the Indonesian Petroleum Association 24th Annual Convention.* Indonesian Petroleum Association, Jakarta: 71-84.

Hall, R. 1996. Reconstructing Cenozoic SE Asia. In: R. Hall and D.J. Blundell (Editors) *Tectonic Evolution of Southeast Asia.* Geological Society Special Publication No. 106: 152-84.

Hall, R. 1998. The Plate Tectonics of Cenozoic SE Asia and the Distribution of Land and Sea. In: Hall, R. and Holloway, J.D. (Editors) *Biogeography and Geological Evolution of SE Asia.* Backhuys, Amsterdam: 99-131.

Harland, W.B., Armstrong, R.L., Cox, A.V., Craig, E.L., Smith, A.G. and Smith, D.G. 1990. *A Geological Time Scale.* Cambridge University Press, Cambridge: 263 pp.

Hickey, L.J. and Doyle, J.A. 1977. Early Cretaceous fossil evidence for angiosperm evolution. *The Botanical Review,* 43: 1-183.

Hill, R.S. 1994. *History of Australian Vegetation, Cretaceous to Recent.* Cambridge University Press, Cambridge: 433 pp.

Morley, R.J. and Flenley, J.R. 1987. Late Cenozoic vegetational and environmental changes in the Malay Archipelago. In: T.C. Whitmore (Editor) *Biogeography of the Malay Archipelago.* Clarendon Press, Oxford: 50-9.

Morley, R.J. 1991. Tertiary stratigraphic palynology in SE Asia: Current Status and new direction, *Geological Society of Malaysia Bulletin,* 28: 1-36.

Morley, R.J. 1998. Palynological Evidence for Tertiary Plant Dispersals in the SE Asia Region in Relation to Plate Tectonics and Climate. In: R. Hall and J. Holloway (Editors) *Biogeography and Geological Evolution of Southeast Asia.* Bakhuys Publishers, Leiden: 7-200.

Morley, R.J. 2000. *Origin and Evolution of Tropical Rain Forests.* J Wiley and Sons, New York: 362 pp.

Morley. R.J. 2001. Why are there so many primitive angiosperms in the rain forests of Asia-Australasia? In: I. Metcalfe, J.M.B. Smith, M. Morwood and I. Davidson (Editors) *Floral and Faunal Migrations and Evolution in SE Asia-Australasia.* Balkema, Lisse.

Muller, J. 1966. Montane pollen from the Tertiary of NW.Borneo. *Blumea,* 14: 231-5.

Muller, J. 1968. Palynology of the Pedawan and Plateau sandstone formations (Cretaceous-Eocene) in Sarawak, Malaysia. *Micropalaeontology,* 14: 1-37.

Muller, J. 1972. Palynological Evidence for Change in Geomorphology, Climate and Vegetation in the Mio-Pliocene of Malesia. In: P.S. Ashton and M. Ashton (Editors) *The Quaternary Era in Malesia.* University of Hull, Geography Department Miscellaneous Serial No. 13: 6-34.

Pribatini, H. and Morley, R.J. 1999. Palynology of the Pliocene Kalibiuk and Kaliglagah Formations, near Bumiayu, Central Java. In: *Tectonics and Sedimentation of Indonesia.* Indonesian Sedimentologists Forum Special Publication No. 1: 53. (Abstract only)

Smith, A.G., Smith, D.G. and Funnell, B.M. 1994. *Atlas of Mesozoic and Cenozoic Coastlines.* Cambridge University Press, Cambridge: 99 pp.

Takhtajan, A. 1969. *Flowering Plants, Origin and Dispersal* (translated by C. Jeffrey). Oliver and Boyd, Edinburgh: 300 pp.

Takhtajan, A. 1987. Flowering Plant Origin and Dispersal: Cradle of the Angiosperms Revisited. In: T.C. Whitmore (Editor) *Biogeographical Evolution of the Malay Archipelago.* Oxford Monographs on Biogeography No. 4: 26-31.

Truswell, E.M., Kershaw, A.P. and Sluiter, A.R. 1987. The Australian – SE Asian Connection: Evidence from the Palaeobotanical Record. In: T.C. Whitmore (Editor), *Biogeographical Evolution of the Malay Archipelago.* Oxford Monographs on Biogeography No. 4: 32-49.

van Aarssen, B.G.K, Cox, H.C., Hoogendoorn, P. and de Leeuw, J.W. 1990. A cadinane bipolymer present in fossil and extant dammar resins as a source for cadinanes and bicadinanes in crude oils from SE Asia. *Geochimica et Cosmochimica Acta,* 54: 3021-31.

van Balgooy, M.M.J. 1987. A plant geographical analysis of Sulawesi. In: T.C. Whitmore (Editor) *Biogeographical Evolution of the Malay Archipelago.* Oxford Monographs on Biogeography No. 4: 94-102.

van der Kaars, W.A. 1991(a). Palynology of eastern Indonesian marine piston-cores: A Late Quaternary vegetational and climatic history for Australasia. *Palaeogeography, Palaeoclimatology, Palaeoecology,* 85: 239-302.

van der Kaars, W.A. 1991(b). Palynological aspects of site 767 in the Celebes Sea. *Proceedings of the Ocean Drilling Program, Scientific Results,* 124: 369-74.

Plant Biogeography of the SE Asian-Australian Region

Trevor Whiffin

Introduction

The SE Asian-Australian region continues to interest and to challenge the student of plant biogeography. This region is the second richest in the world as regards plant biodiversity, being eclipsed only by tropical South America. This floristic richness, however, shows complex patterns, the causes of which appear to relate at least in part to the equally complex patterns of past and present climates and plate tectonic movements. The details of these patterns, and their origins, are gradually being elucidated (Holloway and Hall 1998; Morley 1998).

The geographic setting

The SE Asian-Australian region is here taken as including the area from mainland SE Asia through to New Guinea and northern Australia, including the archipelagoes that now constitute Indonesia and the Philippines. These land areas lie on two main continental shelves, Sunda to the west and Sahul to the east (Figure 5.1). Much of these shelves, generally delimited by the 200 m bathymetric line, would have been exposed during periods of lower sea levels during the Quaternary. The region between the two shelves, consisting of the Philippines, Sulawesi, the Moluccas and the Lesser Sunda Islands, contains islands separated by much deeper seas.

Climatically, there are centres of wet tropical aseasonal climate in the west, centred on Borneo and adjacent parts of Sumatra and peninsular Malaysia, and in the east, centred on the major part of New Guinea. There are smaller areas of similar climate in the eastern Philippines and central Sulawesi. To the north and south there are areas of tropical monsoonal climate with a distinct dry season, based in mainland SE Asia (north of peninsular Malaysia) and in northern Australia. Between these, in the western Philippines, parts of Sulawesi, most of Java, and all of the Lesser Sunda Islands, there are areas of tropical monsoonal climate with a short dry season.

The vegetation generally reflects the climate. There are two core areas of aseasonal tropical rain forest, one in the west and the other in the east. The western core is centred in Borneo and extends to Sumatra, peninsular Malaysia, western Java, central Sulawesi, and parts of the Philippines. The eastern core is centred in New Guinea, with small extensions to neighbouring areas including

ISBN 3-923381-47-6

northern Australia. Between these two core areas of tropical rain forest are 'wedges' of monsoonal forest (Whitmore 1989). The northern wedge extends from mainland SE Asia through to the western Philippines, while the southern wedge extends from northern Australia through most of Java, the whole of the Lesser Sunda Islands, parts of Sulawesi and most of the Moluccas. During drier periods in the Quaternary, these wedges would have been joined to form a definite 'dry corridor' connecting, at least climatically, parts of northern Australia and mainland SE Asia.

Wallace's Line and other biogeographic lines

The naturalist and explorer A.R. Wallace documented the marked differences in mammals and birds either side of a line that passed between Bali and Lombok, north between Sulawesi and Borneo, and then to the east of the Philippines. This line became known as Wallace's Line (Figure 5.1), and was seen as representing the distinction between essentially Asian fauna to the west, and Australian fauna to the east.

Later workers, including Wallace himself, modified this line or suggested other lines, each of which marked a disjunction in the distribution of certain plant or animal groups (George 1981). The increasing number of lines reflects the complexity of the biogeography of the area, where different plant and animal groups show different patterns of distribution. These different distribution patterns are in part due to the different ecological and dispersal characteristics of the organisms themselves, and in part due to the different reactions of the organisms to the complex climatic and geological history of the area. As our knowledge increases then each line, originally derived in biogeographic terms, is seen as reflecting one or a combination of climatic or plate tectonic boundaries.

The concept of Wallacea

Wallace's Line was originally described as the line separating the Asian and Australian faunas, but it is now seen as the eastern boundary of the strictly Asian fauna (George 1981; Moss and Wilson 1998). Lydekker's Line is now seen as the western boundary of the strictly Australian fauna, while Weber's Line is the line of faunal balance, at least for some groups such as the mammals.

Wallace's Line coincides with the eastern edge of the Sunda Shelf, at least in its southern part; in the area north of Sulawesi, Huxley's Line more closely follows the edge of this shelf. Lydekker's Line coincides with the western edge of the Sahul shelf. The area between Wallace's Line and Lydekker's Line consists of islands not on a continental shelf, and thus separated by deeper water.

The area between Wallace's Line and Lydekker's Line is often recognised as the area of Wallacea (Moss and Wilson 1998). This is often seen as a transitional zone between the Asian and Australian regions, but the area does also possess a number of endemic species of its own. Wallacea also coincides with the drier zone, with extensive areas of monsoon forest, from northern Australia through the Lesser Sunda Islands and southern Sulawesi. Thus Wallacea is ecologically and climatically different from the areas to its west and east.

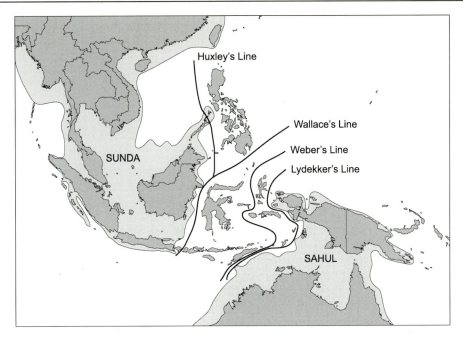

Fig. 5.1 The SE Asian-Australian region, showing the major land areas, the Sunda and Sahul shelves, and the main biogeographic lines.

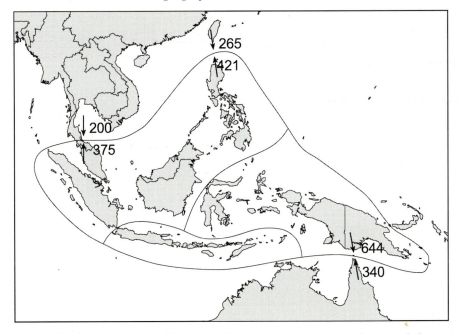

Fig. 5.2. The botanical area of Malesia, showing the demarcation knots and the main subdivisions within the region; redrawn from van Steenis (1950).

Peter Kershaw, Bruno David, Nigel Tapper, Dan Penny and Jonathan Brown (Eds): Bridging Wallace´s Line

Wallacea is sometimes also taken to include the Philippines (except for Palawan), as was done by Simpson (1977) and Brandon-Jones (1998), where the western edge of the region is taken as Huxley's Line rather than Wallace's Line. Climatically, parts at least of the Philippines can be seen as a northward extension of the dry belt that passes through the Lesser Sunda Islands and southern Sulawesi.

The botanical concept of Malesia

To the zoologist, the major line of demarcation between the Asian and Australian biota passes somewhere to the west of New Guinea, and to the east of Borneo. However, to the botanist, New Guinea is almost invariably seen as floristically part of SE Asia, with the line of demarcation passing between New Guinea and Australia.

Following an analysis of the flora of this region, van Steenis (1950) delimited the botanical area of Malesia (Figure 5.2). This essentially comprises Malaysia, Indonesia, the Philippines and the island of New Guinea. These limits were defined by demarcation knots, areas that marked the distributional limits of numerous genera, so that in each case there were many genera that did not cross these demarcation knots from both sides (Figure 5.2).

In this botanical concept of Malesia, New Guinea is placed with SE Asia and separated from Australia. The limits of Malesia coincide in part with climatic and vegetational boundaries. The northwestern boundary, at the Isthmus of Kra, coincides with the change between predominantly tropical wet peninsular Malaysia and the drier, more monsoonal part of mainland SE Asia. The south-eastern boundary, across the Torres Strait, separates predominantly tropical wet New Guinea and the more arid area of Australia.

Plant biogeographic evidence

There are a number of different lines of evidence relating to the plant biogeography of the SE Asian-Australian region, and these will be briefly reviewed here, in each case indicating the type of information that they provide. These studies are placed in an order more or less corresponding with an increased level of detail, and later an increasingly objective method of analysis.

Patterns of floristic richness and endemism

At the most general level, an investigation of the patterns of distribution of numerous species will indicate those geographic areas that are centres of richness or centres of endemism. While both are important in the study of the biogeography of an area, it is possible that the centres of floristic richness are indicative only of more recent environmental conditions. However, it is generally believed that centres of endemism may reflect older events of biogeographic significance.

There is no complete and comprehensive information on the plant species of SE Asia and Australia. The area of Malesia has perhaps 40,000 species of higher plants (vascular plants, including ferns, gymnosperms, and flowering plants), while Australia has up to 25,000. However, within Australia many areas of species richness are outside of the tropics, and of less significance as regards relationships with Malesia, and so the more relevant figure for Australia would be that for the tropical regions. The tropical rain forests of northern Australia have up to 3000 species, while the geographically more widespread monsoonal and semi-arid areas of tropical Australia may have an additional 7000.

Some comparative information on the patterns of species richness and species endemism in Malesia is presented in Table 5.1. Although estimated in different ways, and at different times, the information is broadly consistent.

There are clearly two major centres of species richness and endemism within Malesia. Borneo has the most diverse flora of western Malesia, with up to 15,000 species of flowering plants (Whitmore 1989), of which approximately one third are endemic. New Guinea, in eastern Malesia, is even richer, with quite possibly in excess of 20,000 species (Womersley 1978), of which nearly half are endemic. It is one of the richest floras in the world, particularly in relation to its size (van Balgooy 1976). Other moderately species-rich areas in Malesia include Sumatra, Malaysia and the Philippines, with generally up to 10,000 species each.

A somewhat similar picture can be seen in relation to the genera (Table 5.1). Although these figures may now be incomplete, the picture is unlikely to change. The floristically rich areas in fact have very similar numbers of genera, indicating that they differ in the relative species richness of some of the genera. There are, however, marked differences in the numbers of endemic genera. Here, the major centres of generic endemism are again New Guinea and Borneo, with smaller centres in Malaysia and the Philippines.

The closest approach to comparative data available are those provided by van Balgooy *et al.* (1996). These were derived from the numbers of species, and their distributions, in the published revisions in *Flora Malesiana* up to 1994 (volumes 4 to 11, part 2). These are estimated to comprise perhaps one sixth of the total flora in Malesia. There is some bias in these data, in that in general the smaller and less complex families have been revised, while the larger families have not. Nevertheless, the information should show the general trends, and is consistent with the other information available (Table 5.1).

The floristic information from the published families shows once again major centres of species richness in New Guinea and Borneo, with lesser centres in Malaysia, Sumatra and the Philippines (Table 5.1). Important and significant differences are seen as regards species endemism. New Guinea, Borneo and, to a lesser extent, the Philippines show high levels of species endemism. The levels of endemism for these three areas are greater than expected based on the total number of species in each area, while the levels for the other areas are more or less what would be expected (van Balgooy *et al.* 1996).

Floristic elements and generalised tracks

The pattern of distribution of each species, or group of species, is in some way unique. However, it has long been recognised that many species or groups of species show broadly similar patterns of distribution, and it is to be expected that these general patterns will reflect the same sequence of geological or climatic events.

Peter Kershaw, Bruno David, Nigel Tapper, Dan Penny and Jonathan Brown (Eds): Bridging Wallace´s Line

Table 5.1 Floristic richness and endemism in the major land areas of SE Asia. (a) based on information in Campbell and Hammond (1989); (b) based on information in Whitmore (1989); (c) based on information in van Balgooy (1976); (d) based on information in van Steenis (1987); (e) based on information in van Balgooy et al. (1996).

	Number of species (a) (b)	Number of genera (a) (c)	Endemic genera (d)	Number of species in sample (e)	Endemic species in sample (e)	Percentage species endemic (e)
Sumatra	8,000 - 10,000	1500	13	1650	172	10.4
Malaysia	7900	1500	24	1693	202	11.9
Borneo	10,000 - 15,000	1500	61	2062	639	31.0
Java	4500	1320	5	1093	71	6.5
Lesser Sunda Islands	-	750	1	706	33	4.7
Sulawesi	5000	-	5	973	112	11.5
Philippines	8000	1308	23	1425	349	24.5
Moluccas	3000	-	4	670	36	5.4
New Guinea	20,000	1465	82	2126	1081	50.8

These general patterns allow the recognition of floristic elements – groups of species (or higher taxa) that show the same or similar distribution patterns. These elements may be recognised subjectively, by examination of multiple distribution maps, or more objectively by numerical analysis of distribution patterns. Two aspects of floristic elements should be emphasised, at least as they are understood here. Firstly, these floristic elements are static, reflecting the current distribution of the plants involved. Secondly, they are neutral, in that there is no implied origin or direction to the floristic elements. When taken in this way, floristic elements are essentially the same as generalised tracks, for example as developed recently for Australia by Crisp *et al.* (1999). Generalised tracks relate to areas of endemism. Where a plant group is present in different areas of endemism, it is possible to draw lines on a map (called tracks) connec-ting these areas. Where several tracks, from different plant groups, coincide they are said to represent general tracks. These general tracks indicate relation-ships between the areas, but are not in themselves explanations of the patterns.

For broad scale regional analyses, the most suitable unit of study is the genus (van Balgooy 1976). There is in general more taxonomic agreement about the circumscription of genera than for other levels, including species. In addition, analyses at the generic level are more likely to reveal the older, 'deeper' patterns of biogeography, whereas those at species level are more likely to reflect more recent climatic or ecological events.

For the SE Asian-Australian region there are three relevant studies where the flora of an area has been categorised, on the basis of the distribution patterns of its constituent genera, into floristic elements. These studies cover New Guinea, which is an important part of eastern Malesia, and New Caledonia which, while peripheral to the area, is indicative of an essentially Australian (Gondwanan) derived area. The entire flora of New Guinea was categorised by van Balgooy (1976) into 15 distribution types, which could be grouped into five major categories. In the other studies, Hartley (1986) divided the rain forest flora of New Guinea into 17 distribution types, with five major categories, and Morat *et al*. (1986) divided the rain forest flora of New Caledonia into 12 distribution types, with six major categories.

For the purposes of comparison, the distribution types in these three studies have been grouped into five major floristic elements (Table 5.2). These five major elements are similar to those recognised by van Balgooy (1976), but do not correspond exactly with those, or with the major categories in the other two studies.

New Guinea and New Caledonia both include a reasonably high percentage of widespread genera; this equates to one-third or more of the genera in each case. In New Guinea, the flora is more predominantly Asian or western Malesian than Australian; the ratio is nearly three to one for all genera, and four to one when rain forest genera only are considered. New Caledonia has a more Australian related flora, but even here there is a higher percentage of western Malesian centred genera than Australian centred genera. However, these figures do not include the relationships of the endemic and sub-endemic genera.

A notable distinction between New Guinea and New Caledonia is the extent of generic endemism. New Caledonia has more than one quarter of its rain forest genera endemic or sub-endemic to the island. New Guinea, despite having comparatively high levels of generic endemism (Table 5.1), has much lower levels than New Caledonia (Table 5.2). This emphasises the isolation experienced by New Caledonia over much of the Tertiary.

Most of the endemic and sub-endemic genera in New Caledonia appear to have an Australian or Gondwanan origin, and the same is true for many of the endemic or sub-endemic genera in New Guinea. If these genera are included in the comparison of Asian or western Malesian versus Australian genera (Table 5.2), then New Caledonia has a predominantly Australian flora, the ratio of Asian or western Malesian to Australian genera being approximately one to two. On the other hand, New Guinea still has a predominantly Asian or western Malesian flora, with the ratio of Asian or western Malesian to Australian genera being approximately two to one.

Crisp *et al*. (1999) recognised four general tracks for the Australian flora; these are the South Pacific track, the Equatorial track, the Trans-Indian Ocean track, and the Pan-temperate track. These can be equated in part to the floristic elements recognised in Table 5.2. The Pan-temperate track consisted of widespread temperate plant groups, while the Trans-Indian Ocean track had predominantly southern hemisphere temperate plant groups. The South Pacific track consisted of two sections, the first of which contained further southern temperate plant groups. These three groups are more or less equivalent to the

Peter Kershaw, Bruno David, Nigel Tapper, Dan Penny and Jonathan Brown (Eds): Bridging Wallace´s Line

temperate floristic element in Table 5.2 and, as might be expected, are more important in Australia than in New Guinea. The second section of the South Pacific track contains plant groups centred in Australia, New Guinea and New Caledonia, extending eastwards to Fiji and westwards into western Malesia. This is equivalent to the Australian element in Table 5.2, centred to the south and east. The Equatorial element contains widespread and pan-tropical plant groups, but also some that are centred in Africa, India and Malesia; these latter are equivalent to the western Malesian element in Table 5.2, centred to the north and west.

Table 5.2. Categorisation of the genera of New Guinea and New Caledonia into floristic elements, on the basis of their distribution. (a) based on van Balgooy (1976); (b) based on Hartley (1986); (c) based on Morat et al. (1986).

	New Guinea (all genera) (a)		New Guinea (rain forest genera) (b)		New Caledonia (rain forest genera) (c)	
	Number	Percentage	Number	Percentage	Number	Percentage
Widespread including pantropical	631	43.1	256	35.7	120	32.9
Centred to north and west (western Malesian)	392	26.8	297	41.5	81	22.2
Centred to south and east (Australian)	153	10.4	74	10.3	53	14.5
Temperate	94	6.4	24	3.4	14	3.8
Endemic and subendemic	195	13.3	65	9.1	97	26.6
Total	1465	100.0	716	100.0	365	100.0

As noted above, analyses of a flora at the genus level provide insights into the older patterns of biogeography, while it is expected that analyses at the species level will provide information on more recent events. In an analysis of the plant biogeography of the Lesser Sunda Islands (east Malesia) at the species level, van Steenis (1979) recognised a number of floristic elements (distribution patterns). These generally correspond with the floristic elements described above, in that there were clearly defined and important elements relating to species that are widespread, species that are centred in western or eastern Malesia, and species that are endemic or sub-endemic. However, the Lesser Sunda Islands are in the dry wedge of monsoon forest that stretches from northern Australia to southern Sulawesi and the Philippines (Wallacea). As might therefore be expected, van Steenis (1979) was also able to recognise a number of elements that relate to these more monsoonal areas. Identified here were distribution patterns that included: species restricted to Wallacea; species in the Lesser Sunda Islands and in monsoonal Australia; and species in the Lesser Sunda Islands and then partially or completely disjunct to monsoonal mainland SE Asia. These species are clearly responding to climate, and are to be found predominantly or solely in areas with a marked dry season.

Distribution of major plant groups

It has long been recognised that, within Malesia, many plant groups have apparent centres of diversity in either the western or eastern part (Whitmore 1989), thus corresponding with the Sunda or Sahul shelves. These groups are thus generally seen as having Asian or Australian (Gondwanan) ancestry. The taxonomic rank of these groups varies from genus (or below) to family, perhaps indicating something of their origin in each case but certainly emphasising the complexity of the biogeography. Other groups are seen as more or less evenly distributed throughout Malesia (Whitmore 1989), or as having major centres in both western and eastern Malesia, the bicentric distributions of Dransfield (1987).

In general, those groups showing a general and relatively even distribution across Malesia are found to be widespread or pan-tropical outside of Malesia. Although of biological interest, in terms of how they achieved their widespread or pan-tropical distribution, these groups are generally uninformative as regards the biogeography of the SE Asian-Australian region. Other groups that may appear to be evenly distributed across Malesia turn out to have distinct centres in both eastern and western Malesia (Whitmore 1989), and thus perhaps should be classed with those that show a bicentric distribution. The most informative groups are those with a distinct western or eastern distribution, and these will be discussed first. The concept of bicentric distributions will be considered later. While there are numerous possible examples of plant groups centred in either western or eastern Malesia, one or two examples will be discussed for each pattern. These examples include some of the more important or significant plant groups in the region.

In many ways the prime example of a group centred in western Malesia is the Dipterocarpaceae. This family occurs in three different areas (Asia, Africa and South America) with a distinct subfamily in each area (Ashton 1989). However, the Asian subfamily Dipterocarpoideae is by far the largest, and the only one to achieve ecological and economic importance.

The distribution limits of the various genera (Figure 5.3) show that there is a major concentration of genera in western Malesia, particularly in Borneo, and a minor concentration in Sri Lanka and adjacent southern India. Six of the genera reach the Philippines, while only two reach Sulawesi (*Shorea* and *Vatica*) and three reach New Guinea (*Anisoptera*, *Hopea* and *Vatica*). In all three of these areas in eastern Malesia it is the widespread genera that are to be found, these genera reaching mainland SE Asia and often also India in their westward limits.

Peter Kershaw, Bruno David, Nigel Tapper, Dan Penny and Jonathan Brown (Eds): Bridging Wallace´s Line

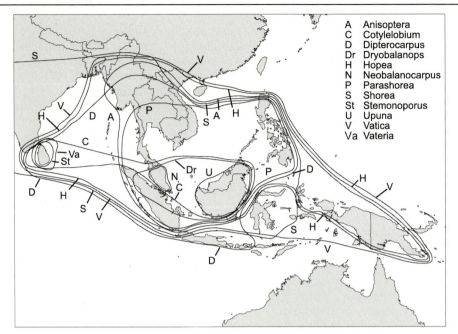

Fig. 5.3   The distribution limits of the genera of Dipterocarpaceae in the SE Asian-Australian region; redrawn from Whitmore (1981).

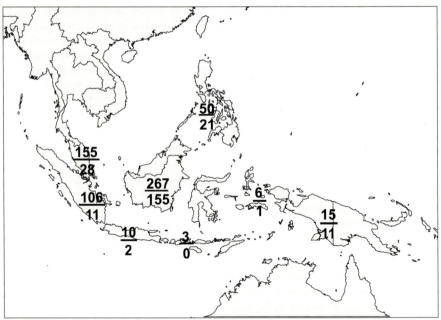

Fig. 5.4   Species richness and species endemism in the Dipterocarpaceae in the SE Asian-Australian region; redrawn from Ashton (1982). For each area, the figure above the line is the total number of species, while the figure below the line is the number of endemic species

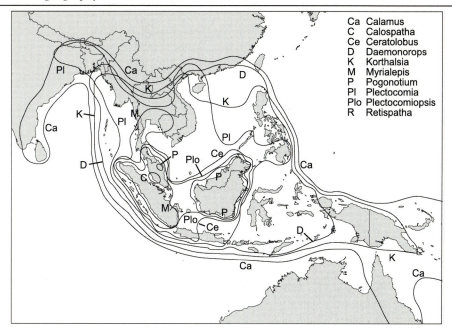

Fig. 5.5   The distribution limits of the genera of the rattan palms (Arecaceae, Calamoideae) in the SE Asian-Australian region; redrawn from Dransfield (1981), additional information from Baker et al. (1998).

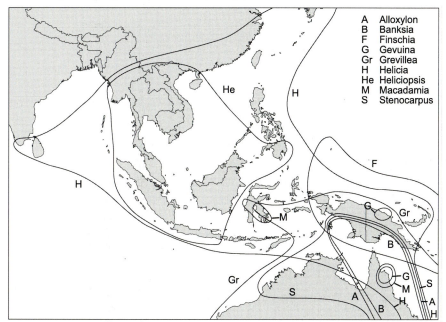

Fig. 5.6 The distribution limits of the genera of Proteaceae in the SE Asian-Australian region; redrawn from Whitmore (1981), additional information from Flora of Australia (1995) and Weston and Crisp (1996).

Peter Kershaw, Bruno David, Nigel Tapper, Dan Penny and Jonathan Brown (Eds): Bridging Wallace´s Line

The patterns of species richness and species endemism (Figure 5.4) similarly show a major centre of species richness in western Malesia, with a major centre of species endemism in Borneo where over half of the species are endemic. There is a minor centre of species richness and endemism in the Philippines, and then the species numbers decrease rapidly eastwards through Sulawesi, Java and the Lesser Sunda Islands, and the Moluccas. There is a very small increase in species richness and, more especially, in species endemism in New Guinea.

The large subfamily Calamoideae of the Arecaceae (palms) is apparently monophyletic, although the relationships of many of the genera within the sub-family require further study (Baker *et al.* 1998). There are ten genera of rattans (climbing palms) within this subfamily that occur in Malesia (Figure 5.5). Although this group of rattans is not necessarily monophyletic, the distribution patterns of the various genera have been well documented and prove to be informative (Dransfield 1981; Baker *et al.* 1998). The genera show a variety of distribution patterns, but taken together indicate that the greatest generic richness is in western Malesia, particularly Borneo, Sumatra and Malaysia, with decreasing numbers of genera in an eastward direction. Four genera reach the southern Philippines, three the northern Philippines, three reach Sulawesi, and two reach the major part of New Guinea. Only one genus reaches Australia, and this genus (*Calamus*) is the most widespread of the group, extending from India and Sri Lanka across SE Asia to the west Pacific and Australia, and is also to be found in Africa.

The family Proteaceae has long been recognised as a southern family, and thus essentially Gondwanan (Weston and Crisp 1996). It has major centres in South Africa and Australia. Of the nearly 80 genera and over 1700 species in the family, 46 genera and approximately 1100 species occur in Australia; of these, 37 genera and nearly all of the species are endemic (Flora of Australia 1995). Many of the genera and species in Australia, such as *Banksia* and *Grevillea*, occur predominantly in woodland and heathland. There are, however, a significant number of small, primitive and relictual genera occurring in the rain forests of NE Queensland.

There are nine genera of Proteaceae that occur in Malesia (Figure 5.6), and all but two of these are also found in Australia; the exceptions are *Finschia* and *Heliciopsis* (Sleumer 1955; Weston and Crisp 1996). The seven Australian genera tend to have major centres of species richness in Australia, with a small number of species extending into Malesia, often only reaching New Guinea, the Moluccas or Sulawesi. Four genera fit this pattern well. *Banksia* and *Grevillea* are large genera, with many species, widely distributed in Australia. On the other hand, *Alloxylon* and *Macadamia* are essentially small, rain forest genera with somewhat limited distribution in Australia and eastern Malesia. *Gevuina* is similar, with a restricted distribution in eastern Australia and New Guinea, but with a third species in southern South America. *Stenocarpus* also fits this general pattern, with nine species in Australia, two of which extend to New Guinea and the Moluccas. In addition, however, there are approximately 16 endemic species in New Caledonia (Sleumer 1955; Weston and Crisp 1996). *Finschia* probably also fits in this group, differing in that it is absent from Australia. It has four species in New Guinea, three of them endemic and one extending from the Moluccas through New Guinea into the west Pacific (Sleumer 1955; Weston and Crisp 1996).

The genus *Helicia* is the most widespread in this region, extending from southern India and Sri Lanka through SE Asia to Japan in the north and Australia in the south (Sleumer 1955; Foreman 1995). It shows centres of richness in both western and eastern Malesia, with 44 species (40 endemic) in New Guinea and 27 species (23 endemic) in mainland SE Asia; there are also minor centres of diversity, with nine species (eight endemic) in Australia, eight species (three endemic) in Borneo, and five species (four endemic) in the Philippines (Sleumer 1955; Weston and Crisp 1996). While the major centre of diversity is in eastern Malesia (New Guinea), the genus is clearly well established in western Malesia and, more particularly, in mainland SE Asia.

The genus *Heliciopsis* is unique in the Proteaceae in that it is distributed entirely in western Malesia and SE Asia (Figure 5.6). There are seven or more species in the genus (Sleumer 1955; Weston and Crisp 1996), with four species (three endemic) in mainland SE Asia, and four species (three endemic) in western Malesia, from Sumatra to the southern Philippines.

A number of families are said to be bicentric, in that they exhibit major centres of species richness in both western and eastern Malesia (Dransfield 1987). For most, if not all, of these it is found that they contain distinct subgroups that individually are centred in either western or eastern Malesia. The relationships between these subgroups need to be studied in detail in each case, in order to obtain a clear picture of the phylogeny and biogeography of the family.

The Arecaceae (palms) are bicentric, but some groups such as the Calamoideae are clearly centred in western Malesia (Figure 5.5) while others such as *Cyrtostachys* are centred in eastern Malesia (Dransfield 1981). Among the other major families that appear to be bicentric in this way are the Elaeocarpaceae (Baker *et al.* 1996), the Myrtaceae and the Rutaceae. There are even apparent examples of bicentric genera, such as *Chisochiton* in the Meliaceae (Whitmore 1989), *Zanthoxylum* in the Rutaceae (Hartley 1986) and perhaps also *Helicia* in the Proteaceae (Figure 5.6). In some cases at least, closer study reveals subgroups within the genus that are either western or eastern in their centre.

Cladistic analysis

A major advance in systematics over the last forty years has been the development of objective methods for the estimation of the phylogeny (evolution) within a group of plants or animals. These techniques go under the general term of cladistic analysis. This involves the detailed, and generally now numerical, analysis of changes in characters within the group, allowing the estimation of the most likely phylogeny of the group.

Initial studies of this type were based on morphological characters. These provided useful information, particularly on the order of the origin of groups and subgroups within the study. More recently, the analysis techniques have been extended to handle molecular (DNA) information. These latter studies have the

Peter Kershaw, Bruno David, Nigel Tapper, Dan Penny and Jonathan Brown (Eds): Bridging Wallace's Line

additional advantage of possibly allowing the determination of the timing of the various phylogenetic events.

There have been several studies involving the cladistic analysis of plant groups within the SE Asian-Australian area. These studies are providing important information on the relationships between phylogeny and biogeography in this region. Three examples will be discussed here, to show the nature of the information that may be provided.

The genus *Guioa* (Sapindaceae) contains 65 species distributed from mainland SE Asia through Malesia to Australia, and extending into the west Pacific (van Welzen 1994). It is clearly centred in Malesia, with 43 species in this region. The major centre of species richness would appear to be New Guinea (25 species), with more minor centres in the Philippines (12 species), western Malesia (six species) and Australia (five species). The genus was revised by van Welzen (1989) who undertook a cladistic analysis based on morphological characters. The analysis supported the hypothesis that the genus was widely distributed in eastern Malesia and surrounding areas (Australia, New Caledonia and New Guinea), and that it migrated from there into western Malesia. While the timing of this migration cannot be determined from this analysis, it would have had to be sufficiently early to have allowed the diversification now seen in western Malesia and mainland SE Asia.

The genus *Caryota* (Arecaceae) is one of the genera of palms that spans Wallace's Line, occurring in both western and eastern Malesia. It is a small genus of eleven species. Most species are found in western Malesia, with only two or three extending into eastern Malesia and as far eastwards as north eastern Australia. A phylogenetic analysis using molecular data (chloroplast DNA variation) was undertaken by Hahn and Sytsma (1999). Three main clades were found, two of which were found predominantly in western Malesia, with the third predominantly in eastern Malesia. Two of these clades show distributions across Wallace's Line, but in each case it involves those islands adjacent to the line, that is Borneo, Sulawesi and the Philippines. Thus while the genus itself occurs across Malesia, there are subgroups centred in each case either in western or in eastern Malesia. When species in a subgroup cross Wallace's Line, it is only as far as the neighbouring islands, indicating a relatively recent dispersal event.

The genus *Cyrtandra* (Gesneriaceae) contains some 500-600 species, and is found from mainland SE Asia through Malesia to Taiwan in the north, Australia in the south, and the Pacific islands and Hawaii in the west (Atkins *et al.* 2001). The centres of species richness are New Guinea and Borneo, each with more than 150 species, and the Philippines, with more than 80 species. It is thus similar to many other Malesian genera. The species in western Malesia and the Philippines allow a study of the relationships between the floras of the Philippines and Borneo. In particular, they allow a study of the floristic position of Palawan. This is the most westerly of the islands of the Philippines, and it is now geographically closer to Borneo. A study of 26 species from this region using molecular data (nuclear ribosomal ITS) was undertaken by Atkins *et al.* (2001). The analysis showed a major division between the *Cyrtandra* species of Malaysia and Borneo on the one hand and those of the Philippines on the other. The island of Palawan showed some links with Borneo at the species level.

However, the main Palawan species formed a clade that linked back to the Philippine species. Thus while there is evidence for recent dispersal from Borneo, the flora of Palawan has older and stronger links with the other islands of the Philippines.

Cladistic biogeography

To a certain extent, each group of plants is unique, and has a unique phylogeny. Neverthless, different groups within an area will respond in a similar manner to major events, such as vicariance caused by geological (plate tectonic and other) events. Thus it is expected that the areas of agreement (congruence) between the phylogenies of different groups of plants from the same geographic area will reflect these shared, major events. The areas of disagreement will reflect unique, usually minor stochastic events such as dispersal or local extinction.

Cladistic biogeography involves the search for congruent patterns in phylogenies (cladograms) of several, often unrelated, groups of organisms from the same geographic area. The area of distribution of a species is used to replace that species within the relevant cladogram. Then points of congruence are sought in these area cladograms, and it is expected that these will reflect the major shared biogeographic events that affected the flora as a whole. This is proving most informative, especially where these major biogeographic events can be correlated with the known geological history of the area.

The most informative study of cladistic biogeography for the SE Asian-Australian region is that of selected genera of Sapindaceae by Turner (1995, 1996). The eight genera studied fell into three distribution patterns. Three genera (*Arytera*, *Cupaniopsis* and *Guioa*) have a widespread distribution in the region, from mainland SE Asia or western Malesia across Malesia to New Guinea and eastern Australia, and extending into the west Pacific. Two genera (*Lepidopetalum* and *Rhysotoechia*) have a more westerly distribution that is similar to the widespread pattern except that it does not extend into the Pacific. Three genera (*Cnesmocarpon*, *Jagera* and *Mischarytera*) have a more restricted, and eastern, distribution in New Guinea and eastern Australia, with occasional marginal extensions into nearby western Malesia. Cladistic analyses of the individual genera are indicative of the ancestral species being in Australia and New Guinea, with minor centres also in New Caledonia and Fiji (van Balgooy *et al.* 1996).

Turner (1995, 1996) combined the cladograms of the eight genera, to derive a broad summary pattern (Figure 5.7). This can be interpreted as showing a Gondwanan origin, with two major vicariance events. The first was the separation of New Caledonia from Australia, and the second the separation of New Guinea from Australia. The west Pacific Sapindaceae were derived by dispersal from both New Caledonia and New Guinea. Similarly, the western Malesian Sapindaceae were derived by dispersal from New Guinea. That there are differences in the various separate cladograms indicates that there were probably different source areas for the migration westwards. Finally, some Sapindaceae may have re-invaded Australia. This scenario is in broad agreement with the known plate tectonic and climatic history.

Peter Kershaw, Bruno David, Nigel Tapper, Dan Penny and Jonathan Brown (Eds): Bridging Wallace's Line

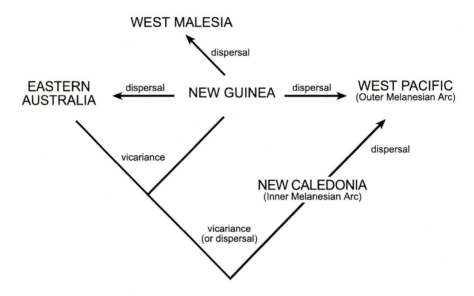

Fig. 5.7 Summary of the cladistic biogeography of Sapindaceae in the SE Asian-Australian region; redrawn from Turner (1996).

The biogeographic relationships of New Guinea and Australia were studied by Crisp *et al.* (1995) using congruence among eleven groups of flowering plants. Of importance here is that there were two separate and distinct biogeographic connections between New Guinea and Australia. There was a clade relating areas with a monsoonal to arid climate, and a separate clade relating the wet tropical areas. This latter showed successive differentiation down the east coast of Australia. These two clades were not closely related, suggesting an ancient division between them. However, some individual cladograms showed a closer relationship between the two clades. This was due in each case to the effect of one or more widespread species, showing that there had also been recent connections between the floras of the wet tropical and monsoonal areas.

Discussion

The plant biogeography of the SE Asian-Australian region is complex, and it is certain that no single factor will account for the distribution of plants in the region (Hall 1998). Climate, both past and present, plate tectonics, and the biological characteristics of the plants themselves will all influence the patterns of distribution. Different groups of plants will generally react differently to the various changes, but general patterns do emerge, and these patterns are supported by a number of different types of biogeographic evidence, as discussed above.

The major patterns emerging are firstly of high floristic richness and endemism in the core areas of Borneo and New Guinea. These areas correspond both with central positions on the respective plates, Sunda and Sahul, and with the current centres of wet and relatively aseasonal climates.

Borneo has the richest flora of western Malesia, as can be seen in total species richness (Table 5.1). There are numerous groups of plants that are clearly centred in Borneo, such as the dipterocarps (Figure 5.3) and the rattan palms (Figure 5.5). These groups show a reduction in diversity and richness in both a northwesterly and a southeasterly direction. Within Borneo, there is a concentration of endemic species in the topographically more diverse areas of northern Borneo, rather than in the southern half of the island (van Balgooy *et al*. 1996). The areas immediately adjoining Borneo, specifically Sumatra and peninsular Malaysia, are in many ways similar in their flora, but differ from Borneo most noticeably in the lower number of endemic genera and species. Borneo has a predominance of groups that are centred in western Malesia, but there are also many groups that are apparently of Australian origin. These are generally readily recognisable where they are adapted to monsoonal climates, or poor soil conditions, such as much of the sclerophyll flora of monsoonal tropical Australia. However, there are also rain forest groups, such as some members of the Myrtaceae, Proteaceae, Rutaceae and Sapindaceae, that also appear to be of Australian origin (van Steenis 1985, 1987)

New Guinea has the richest flora of eastern Malesia. In part this statement merely reflects the fact that New Guinea is the only major centre of wet, more or less aseasonal tropical rain forest in eastern Malesia. The areas to the north (Philippines), west (Sulawesi) and south (tropical Australia) have predominantly more monsoonal climates, and a lower floristic richness. However, New Guinea apparently has the richest flora in the whole of Malesia, and the highest level of endemism.

New Guinea emerges in most analyses as having the largest proportion of endemic genera and species (Tables 5.1 and 5.2). Many of these endemic species are in the northern part of the island, where there are significantly more endemic species than in the southern half (van Balgooy *et al*. 1996). New Guinea also emerges as having a predominantly Asian or western Malesian flora, rather than an Australian flora (Table 5.2). Genera and species from both floras are present, but often with different distributions within New Guinea. The lowland rain forest flora is generally dominated by plants with an Asian or western Malesian relationship, whereas the montane rain forest flora has a notably higher proportion of plants with an Australian relationship (van Balgooy 1976; Hartley 1986). Most endemic species in New Guinea are found at higher altitudes (van Balgooy *et al*. 1996), and many show an Australian relationship.

Geographically between the core rain forest areas of Borneo in western Malesia and New Guinea in eastern Malesia is the 'dry corridor' of Wallacea, comprising the Lesser Sunda Islands, parts of Sulawesi, and parts of the Philippines. The floristic richness is lower and, in general, the level and proportion of endemism is lower (Table 5.1). The plate tectonic history of this whole area of eastern Malesia, in particular Sulawesi, the Philippines and New Guinea, indicates that these islands have essentially taken their current configuration and position only during the later part of the Tertiary.

Peter Kershaw, Bruno David, Nigel Tapper, Dan Penny and Jonathan Brown (Eds): Bridging Wallace´s Line

The flora of Wallacea shows a mixture of essentially western Malesian and eastern Malesian groups, with a higher proportion of monsoonal species showing relationships directly with mainland SE Asia on the one hand or with northern Australia on the other (van Steenis 1979, 1985). The strongest relationships of the Sulawesi flora are with the Philippines, the Moluccas and the Lesser Sunda Islands (van Balgooy 1987). These main relationships are therefore with other islands in the dry belt from northern Australia to the Philippines, and it is likely that climate plays a significant role. The relationships with the flora of Borneo are much lower, indicating that the Makassar Straits have remained a barrier to the movement of plants. The relationships of the Sulawesi flora with Borneo tend to be via either the Philippines or via the Lesser Sunda Islands, rather than directly (van Balgooy 1987).

The importance of climate in the plant biogeography of the SE Asian-Australian region has been emphasised by Morley (1998); this is true both currently and historically. The present centres of floristic richness and endemism, Borneo and New Guinea, reflect the climatic optima for tropical rain forest. However, it should be noted that Borneo has undergone cycles of aridity during the Tertiary (Morley 1998), and that New Guinea as such has only existed since the later part of the Tertiary. New Guinea is similar to much of eastern Malesia in this last respect. However, New Guinea is under a wetter and more aseasonal climate, and shows high floristic richness and endemism, whereas Sulawesi and the Philippines are more monsoonal, and with a lower floristic richness and endemism. Australia, once covered with rain forest, is now predominantly monsoonal to semi-arid in the tropical regions. The rain forest flora has become restricted to NE Queensland, where the levels of floristic richness and endemism are much higher than in the monsoonal Australian tropics. A somewhat similar picture may well apply to India, on the other side of the SE Asian-Australian region (Morley 1998).

Care must be taken when considering areas of high species richness and endemism. These do not necessarily equate to areas of origin, and there is increasing evidence that many plant groups evolved in areas where they are now either extinct or represented by relatively low numbers of species. Morley (1998) shows that many plant groups now centred in western Malesia may well have arrived in this region from India, and that their origin may have been in Gondwana, in Africa or in India. While these groups are now well represented in western Malesia, they are poorly or not represented in their apparent area of origin, due mainly to climatic changes in those areas. In a similar manner, the flora centred in eastern Malesia appears to have originated in Australia (when it was part of Gondwana), but in many cases this flora is much richer in areas such as eastern Malesia that it is in Australia, due again to climatic changes in Australia.

Vegetation appears to have a remarkable ability to track climatic changes. There have been marked changes in climate over the late Tertiary and the Quaternary, and for the most part the vegetation has responded by showing expansion and contraction of range. For example, studies in NE Australia show that there have been rapid changes in the local flora over relatively short intervals of time, responding to rapid climatic changes (Kershaw 1994). However, analysis of the actual versus potential distribution of rain forest vegetation indicates that for the most part the present distribution of rain forest types reflects the present distribution of suitable climate (Webb and Tracey 1981; Nix 1991). There is, however, not an exact match of suitable climate (and other environmental conditions) and the present distribution of rain forest types. This is a reflection of the effects of stochastic events such as dispersal and local extinction. The rain forest flora is also not the same in all areas, showing centres of high species richness and endemism, with other areas of much lower species richness and (especially) endemism (Whiffin and Hyland 1986). Species with poorer dispersal mechanisms, or with more tightly circumscribed climatic requirements, tend to remain in areas with greater persistence of the climatic conditions to which they are adapted. These species appear as the local or regional endemics, and the areas in which they are found appear as the long term refugial areas (Whiffin and Hyland 1986). However, for the most part, these species are identifying areas of long-term suitability of climate, rather than areas of origin as such.

In this light, the current centres of floristic richness and endemism, in Borneo and New Guinea, should be seen as reflecting the current optimal climate for aseasonal tropical rain forest. The topographic diversity, particularly in northern Borneo and northern New Guinea, has presumably ensured the persistence of small areas of suitable climate over the period of the late Tertiary and Quaternary.

The traditional view of an Asian (Laurasian) rain forest flora extending over western Malesia, and then invading eastern Malesia and Australia, has long been overturned. However, the more current view of a western Malesian (Asian) flora and an eastern Malesian (Australian) flora intermingling in the late Tertiary may also need refinement. Morley (1998) demonstrates that the western Malesian flora is for the most part not Laurasian, but Gondwanan, having arrived via India. There has also been the opportunity for the exchange of floras across Wallace's Line on a number of different occasions during the late Tertiary and Quaternary, with climate perhaps playing the limiting role. In addition, part of the western Malesian flora may have arrived in eastern Malesia before the formation of the Makassar Straits, via Sulawesi (Morley 1998).

At the risk of oversimplification, it is possible to recognise three main ecological elements in the flora of Malesia: wet tropical; dry tropical (monsoonal); and montane. Continuing the oversimplification, it is also possible to recognise three biogeographic elements: Sundaic (or western Malesian); Australian (or eastern Malesian); and temperate. However, there is not a one to one correspondence between these biogeographic elements and the ecological elements, as will be discussed.

It is also possible to recognise two other biogeographic elements (Table 5.2), being the widespread (including pan-tropical) and endemic elements. However, while of biological (and biogeographical) significance and interest, these are not informative in the current situation.

The term Gondwanan is often used for the Australian element, but this may be misleading. As noted above, it is likely that at least the major part of the Sundaic element was also Gondwanan in origin.

Peter Kershaw, Bruno David, Nigel Tapper, Dan Penny and Jonathan Brown (Eds): Bridging Wallace's Line

The wet tropical element in Malesia appears to contain plant groups that have been derived predominantly from a Gondwanan origin, and that are now associated either with the Sundaic or the Australian biogeographic elements. In other words, both the major elements (Sundaic and Australian) have contributed to the wet tropical element. There may be some differentiation between the two, in that many of the more clearly Australian groups are subtropical (mesothermal) in their climatic requirements (Nix 1991), while many of the Sundaic groups are tropical (megathermal). This is apparent in New Guinea, where the Sundaic groups dominate the lowland rain forest, with the Australian groups being much more common in the montane rain forest. However, there is probably not a clear division along these lines, with plant groups or even species within groups moving across these ecological boundaries. Particularly in the case of the Australian element, these rain forest groups are recognised as relictual Gondwanan within Australia, restricted to small areas of rain forest mostly in NE Australia.

The dry tropical (monsoonal) element in Malesia predominates in the dry corridor from northern Australia through southern Sulawesi to parts of the Philippines. At times of increased aridity during the late Tertiary and Quaternary, the distribution of this element would have been wider, and this is now reflected in the disjunct distribution of certain species. This biogeographic element has also been derived from both mainland SE Asia and Australia, where these two areas have a wider distribution of monsoonal climate than does Malesia. In Australia, at least, this element is part of the derived (sclerophyllous) Gondwanan flora.

The temperate flora in Malesia is very restricted, occurring in the high montane areas of the higher mountains. It will have achieved its present distribution, sometimes crossing between the two hemispheres, by dispersal along the mountain chains (Morley 1998). Plant groups within this element can often be assigned to Gondwanan or Laurasian origin, but some are now widespread and difficult to assign as to origin.

In summary, the plant biogeography of the SE Asian-Australian region is complex, reflecting in part the complex geological history of the area. While the plate tectonic history of the area is gradually being elucidated, our increasing knowledge of the systematics, biogeography and phylogeny of important plant groups is also increasing, and is aiding in a better understanding of the plant biogeography of this region.

References

Atkins, H., Preston, J. and Cronk, Q.C.B. 2001. A molecular test of Huxley's line: *Cyrtandra* (Gesneriaceae) in Borneo and the Philippines. *Biological Journal of the Linnean Society* 72: 143-159.
Ashton, P.S. 1982. Dipterocarpaceae. *Flora Malesiana* 9: 237-552.
Ashton, P.S. 1989. Dipterocarp reproductive biology. In: H. Lieth and M.J.A. Werger (Editors), *Tropical Rain Forest Ecosystems: Biogeographical and Ecological Studies*. Elsevier, Amsterdam, pp. 219-240.

Baker, W.J., Coode, M.J.E., Dransfield, J., Dransfield, S., Harley, M.M., Hoffmann, P. and Johns, R.J. 1998. Patterns of distribution of Malesian vascular plants. In: R. Hall and J.D. Holloway (Editors), *Biogeography and Geological Evolution of SE Asia*. Backhuys, Leiden, pp. 243-258.

Brandon-Jones, D. 1998. Pre-glacial Bornean primate impoverishment and Wallace's Line. In: R. Hall and J.D. Holloway (Editors), *Biogeography and Geological Evolution of SE Asia*. Backhuys, Leiden, pp. 393-403

Campbell, D.G. and Hammond, H.D. (Editors). 1989. *Floristic Inventory of Tropical Countries*. New York Botanical Garden, New York, 545 pp.

Crisp, M.D., Linder, H.P. and Weston, P.H. 1995. Cladistic biogeography of plants in Australia and New Guinea: congruent pattern reveals two endemic tropical tracks. *Systematic Biology* 44: 457-473.

Crisp, M.D., West, J.G. and Linder, H.P. 1999. Biogeography of the terrestrial flora. *Flora of Australia*, 2nd edn, 1: 321-367.

Dransfield, J. 1981. Palms and Wallace's Line. In: T.C. Whitmore (Editor), *Wallace's Line and Plate Tectonics*. Clarendon Press, Oxford, pp. 43-56.

Dransfield, J. 1987. Bicentric distribution in Malesia as exemplified by palms. In: T.C. Whitmore (Editor), *Biogeographical Evolution of the Malay Archipelago*. Clarendon Press, Oxford, pp. 60-72.

Flora of Australia. 1995. Volume 16, *Elaeagnaceae, Proteaceae* 1. CSIRO Australia, Melbourne, 522 pp.

Foreman, D.B. 1995. *Helicia*. Flora of Australia 16: 393-399.

George, W. 1981. Wallace and his line. In: T.C. Whitmore (Editor), *Wallace's Line and Plate Tectonics*. Clarendon Press, Oxford, pp. 3-8.

Hall, R. 1998. The plate tectonics of Cenozoic SE Asia and the distribution of land and sea. In: R. Hall and J.D. Holloway (Editors), *Biogeography and Geological Evolution of SE Asia*. Backhuys, Leiden, pp. 99-131.

Hahn, W.J. and Sytsma, K.J. 1999. Molecular systematics and biogeography of the southeast Asian genus *Caryota* (Palmae). *Systematic Botany* 24: 558-580.

Hartley, T.G. 1986. Floristic relationships of the rainforest flora of New Guinea. *Telopea* 2: 616-630.

Holloway, J.D. and Hall, R. 1998. SE Asian geology and biogeography: an introduction. In: R. Hall and J.D. Holloway (Editors), *Biogeography and Geological Evolution of SE Asia*. Backhuys, Leiden, pp. 1-23.

Kershaw, A.P. 1994. Pleistocene vegetation of the humid tropics of northeastern Queensland, Australia. *Palaeogeography, Palaeoclimatology and Palaeoecology* 109: 399-412.

Kershaw, A.P., Martin, H.A. and McEwen Mason, J.R.C. 1994. The Neogene: a period of transition. In: R.S. Hill (Editor), *History of the Australian Vegetation: Creataceous to Recent*. Cambridge University Press, Cambridge, pp. 299-327.

Morley, R.J. 1998. Palynological evidence for Tertiary plant dispersals in the SE Asian region in relation to plate tectonics and climate. In: R. Hall and J.D. Holloway (Editors), *Biogeography and Geological Evolution of SE Asia*. Backhuys, Leiden, pp. 211-234.

Morat, Ph., Veillon, J.-M. and MacKee, H.S. 1986. Floristic relationships of New Caledonian rainforest phanerogams. *Telopea* 2: 631-679.

Moss, S.J. and Wilson, M.E.J. 1998. Biogeographic implications of the Tertiary palaeogeographic evolution of Sulawesi and Borneo. In: R. Hall and J.D. Holloway (Editors), *Biogeography and Geological Evolution of SE Asia*. Backhuys, Leiden, pp. 133-163.

Nix, H.A. 1991. An environmental analysis of Australian rainforests. In: G. Werren and P. Kershaw (Editors), *The Rainforest Legacy, Australian National Rainforests Study*, volume 2. Canberra, Australian Government Publishing Service, pp. 1-26.

Sleumer, H. 1955. Proteaceae. *Flora Malesiana* 5: 147-206.

Simpson, G.G. 1977. Too many lines: the limits of the Oriental and Australian zoogeographic regions. *Proceedings of the American Philosophical Society* 121: 107-120.

Turner, H. 1995. Cladistic and biogeographic analyses of *Arytera* Blume and *Mischarytera* gen. nov. (Sapindaceae) with notes on methodology and a full taxonomic revision. *Blumea Supplement* 9: 1-230.

Turner, H. 1996. Sapindaceae and the biogeography of eastern Australia. *Australian Systematic Botany* 9: 133-167.

van Balgooy, M.M.J. 1976. Phytogeography. In: K. Paijmans (Editor), *New Guinea Vegetation*. Canberra, Australian National University Press, pp. 1-22.

van Balgooy, M.M.J. 1987. A plant geographical analysis of Sulawesi. In: T.C. Whitmore (Editor), *Biogeographical Evolution of the Malay Archipelago*. Clarendon Press, Oxford, pp. 94-102.

van Balgooy, M.M.J., Hovenkamp, P.H. and van Welzen, P.C. 1996. Phytogeography of the Pacific – floristic and historical distribution patterns in plants. In: A. Keast and S.E. Miller (Editors), *The Origin and Evolution of Pacific Island Biotas, New Guinea to Eastern Polynesia: Patterns and Processes*. SPB Academic Publishing, Amsterdam, pp. 191-213.

van Steenis, C.G.G.J. 1950. The delimitation of Malaysia and its main plant geographical divisions. *Flora Malesiana* 1: 70-75.

van Steenis, C.G.G.J. 1979. Plant-geography of east Malesia. *Botanical Journal of the Linnean Society* 79: 97-178.

van Steenis, C.G.G.J. 1985 [1986]. The Australasian generic element in Malesia. *Brunonia* 8: 349-372.

van Steenis, C.G.G.J. 1987. *Checklist of Generic Names in Malesian Botany*. Flora Malesiana Foundation, Leiden, 162 pp.

van Welzen, P.C. 1989. *Guioa* Cav. (Sapindaceae): taxonomy, phylogeny and historical biogeography. *Leiden Botanical Series* 12: 1-315.

van Welzen 1994. *Guioa. Flora Malesiana* 11: 548-598.

Webb, L.J. and Tracey, J.G. 1981. Australian rainforests: patterns and change. In: A. Keast (Editor), *Ecological Biogeography of Australia*. W. Junk, The Hague, pp. 607-694.

Weston, P.H and Crisp, M.D. 1996. Trans-Pacific biogeographic patterns in the Proteaceae. In: A. Keast and S.E. Miller (Editors), *The Origin and Evolution of Pacific Island Biotas, New Guinea to Eastern Polynesia: Patterns and Processes*. SPB Academic Publishing, Amsterdam, pp. 215-232.

Whiffin, T. and Hyland, B.P.M. 1986. Taxonomic and biogeographic evidence on the relationships of Australian rainforest plants. *Telopea* 2: 591-610

Whitmore, T.C. 1981. Wallace's Line and some other plants. In: T.C. Whitmore (Editor), *Wallace's Line and Plate Tectonics*. Clarendon Press, Oxford, pp. 70-80.

Whitmore, T.C. 1989. Southeast Asian tropical forests. In: H. Lieth and M.J.A. Werger (Editors), *Tropical Rain Forest Ecosystems: Biogeographical and Ecological Studies*. Elsevier, Amsterdam, pp. 195-218.

Womersley, J.S. (Editor). 1978. *Handbooks of the Flora of Papua New Guinea*, volume 1. Melbourne, Melbourne University Press, 278 pp.

PART II

QUATERNARY ENVIRONMENTS

Quaternary Flora and Vegetation of Java

A.A. Polhaupessy

Introduction

Pollen analysis, initially developed for the study of Quaternary vegetation in temperate areas, has since proved to be equally applicable in the study of species-rich equatorial vegetation, despite the much greater diversity of its pollen rain, and predominance of zoophily over anemophily (Flenley 1979). The application of pollen analysis in the Indonesian region is still very much in its infancy, although some progress has been made with respect to the younger Quaternary (e.g. Stuijts *et al.* 1988). This paper examines palynological evidence for the history of vegetation, palaeoenvironments and climates during the Quaternary and latest Tertiary for the island of Java.

Javanese fossil pollen was first observed by Polak (1933) from the Rawa Lakbok, a large coastal swamp in Central Java, although no actual counts showing the relative abundances of the pollen types recorded from this site were made. The first pollen diagrams to be produced for the island of Java were from Holocene sediments from two small lakes, Situ Gunung and Telaga Patenggang, within the lower montane rain forest formation of West Java (van Zeist *et al.* 1979), and subsequently, Stuijts (1984) published a diagram from nearby Situ Bayongbong, which was the first Quaternary profile from Java to extend beyond the Holocene into the period represented by the last glacial maximum (Figure 6.1).

No late Quaternary pollen profiles have yet been published from lowland settings in Java, although several palynological studies have been performed from fluvial and shallow marine Early Quaternary and latest Tertiary sediments from Central and East Java, in particular, the sections from Trinil and Sangiran which have become famous for their fossil skulls of *Homo erectus*. These studies do not reveal detailed vegetational histories in the manner of Late Quaternary profiles, but provide more generalised information about the nature of the vegetation which grew on lowland flood plains at the time, and also important data regarding early Quaternary and latest Tertiary climates. In addition to the hominid localities mentioned above, palynological studies are reviewed from the Solo-Madiun area in East Java, and Bumiayu in Central Java (Figure 6.1). In addition to providing information on vegetation and climate, the presence of

ISBN 3-923381-47-6
© 2002 by CATENA VERLAG, 35447 Reiskirchen

stratigraphically restricted pollen and spores within these sections helps to establish their relative age.

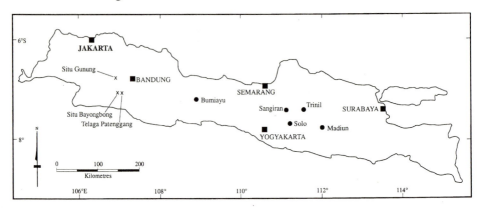

Fig. 6.1 Java: study area.

Late Quaternary Pollen Diagrams

Situ Gunung

This lake is situated on the southwest facing slope of the extinct volcano Gunung
Gede-Prangrango, at an elevation of about 1000 m, 10 km north of Sukabumi
(Figure 6.1), and is surrounded by a belt of swamp vegetation. The section
analysed covers a time-span of nearly 8000 years, based on a radiocarbon date
from 7.45-7.57 m, and may date from the last major eruption of the Gede-
Pangrango volcano. In the original study of van Zeist *et al.* (1979), pollen
diagrams were provided for pollen and spores produced by the local swamp and
aquatic vegetation, and also the surrounding dry land vegetation, with a minor
proportion of the pollen being been blown in from a long distance. Only pollen
from trees and shrubs is presented here, since this provides the most information
about the character of the surrounding forests (Figure 6.2), although some of the
pollen representing woody taxa may have been produced by plants growing in
swamp vegetation communities on or around the lake.

The pollen profile is dominated by *Altingia* (Hamamelidaceae) pollen, which
suggests that *Altingia* grew abundantly around the lake throughout the Holocene,
but pollen of Fagaceae (*Quercus* and *Castanopsis* type), Euphorbiaceae *(Bischofia, Macaranga, Trewia)*, Moraceae *(Ficus)*, Myricaceae *(Myrica)*, Ulmaceae
(*Trema*), Juglandaceae *(Engelhardia)* and *Stephania* and Podocarpaceae *(Podocarpus)* are also well represented, and these taxa must have all been important
elements of the surrounding vegetation.

The pollen diagram shows a single major break at about 4.50 m, dated by
radiocarbon at about 5000 years BP, below which depth *Altingia* pollen is
particularly common, and *Myrica* and *Trewia* significantly represented, whereas
above 4.50 m, *Castanopsis* is common, and *Pandanus* (Pandanaceae) abundant,
and pollen of Myrtaceae and Melastomataceae well represented. The precise

meaning of this change in terms of regional vegetation is unclear, but the increased representation of *Pandanus* pollen most probably relates to the expansion of *Pandanus* within the surrounding swamp vegetation, especially since the maximum of *Pandanus* pollen coincides with increased fern spores and herbaceous swamp elements, suggesting the widespread development of herbaceous marsh vegetation at the site. The diagram therefore suggests that the regional vegetation surrounding Situ Gunung changed little during the major part of the Holocene, but any changes which did occur appear closely related to the character of vegetation growing on and around the site.

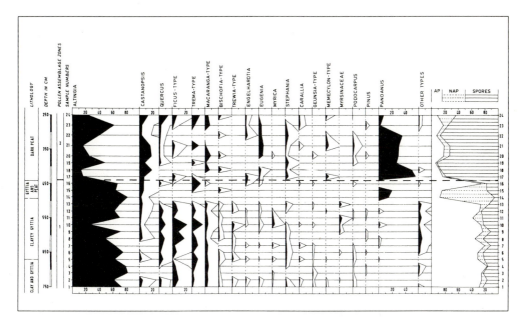

Fig. 6.2 Pollen diagram of Situ Gunung (after Polhaupessy, in Zeist et al. 1979).

Telaga Patenggang

This lake is situated at 1575 m, about 55 km SW of Bandung (Figure 6.1), and consists of a crater lake (Verbeek and Fenema 1896). The lake lies at the east side of a large marshy depression, now covered with tea plantations. A 3.8 m core was retrieved, but no radiocarbon dates were obtained, and so the age is unclear, although a Holocene age is assumed. The pollen diagram (Figure 6.3) shows high frequencies for *Altingia* pollen (although less than for Situ Gunung), *Quercus, Castanopsis* type, and *Podocarpus*, and the fair representation of *Ilex, Schima* (Theaceae), *Ficus* and *Elaeocarpus*. Other taxa which are reasonably represented are Myrtaceae, *Trema*, Melastomataceae, and Euphorbiaceae, including *Bischofia, Macaranga* and *Trewia*. All the above taxa must have been prominent in the vegetation surrounding the lake during the Holocene.

Peter Kershaw, Bruno David, Nigel Tapper, Dan Penny and Jonathan Brown (Eds): Bridging Wallace´s Line

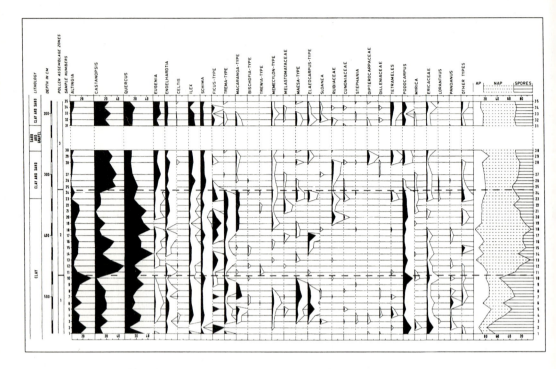

Fig. 6.3 Pollen diagram of Telaga Patenggang (after Stuijts, in Zeist et al. 1979).

The main changes within this profile are in the abundance of *Altingia* pollen relative to *Quercus* and *Castanopsis*. *Altingia* shows a clear abundance maximum below 4.65 m, whereas Fagaceae pollen shows a sharp increase in abundance above this depth, although whereas *Quercus* pollen remains abundant up to 3.20 m, *Castanopsis* pollen declines in abundance. Above 3.20 m, *Castanopsis* pollen shows a sharp increase in abundance, whereas *Altingia* pollen shows a sudden decline, and remains at low levels to the top of the profile.

It is difficult to make a detailed interpretation of this pollen diagram in terms of regional vegetation. As with Situ Gunung, the main assemblage changes are closely associated with features of the hydrosere; the lower, *Altingia*-dominated interval corresponds with high frequencies of spores, possibly reflecting fern-dominated marsh around the lake, whereas the upper interval, with very low percentages of *Altingia*, corresponds to a change from clay to sandy and gravel lithologies.

The successive reduction of *Altingia* in both Telaga Patengan and Situ Gunung may relate to similar ecological successions in both areas, although it cannot be suggested that any of the events seen in the two sections are synchronous, since no radiocarbon dates were obtained for the Telaga Patenggang section. Possible explanations are: a) direct relationship with the hydroseral succession, since the main pollen assemblage changes seen in both profiles coincide closely with hydroseral changes, or b) a relationship with volcanic activity, since pollen of the regrowth forest taxa *Macaranga, Trema* and *Ficus* are often common close to the main assemblage changes. A third possibility is human activity, since *Altingia*, which shows stepwise reductions in abundance in both profiles, is a highly favoured timber.

The main difference between the Situ Gunung and Telaga Patenggang profiles is in the overall abundance of *Altingia* and Fagaceae pollen, with *Altingia* being most abundantly represented in the Situ Gunung profile, whereas Fagaceae are most common at Telaga Patengan. This simply reflects the higher altitude of Telaga Patengan compared to Situ Gunung.

Situ Bayongbong

This small swamp is situated at an elevation of 1300 m, about 55 km SW of Bandung (Figure 6.1). The swamp has an outlet, on its northern side, and is covered with herbaceous marsh. The site was investigated by Stuijts (1984).

Situ Bayongbong has yielded an 8 m core consisting of lake muds overlain by peat, and the oldest sediments have been dated at 16,800 years BP. The pollen profile which was produced from this site (Figure 6.4) is therefore particularly important, since it is the first profile from Java to penetrate the period of the last glacial maximum (more recently, van der Kaars and Dam (1996) have published a long Late Quaternary profile from the Bandung Lake itself, but this study is excluded from this review).

The pollen profile for this site shows that during the time of the last glacial maximum, from 16,800 to about 12,000 years BP, the podocarp *Dacrycarpus imbricatus* was extensively represented around the site, and was associated with *Engelhardia* and Fagaceae (both *Quercus* and *Lithocarpus*), and intermittently with *Myrsine* (Myrsinaceae). The abundance of *D. imbricatus* with *Myrsine* suggests that the climate at this time was considerably cooler than at present.

Between about 12,000 and 9000 years BP, *Dacrycarpus imbricatus* pollen shows a gradual decline in abundance, whereas pollen of *Engelhardia* shows a sharp increase, peaking at about 11,000 years BP. It is likely that the decline of *D. imbricatus* reflects a climatic change to warmer climates, but the expansion of *Engelhardia*, coupled with the presence of pollen of *Myrica* at this time, may also suggest some kind of volcanic disturbance (Stuijts *et al.* 1988).

After about 9000 years BP, pollen of *Engelhardia* declines rapidly, *Myrsine* pollen is virtually absent, and *Dacrycarpus imbricatus* falls to low values. On the other hand, pollen of *Altingia, Castanopsis* and *Quercus* increase in abundance, suggesting that from the earliest Holocene onward, the *D. imbricatus* and *Engelhardia* dominated forests were replaced by forests with abundant oaks and *Altingia*, which are presently the main dominants in undisturbed forests surrounding this site. This change is thought to reflect a change to warmer climates at the beginning of the Holocene.

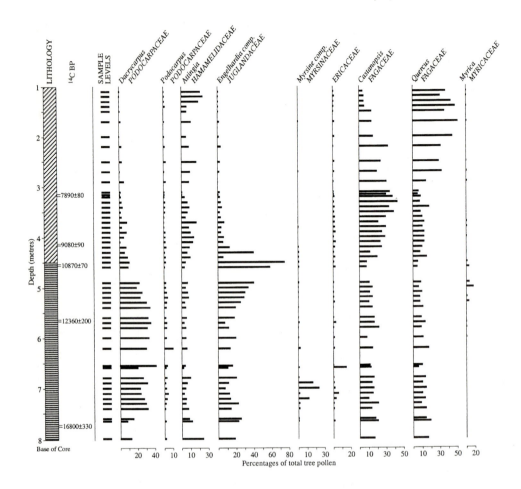

Fig. 6.4 Pollen diagram of Situ Bayongbong (after Stuijts et al. 1988).

Studies of Late Pliocene and Early Quaternary sediments

Late Pliocene and Early Quaternary sediments are widespread in Central and East Java, locally attaining considerable thickness, and consist mainly of gently folded shallow marine, coastal and flood plain deposits, which are particularly well known for their content of hominid skulls and also of fossils of other large vertebrates.

Bumiayu

The region of Bumiayu lies within the western part of Central Java, equidistant from the Java Sea to the north, and Indian Ocean to the south. The data summarised here was generated from a suite of samples collected from outcrops of the predominantly claystone Kalibiuk Formation and the overlying sandy and conglomeratic Kaliglagah Formation, from along the Cisaat River. A more detailed presentation of results from this location is presented in Polhaupessy (1990).

A total of 30 samples were analysed for their pollen content, and all yielded moderate recovery of well-preserved pollen. The assemblages recovered principally reflect the broad depositional environment within a brackish to freshwater coastal plain, but also provide some indications of the character of more regional vegetation. The most characteristic feature of the pollen diagram is the dominance of pollen of Gramineae and Cyperaceae, and of mangrove pollen, with the result that the diagram contrasts markedly with those from the Late Quaternary lakes discussed earlier. For the convenience of description, the section is divided into six pollen zones (Figure 6.5), which are discussed below.

Pollen zones 1-3 are associated with clay lithologies, which suggest deposition within low energy settings. The oldest interval, zone 1, is characterised by pollen of mangrove. and swamp forest trees, as well as abundant Gramineae pollen. Podocarp pollen is also present in low frequencies. The presence of mangrove pollen suggests a coastal brackish setting, whereas the presence of pollen of freshwater swamp trees and abundant Gramineae suggests slow moving rivers and small lakes surrounded by grass-dominated swamps within the flood plain. Podocarp pollen was probably derived from a nearby upland source. Within zone 2, mangrove pollen is much more common, and Gramineae pollen abundant, whereas pollen of freshwater swamp trees is less well represented. A depositional setting is suggested with extensive freshwater lakes or brackish lagoons, with widespread grass-dominated swamps, and with mangrove swamps occurring widely in the vicinity. Pollen zone 3 is again characterised by mangrove pollen and pollen of Gramineae, and a depositional setting similar to that for zone 2 is suggested, but with brackish conditions being predominant. The *Turritella* bed (Figure 6.5) coincides with a maximum of *Rhizophora* type pollen, and may reflect a sudden sea level rise. The presence of *Podocarpus* pollen through this zone may reflect long distance transport from a nearby upland source.

Pollen zone 4 represents a transitional sequence, both in terms of lithologies, which include sand, clay, and also a thin lignite bed, indicating peat accumulation, and pollen content. Mangrove pollen is reduced, but Gramineae pollen remains common, suggesting a mainly freshwater setting, but still with some brackish influence. The presence of caliche nodules within this interval suggests a strongly seasonal climate. Zone 5, which coincides with an interval of sands and gravels, is mainly characterised by pollen of Gramineae, but with some pollen of freshwater swamp trees. Caliche nodules are common, again suggesting a strongly seasonal climate. It is possible that the Gramineae pollen within this interval is derived from grassy savannas as well as grass-dominated freshwater swamps on the floodplain. *Podocarpus* pollen is present within both zones 4 and 5, and its presence suggests that montane forests were present on adjacent uplands throughout the period of deposition.

Peter Kershaw, Bruno David, Nigel Tapper, Dan Penny and Jonathan Brown (Eds): Bridging Wallace's Line

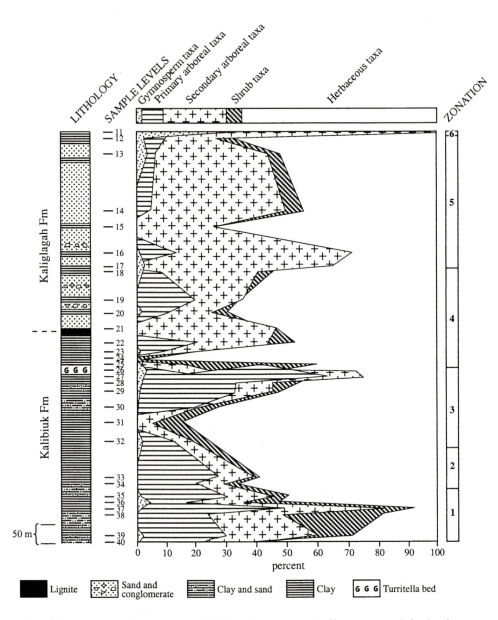

Fig. 6.5 Summary pollen diagram of CISAAT, Bumiayu. Pollen sum: total dry land taxa.

Within pollen zone 6, there is a return of mangrove pollen, suggesting a final phase of transgression. Deposition within a coastal brackish setting is suggested.

The Bumiayu succession yielded two pollen types which determine the approximate age of the sequence. *Dacrycarpus imbricatus* pollen occurs throughout the succession, and this first appears in the region of Java at the beginning of the Late Pliocene, about 3.5 million years ago (Morley 1991, 1998). Spores of the climbing fern *Stenochlaena milnei* and also present through the succession, and this species became extinct in Java in the latest Pliocene, or possibly very earliest Quaternary (Morley 1991, 1998). The co-occurrence of these two palynomorphs therefore suggests a Late Pliocene to earliest Quaternary age, and confirms earlier conclusions regarding age by von Koenigswald (1934).

In summary, the pollen record from the Kalibiuk and Kaliglagah formations at Bumiayu suggests that deposition occurred during the Late Pliocene to earliest Quaternary, initially within mangrove, lagoonal and freshwater lacustrine environments (zones 1-4), again confirming the earlier interpretations based on geological observations by von Koenigswald (1934), while zone 5 reflects a high energy floodplain depositional setting, with zone 6 relating to a phase of transgression and the return of brackish mangrove swamps. The most distinctive feature of the Bumiayu pollen diagram, however, is the dominance of pollen of Gramineae, reflecting the widespread occurrence of grass-dominated swamps, and possibly also of savanna. This suggests that during the Late Pliocene, the climate of the Bumiayu region was much drier, and more strongly seasonal than at present, and this would be consistent with the rich vertebrate faunas found in this, and laterally equivalent stratigraphic successions in Central and East Java.

Sangiran

The stratigraphic succession at Sangiran, where many hominoid fossils have been found, is composed of two formations, with the Kabuh Formation overlying the Pucangan Formation. Palynological examination was undertaken by Tokunaga *et al.* (1985) and Semah (1984) from several localities. The Pucangan Formation is generally characterised by low pollen concentrations, with fern spores and *Sonnaratia caseolaris* pollen, whereas analyses of the Kabuh Formation indicate a predominance of non-arboreal pollen, with Gramineae, Cyperaceae and abundant fern spores. Tokunaga *et al.* (1985) and Semah (1984) suggest that sediments of the Pucangan Formation accumulated within a freshwater swamp to brackish setting, whereas the deposition of the Kabuh Formation occurred in a setting characterised by freshwater swamps. With the common occurrence of Gramineae pollen throughout the sections studied, several parallels can be seen with the Kalibiuk and Kaliglagah Formations at Bumiayu, and a strongly seasonal climatic setting is suggested.

Trinil

The Pucangan and Kabuh formations within the Trinil area, which is also well known for its hominid fossils, were studied palynologically by Polhaupessy (1990), who divided the succession into two palynological zones (Figure 6.6). Pollen zone 1 is dominated by herbaceous pollen, with abundant pollen of Gramineae, suggesting the widespread occurrence of grass-dominated swamps

Peter Kershaw, Bruno David, Nigel Tapper, Dan Penny and Jonathan Brown (Eds): Bridging Wallace´s Line

within a fluvial setting, and also possibly the occurrence of savanna grassland, and suggests a markedly seasonal climate. Only low frequencies of arboreal pollen were noted, mainly consisting of *Podocarpus*, and it is likely that these were derived from nearby upland regions with montane forest.

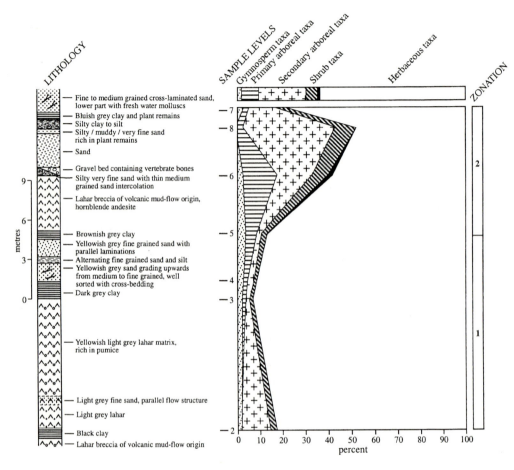

Fig. 6.6 Summary pollen diagram of Trinil, East Java. Pollen sum: total dry land taxa.

Pollen zone 2 is characterised by pollen of rain forest and riverine trees, and reduced frequencies of herbaceous pollen, although pollen of Gramineae remain very abundant, and low frequencies of podocarp pollen are also present. These assemblages suggest both freshwater swamp forest and grass-dominated swamps within the floodplain, and possibly a mosaic of rain forest and grass-dominated savanna on well-drained areas. *Podocarpus* and *Dacrycarpus imbricatus* were probably well represented in montane forest on adjacent upland areas. The climate for zone 2 was still considered seasonal, to account for the abundance of Gramineae, but probably considerably more moist than for zone 1.

It is noteworthy that no specimens of *Stenochlaena milnei* were observed from the Trinil section, and since *Dacrycarpus imbricatus* is regularly present, following Morley (1991), a Pleistocene age is suggested on the basis of age restricted palynomorphs. This suggestion is consistent with results of a recent magnetostratigraphic study by Leinders *et al.* (1985) which suggests that the Trinil fauna is about 1.0 million year old (Early Pleistocene).

Solo-Madiun area

During 1976-9, JICA (Japanese aid programme) and the Geological Research and Development Centre in Bandung undertook a study with the aim of establishing a detailed chronostratigraphy for the Quaternary of Java. Palynology was used primarily to reconstruct past vegetation. Samples were collected from the Pucangan, Kabuh and overlying Setri Formation along the Solo and Madiun Rivers. Results of these analyses are presented in Polhaupessy and Sudijono (1985).

Pollen assemblages from the Pucangan Formation were similar to those recorded from the same formation elsewhere, with Gramineae and Cyperaceae pollen, and pteridophyte spores dominant, but with some pollen of mangrove taxa, suggesting a brackish to freshwater, lagoonal setting. Mangrove pollen was not recorded from the overlying Kabuh Formation, assemblages here being dominated by Gramineae pollen and pteridophyte spores but with reduced Cyperaceae compared to the Pucangan Formation. Deposition on a fluvial plain is suggested in these regions, possibly with grass-dominated savanna growing on well-drained areas.

Palynomorph assemblages from the overlying Setri Formation are dominated by pollen of herbs, and fern spores, with Gramineae pollen dominating. Cyperaceae pollen proved to be absent from the localities studied. Several gymnosperm pollen types were noted, including several types of *Podocarpus*, *Dacrycarpus imbricatus* and *Dacrydium* pollen. The palynomorph assemblages suggest deposition in a freshwater setting, probably within a fluvial plain, traversed by rivers and with small, ponds and lakes, surrounded by grassy swamps. Gymnosperm pollen was probably derived from montane forest growing on adjacent upland areas. The presence of *Dacrydium* pollen is noteworthy, since *Dacrydium* is now absent from Java (the youngest records are from the early Holocene of Bandung Lake (Polhaupessy unpublished data)).

The presence of *Dacrycarpus imbricatus* pollen within samples studied from the Solo-Madiun region, but absence of *Stenochlaena milnei* spores suggests a Quaternary age for these formations. Also, the predominance of Gramineae pollen in all formations suggests the presence of widespread grass-dominated swamps, and possibly the presence of savanna on well drained areas. This suggests that, as for other Pleistocene sediments in East and Central Java which are rich in vertebrate fossils, the climate was markedly more seasonal than that experienced in these regions today.

Peter Kershaw, Bruno David, Nigel Tapper, Dan Penny and Jonathan Brown (Eds): Bridging Wallace´s Line

Discussion

This paper reviews Quaternary and Late Pliocene palynological evidence for the history of vegetation and climate for the island of Java. Data is available from two strongly contrasting sources; cores collected from lake and swamp sediments from the mountains of West Java provide relatively detailed histories of montane vegetation and climate from the time of the last glacial maximum and the Holocene, whereas studies of the vertebrate-bearing (and hominid-bearing) Early Quaternary and Late Pliocene sediments of Central and East Java provide information on their depositional setting, with rather generalised conclusions regarding palaeoclimate. The presence of age-restricted palynomorphs within these formations also provide a broad age framework.

The oldest sediments for which data is available come from the Kalibiuk and Kaliglagah formations at Bumiayu in western Central Java. Palynological data confirms a Late Pliocene or earliest Quaternary age and suggests that deposition occurred within a basically regressive regime, initially within a brachish lagoonal setting, for the Kalibiuk Formation, later occurring within a fluvial setting for the Kaliglagah Formation. A characteristic feature of the whole succession is the abundant occurrence of Gramineae pollen, suggesting widespread grass-dominated swamps associated with both mangrove swamps and freshwater lakes, and also the possibility of widespread grass-dominated savanna on well drained soils. This suggests that the climate during the Late Pliocene was more strongly seasonal than that experienced in the region today, and the presence of widespread open vegetation, rather than forest, helps explain the occurrence of fossils of large browsing and grazing vertebrates within the Kaliglagah Formation, such as mastodon, elephant, hippopotamus, deer and antelope.

For the Early Quaternary Pucangan, Kabuh and Setri formations, which bear hominid fossils as well as diverse large vertebrates, palynology confirms that these formations are not older than Quaternary, but cannot at present provide a more precise age interpretation. For instance, magnetostratigraphic studies suggest that the hominid-bearing Trinil fauna is about 1.0 million years old. For each of the Early Quaternary formations, palynological data suggests the much more widespread occurrence of Gramineae that might be expected under natural conditions within Central and East Java today, with grass-dominated swamps occurring within brackish settings, and associated with lakes and rivers in fluvial settings, in a similar manner as for the older Kalibiuk and Kaliglagah formations. Also the possibility needs to be raised that open, grass-dominated savanna was also widespread in areas of well drained soils, although there is evidence for a mosiac of both forest and grassland within the Trinil succession. The presence of podocarp pollen suggests the perhaps continuous presence of moist montane forest in nearby upland areas. This scenario again helps to explain the presence of diverse large vertebrates, which in addition to *Homo˙ erectus*, include elephant, stegodon, hippopotamus and antelope.

There is no palynological record for the Middle and most of the Late Quaternary. The record recommences within the latter part of the last glacial period, about 16,000 years BP with the core from the Situ Bayongbong swamp in West Java. This pollen profile suggests that during the time of the last glacial maximum, from 16,800 to about 12,000 years BP, the podocarp *Dacrycarpus imbricatus, Engelhardia*, Fagaceae and *Myrsine* dominated the surrounding dry land vegetation, suggesting a climate several degrees cooler than at present. However, between about 12,000 and 9000 years BP, *Dacrycarpus imbricatus* gradually declined in abundance, whereas pollen of *Engelhardia* suddenly increased, peaking at about 11,000 years BP, clearly relating to climatic amelioration, but possibly also suggesting volcanic disturbance of the vegetation. After about 9000 years BP, *Engelhardia* declined rapidly, and *Dacrycarpus imbricatus* became rare, whereas Fagaceae and *Altingia* increased in abundance, reflecting early Holocene climatic amelioration.

The palynological succession from Situ Bayongbong suggests a very similar pattern of vegetational and climatic change to that suggested for nearby Sumatra by Morley (1982) and Stuijts *et al.* (1988).

Palynological records for Situ Gunung and Telaga Patenggang cover the Holocene only. Both diagrams show some changes, but the major elements of the surrounding dry land vegetation can mostly be explained in terms of changes within the forests which surround the sites at the present time. The main changes within both sites are the successive reduction in representation of *Altingia* through time. Three possible explanations for these successive declines are proposed; firstly, they may relate to changes relating to drainage around the sites, since several of the horizons at which *Altingia* pollen shows sudden reductions in abundance coincide with hydroseral changes within the lake or swamp; secondly they may relate to volcanic activity, and thirdly, human factors may be involved, since *Altingia* is prized in Java for its wood.

The main differences in the pollen record from Situ Gunung and Telaga Patenggang relate to the different altitudes of the sites, since Telaga Patenggang is located some 300 m above Situ Gunung. For this reason, Telaga Patenggang has yielded more abundant *Schima* and Fagaceae pollen, whereas Situ Gunung has yielded more abundant *Altingia* pollen.

References

Flenley, J.R. 1979. *The Equatorial Rain Forest: A Geological History.* Butterworths, London: 162 pp.

Leinders, J.J.M., Aziz, F, Sondaar, P.Y. and de Vos, J. 1985. The age of the hominid-bearing deposits of Java. State of the Art. *Geologie en Mijnbouw*, 64: 167-73.

Morley, R.J. 1982. A palaeoecological interpretation of a 10,000 year pollen record from Danau Padang, Central Sumatra, Indonesia. *Journal of Biogeography*, 9: 151-90.

Morley R.J. 1991. Tertiary stratigraphic palynology in South-East Asia; current status and new directions. *Proceedings of the Geological Society of Malaysia*, 28: 1-36.

Morley, R.J. 1998. Palynological Evidence for Tertiary Plant Dispersals in the SE Asia Region in Relation to Plate Tectonics and Climate. In: R. Hall and J. Holloway (Editors) *Biogeography and Geological Evolution of Southeast Asia.* Bakhuys, Leiden: 177-200.

Polak, E. 1933. Uber Torf und Moor in Niederlandisch Indiën. *Verhandelingen der Koninklijke Akademie van Wetenschappen Amsterdam*,Tweede Serie 30: 1-85.

Polhaupessy, A.A. 1990. *Late Cenozoic Palynological Studies on Java.* Unpublished Ph.D.Thesis, University of Hull, Hull.

Polhaupessy, A.A. and Sudijono. 1985. *Palynological study of Quaternary Formations in the Solo and Madiun Areas.* Geological Research and Development Centre Special Publication No. 17: 109-17.

Semah, M. 1984. Palynology and Javanese Pithecanthropus Palaeoenvironment. *Courier Forschungsinstitut Senckenberg*, 69: 237-43.

Stuijts, I. 1984. Palynological study of Situ Bayongrong, West Java. *Modern Quaternary Research in Southeast Asia,* 8: 17-28.

Stuijts, I., Newsome, J.C. and Flenley, J.R. 1988. Evidence for Late Quaternary Vegetational Change in the Sumatran and Javan Highlands. *Review of Palaeobotany and Palynology.* 55: 207-16.

Tokunaga, S., Oshima, H., Polhaupessy, A.A. and Ito, Y. 1985. *A Palynological Study of the Pucangan and Kabuh Formations in the Sangiran Area.* Geological Research and Development Center Special Publication No. 4: 199-217.

van der Kaars, W.A. and Dam, M.A.C. 1996. A 135,000-year record of vegetational and climatic change from the Bandung area, West-Java, Indonesia. *Palaeogeography, Palaeoclimatology, Palaeoecology,* 117: 55-72.

van Zeist,W., Polhaupessy, A.A. and Stuijts, I. 1979. Two pollen diagrams from West Java, a preliminary report. *Modern Quaternary Research in Southeast Asia*, 5: 43-56.

Verbeek, R.D.M. and Fenema, R. 1896. *Geologische beschrijving van Java en Madoera.* Stemler, Amsterdam: 2 Volumes.

von Koenigswald, G.H.R. 1934. Zur Stratigraphie des Javanischen, Pleistozans. *De Ingenieur in Nederlandsch, Indie (iv)*: 185-200.

Quaternary Records of Vegetation, Biomass Burning, Climate and Possible Human Impact in the Indonesian-Northern Australian Region

Peter Kershaw, Sander van der Kaars, Patrick Moss and Sue Wang

Introduction

The Quaternary history of terrestrial environments within the tropics is poorly known despite their importance to an understanding of changing precipitation and temperature patterns over much of the earth's surface (Sturman and Tapper 1996) and their central position in the evolution and pattern of migration of modern people (e.g. Harding *et al.* 1997). A major reason for this lack of knowledge is the relative inaccessibility of tropical areas to major research institutions and, in tropical rainforest environments in particular, the difficulty of locating suitable sites for analysis. Even when sites are found, the incompletely known nature of the biota inhibits clear interpretation of past environments from any fossil evidence.

In recent years, the ability to collect cores from ocean sediments has provided a means of constructing long palaeoenvironmental records which reflect broad regional patterns of change. Much of this evidence is of marine conditions, and biogeographic and evolutionary changes within marine fauna and flora combined with oxygen isotope analysis of the tests of foraminifera has allowed the construction of a detailed and well-dated history of global environments through the last few million years (e.g. Shackleton *et al.* 1995). However, marine cores, located relatively close to land, also contain evidence of terrestrial environments through material that is transported by rivers or wind into ocean environments. The most prevalent terrestrial material is pollen. Although plants exhibit differential pollen production and dispersal features, assemblages have been usually found to be dominated by pollen from taxa that characterise major community types. Charcoal is another constituent, and this provides some measure of the degree of biomass burning.

The Indonesian/northern Australian region is proving very suitable for marine palynology as deep water exists in close proximity to a number of land masses while interest in the complex geological history of this area has resulted in a number of ocean cruises aimed at the collection and analysis of core sediments. The region is also unusual for tropical environments in that long

ISBN 3-923381-47-6
© 2002 by CATENA VERLAG, 35447 Reiskirchen

palynological records have also been able to be constructed from adjacent terrestrial environments due to the existence of depositional basins formed by volcanic and tectonic activity.

This paper summarises palynological results from the first three long marine records to be produced within the region and from two terrestrial sequences. Pollen and charcoal comparisons between the marine and terrestrial records help provide chronologies for the terrestrial records where direct dating has generally proved difficult beyond the limit of radiocarbon.

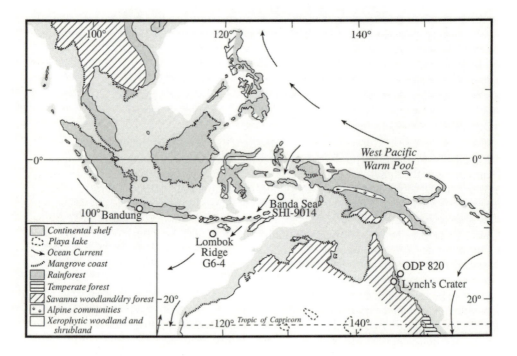

Fig. 7.1 Location of the pollen records in relation to major geographical features of the SE Asian-northern Australian region.

The sites and their regional setting

Selected features of the five records provide a framework for examination of climate, vegetation and biomass burning over the last one to three glacial cycles. All sites, shown in Figure 7.1, are situated along, or close to, substantial gradients in precipitation and/or temperature which should make the records sensitive to past environmental changes. The sites also span the major biogeographical divides between Asian and Australian biotas. The terrestrial records of Bandung (van der Kaars and Dam 1995) and Lynch's Crater (Kershaw 1986) illustrate patterns of change in the humid tropics of Indonesia and Australia respectively while the deep sea records from the Lombok Ridge (van der Kaars

1991; Wang *et al.* 1999) and Banda Sea (van der Kaars *et al.* 2000) give a picture of humid to seasonally dry environments over a broader Indonesian-northern Australian region. The ODP 820 record (Moss and Kershaw 2000), from shallow water on the continental slope, embraces a range of environments in NE Australia. Age control on the marine sequences is provided primarily by oxygen isotope records produced for each core, and on the terrestrial records by limited radiometric dates. Preliminary comparisons between the Bandung and Banda Sea records (van der Kaars 1998) and between the Lynch's Crater and ODP 820 cores (Moss and Kershaw 2000) have generally confirmed the age models proposed for the terrestrial records and allowed the tentative application of the marine isotope stratigraphy to all records. Dates on isotope stages (IS) are derived from Martinson *et al.* (1987). Due to the focus of the paper on ages beyond the calibrated part of the radiocarbon timescale, no attempt has been made to make corrections for possible differences in age models between radio-carbon and the marine isotope record.

Bandung Basin

The Bandung Basin of West Java (7°S, 108°N, altitude 665 m) receives a mean annual rainfall of about 1700 mm which is concentrated in the southern hemisphere summer due to the influence of the north-west Australian monsoon. Surrounding highlands, up to 2400 m, are perhumid with rainfall totals exceeding 3000 mm. The site presently provides the only late Quaternary, terrestrial record from SE Asia to reach beyond 40,000 years BP. Accumulation of lake and swamp sediments began after 135,000 years BP as a result of volcanic and tectonic activity in the Bandung Basin and achievement of suitable climatic conditions, and continued to present although pollen is not preserved from about the Last Glacial Maximum (van der Kaars and Dam 1995). Shorter records have been constructed from elsewhere in the area to provide a picture of Holocene vegetation (van der Kaars and Dam 1995). Selected features of the Bandung record are shown on Figure 7.2.

A tentative timescale for the Bandung record has been established from radiocarbon dates on the later part and linear extrapolation beyond the limit of radiocarbon for the earlier part of the sequence. The derived timescale is consistent with a uranium/thorium date of 135,000 years BP on basal pedogenic carbonate concretions. Palaeosol formation at this time indicates strong seasonality and an annual rainfall of between 750-1000 mm, about half that of the present day (van der Kaars and Dam 1995). Pollen evidence for the development of extensive fresh water swamp forest with lowland components within the Basin, together with the dominance of lowland to lower montane forest types on the Bandung Plain and in the surrounding highlands, during the early part of IS-5 (*c*.126,000 to 107,000 years BP), suggests that temperatures may have been 1-2°C higher than present and that there was a less pronounced dry season. Conditions from about 107,000 to 81,000 years BP may have been marginally drier and cooler with some contraction of swamp forest and a slight increase in representation of higher altitude taxa (e.g. *Dacrycarpus* and *Engelhardia*). Around 81,000 years BP the appearance of the Basin changed

Peter Kershaw, Bruno David, Nigel Tapper, Dan Penny and Jonathan Brown (Eds): Bridging Wallace's Line

Bandung basin, core DPDR-II

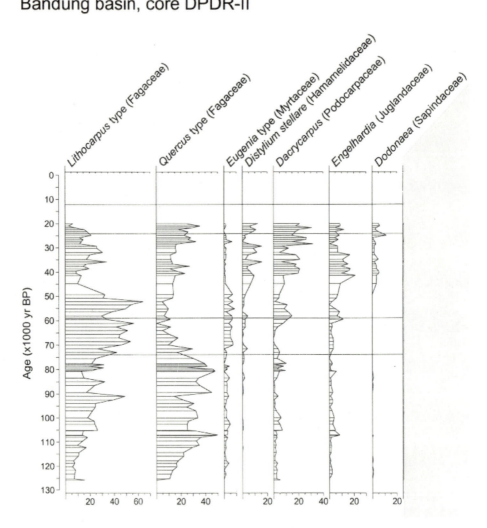

Fig. 7.2 Selected features of the pollen diagram from Bandung basin core DPDR-11, western Java. All predominantly dry land plus swamp forest taxa, which excludes the swamp herbs, compose the pollen sum for each sample upon which all pollen percent-ages are based. Lithocarpus, Quercus, and Eugenia type pollen are major components of lower montane forest while Distylium stellare, Dacrycarpus imbricatus, Engelhardia cf. spicata and Dodonaea cf. viscosa are prominent members of mid-upper montane forest. Charcoal values, indicated on a five point scale (absent, rare, common, very common and abundant) include values in the top part of the core where pollen is not preserved. Data from van der Kaars and Dam (1995).

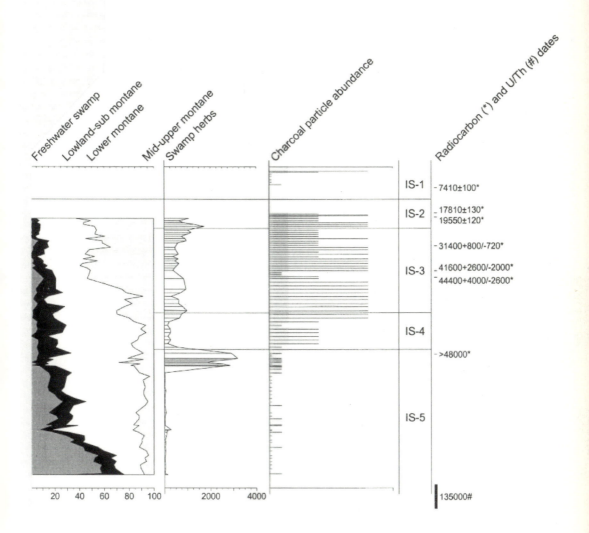

Fig. 7.2 Continuation, right part

dramatically with a substantial replacement of swamp forest by herbaceous swamp vegetation, while there may have been some slight altitudinal depression of the montane forest belts. Precipitation fell to an estimated 1000 mm but temperatures were only marginally cooler. Although dated, by interpolation between the radiocarbon and uranium/thorium dates to the end of IS-5, the fall in moisture could relate to a marked reduction in sea level at the beginning of IS-4 where the exposure of continental shelves would have reduced the moisture

Lombok Ridge, core G6-4

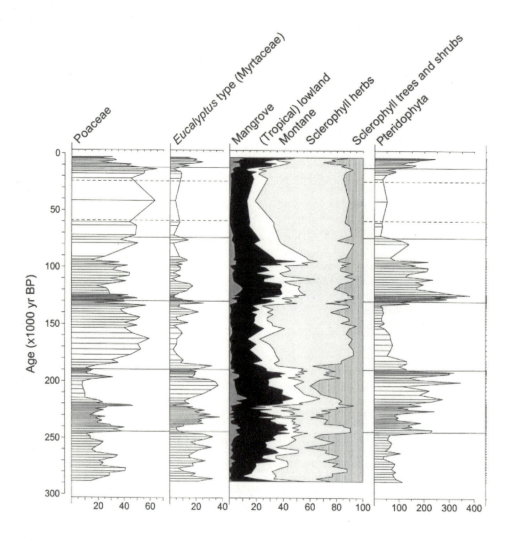

Fig. 7.3 Selected features of the pollen diagram from Lombok Ridge core G6-4. All predominantly dry land pollen taxa, which excludes mangroves and Pteridophyta spores, make up the pollen sum for each sample upon which all pollen percentages are based. Charcoal values are expressed as number of particles per cm^3. Data from Wang et al. (1999).

content of the NW monsoon. The question over the relationship of the record to the marine isotope stratigraphy is not resolved from the subsequent period, indicated as IS-4, which, although different, provides no clear picture of climatic

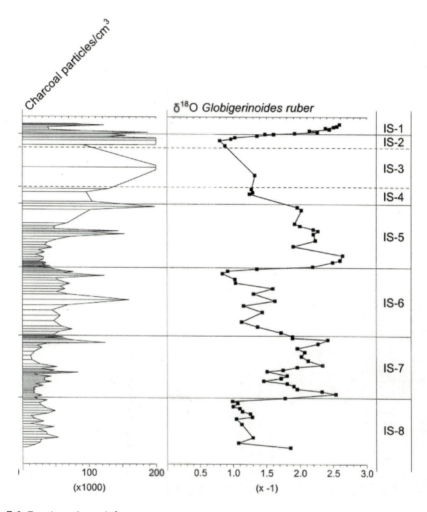

Fig. 7.3 Continuation, right part

conditions. The increase in representation of higher altitude taxa, particularly *Dacrycarpus imbricatus* and *Engelhardia* from an estimated date consistent with that of the beginning of IS-3 indicates further cooling with maximum altitudinal depression of higher montane forests (including the taxa *Distylium stellare, Dacrycarpus imbricatus* and *Engelhardia* cf. *spicata*) and compression of lower montane types (including *Lithocarpus* and *Eugenia*) between about 47,000 and 18,000 years BP. It is estimated from this vegetation configuration combined with evidence from short cores at higher altitude that temperatures during the Last Glacial Maximum were between 4°C and 7°C lower than the 23.7°C presently experienced (van der Kaars and Dam 1995).

Preliminary estimates of biomass burning from a charcoal abundance scale suggest that periodic fires were a feature of the landscape throughout the recorded period. There was a substantial increase in burning from within or subsequent to IS-4, which might be expected under precipitation levels estimated to have been at least 30% lower than those of today (van der Kaars and Dam 1995). However, there was some delay between precipitation reduction at the end of IS-5 and achievement of highest burning levels. The entry of *Dodonaea* into the record about 47,000 years BP has interesting implications for environmental change. Its occurrence was considered by van der Kaars and Dam (1995) to provide additional evidence for temperature lowering. However, it is also an important pioneer species and could indicate a greater effectiveness of fire on the vegetation from this time. Relatively low burning levels are indicated from preservation of charcoal in the Holocene with a substantial increase close to present, presumably because of intensive human impact.

Lombok Ridge

Core G6-4 was taken by the Indonesian-Dutch Snellius-II expedition in 1985 at 10°47'S, 118°04'E, approximately 170 km southwest of the island of Sumba in the Lesser Sunda Islands and 850 km off northern Australia at a water depth of 3510 m. The climate of the general region is dominated by the northwest monsoons which result in a highly seasonal distribution of rainfall between 500 and 1500 mm in Australia and 750-3000 mm in the Lesser Sunda Islands. The vegetation of adjacent Australia is dominated by open eucalypt forests and woodlands with extensive grasslands and small mangrove communities within major river estuaries and small patches of seasonal rainforest near the coast. By contrast, the Lesser Sunda Islands support mainly seasonal rainforest with evergreen types restricted to isolated patches and the mountains.

The nine metre core was analysed for pollen in relation to a marine isotope record by van der Kaars (1991). Subsequently data on biomass burning and a revised oxygen isotope record were added (Wang *et al.* 1999). The record covers the last three glacial cycles and pollen appears to have been derived from both adjacent land areas with a predominance of tropical lowland and montane rainforest pollen and probably pteridophyte spores from Indonesia and eucalypts and grasses from Australia (Figure 7.3). The higher values of tropical lowland rainforest and substantially higher values of pteridophyte spores during 'interglacial' periods suggest that conditions were more humid than in the intervening glacials within the Indonesian region. The correspondence also between pteridophyte peaks and high values of ^{16}O in the associated isotope record suggests that isotope substages 5e, c and a experienced wetter conditions than substages 5d and b. Generally higher values of Poaceae, relative to eucalypts, during the glacial periods suggest also that these periods were less humid than the interglacials within Australia. The mangroves may provide support for generally higher precipitation levels during interglacials in the Australian region although their representation is also influenced by coastal processes. Major peaks are recorded during major marine transgression phases when conditions were most suitable for extensive colonisation (see Grindrod *et al.* this volume).

Quantification of the degree of vegetation and climate change within the region is made difficult by uncertainties over the degree to which changes in wind patterns and ocean currents may have influenced pollen transport from different sources, the substantial expansion of the Australian continent, over the whole of the continental shelf, during glacial low sea levels which would probably have favoured grasses in particular, and by a sustained increase in grasses and concomitant decline in eucalypts around 185,000 years BP. An explanation for the switch from eucalypt to grass domination may relate to a general increase through the record in charcoal and hence biomass burning although there is no correspondence between the increase in eucalypts and any peak in charcoal. Another possibility is that the peak in burning late in IS-7 heralded the onset of more frequent fires which impacted on the vegetation when conditions became drier at the beginning of IS-6. Highest charcoal levels are recorded within the last glacial period, apart from a single high value at the IS-5-6 boundary, but the presence of a hiatus within the record prevents accurate dating of the achievement of these levels. There are reduced charcoal values in the early part of IS-1 before an increase to present.

Banda Sea

The seven metre long Core SHI-9014, collected as part of a French-Indonesian marine geological research programme in 1990, was taken at a water depth of 3163 m in the Banda Basin at 5°46'S, 126°58'E. The climate of the area is very variable being controlled by the monsoon circulation, the migration of the Intertropical Convergence Zone and land-sea distributions. Precipitation in the adjacent areas of northern Australia, southern New Guinea and the Lesser Sunda Islands is relatively low (500-2500 mm) and highly seasonal but seasonality is reduced and rainfall heavy (greater than 2500 mm) in the northern localities of Sulawesi, the Moluccas and northern New Guinea. These northern areas support predominantly tropical lowland and montane rainforests in contrast to the open eucalypt woodlands of Australia and southern New Guinea and seasonal rainforests of the Lesser Sunda Islands.

The record of van der Kaars *et al.* (2000) extends though approximately the last 175,000 years or two glacial cycles (Figure 7.4). As in the Lombok Ridge core, 'interglacials' are characterised by high levels of pteridophyte spores and mangroves indicating higher moisture levels than during glacial periods. Marine transgressions are also marked by mangrove peaks. The high levels of Poaceae and *Eucalyptus* indicate the importance of an Australian pollen source and suggest a continuity of NW to SE wind patterns. The almost total absence of *Nothofagus* – a large pollen producer within the region but confined to New Guinea – is consistent with this interpretation and further suggests that much of New Guinea is likely to have contributed little to the pollen influx. There is little consistent variation in percentage representation of the grasses and eucalypts as percentaged outside of the pollen sum but the relative importance of both, when

Peter Kershaw, Bruno David, Nigel Tapper, Dan Penny and Jonathan Brown (Eds): Bridging Wallace´s Line

Banda Sea, core SHI-9014

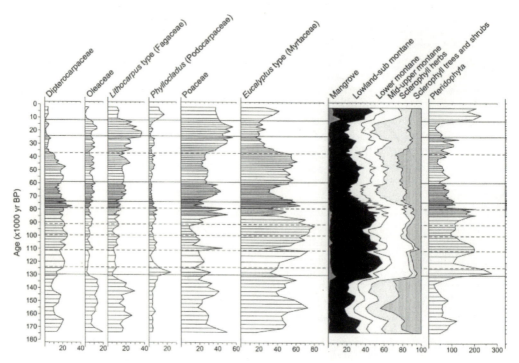

Fig. 7.4 Selected features of the pollen record from Banda Sea core SHI-9014. All rainforest taxa, presumed to have derived from the Indonesian region, make up the pollen sum of each sample on which all percentages are based, except for sclerophyll herbs and sclerophyll trees and shrubs, predominantly Poaceae and Eucalyptus, which are considered to have a mainly Australian origin and form an alternative pollen sum. Elemental carbon values indicate the ratio between ^{13}C and ^{12}C with $\delta^{13}C$ values of around -27 per mil indicative of plants using the C3 photosynthetic pathway and values of around -12 per mil indicative of plants using the C4 photosynthetic pathway. Data from van der Kaars et al. (2000).

expressed as ratios of total pollen, is substantially greater during the glacials than during IS-1 and IS-5. It could be that grassland with a significant eucalypt cover extended over the exposed continental shelf during glacial periods. The general pattern of rainforest pollen, presumably derived largely from Sulawesi, Halmahera and the southern part of the Bird's Head Peninsula of New Guinea which lacks *Nothofagus,* indicates more extensive areas of lowland rainforest (illustrated by Dipterocarpaceae) during interglacials despite reduced lowland area. It is likely that this reduction was due to a temperature decrease as well as a reduction in moisture. During glacial periods, lower montane forest, including *Lithocarpus,* expanded but was not accompanied by an increase in upper montane taxa which show greater representation during interglacials. It is pos-

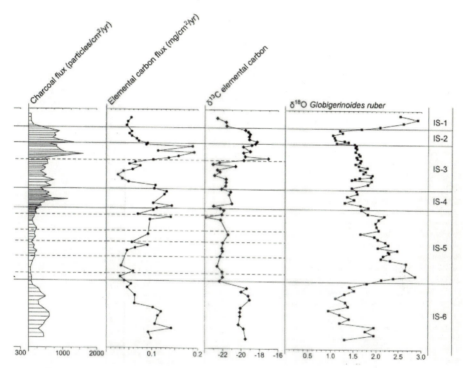

Fig. 7.4 Continuation, right part

sible that glacial lowering of the treeline eliminated much of the upper montane forest but conditions were not conducive for the replacement of lower montane by upper montane forest. However, a similar temperature reduction to that postulated from the Bandung Basin is proposed for lower montane environments during glacial periods. The peaks in *Phyllocladus* at the beginning of the interglacials are characteristic of the expansion phase in upper montane forests within a number of late Quaternary records in the region.

As in the Lombok Ridge record, there is some vegetation response to the causes of isotopic change within IS 5. Lowland rainforest in particular expanded during isotope substages 5e, 5c, and 5a. The pteridophytes also show three distinct peaks but in this record they are out of phase with the isotope record. Clearly relationships within this period are complex.

There is evidence for the continuous presence of fire from amounts of both charcoal and elemental carbon records. Values are generally higher during the drier glacial periods, with this pattern being more consistent for charcoal than elemental carbon. Differences between the two curves may result from a number of factors including the use of a larger size fraction (5-200 µm) for microscopic charcoal than elemental carbon (<100 µm). Smaller particles may have derived

ODP 820A

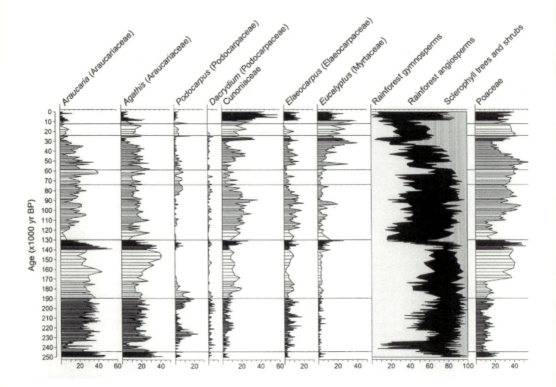

Fig. 7.5 Selected features of the pollen record from ODP Site 820A. All predominantly dry land woody taxa, which excludes Poaceae and mangroves, compose the pollen sum on which all percentages are based. Araucaria, Agathis, Podocarpus and Dacrydium compose the rainforest gymnosperms, Cunoniaceae and Elaeocarpus are the major rainforest angiosperms while Eucalyptus is the dominant sclerophyll tree and shrub taxon. Data from Moss (1999).

from a much broader area and contain a greater wind-borne component (Garstang *et al.* 1997). The carbon isotope record suggests that charcoal was derived mainly from forest vegetation during interglacials and from C4 grasses during glacials. However, this relationship breaks down to some degree during IS-4 and IS-3. There is a marked increase in all three burning-related curves about 37,000 years BP which is related to sustained reductions in Dipterocarpaceae and *Eucalyptus,* and an increase in grasses. It is possible that there was a regional increase in burning which influenced the vegetation of both Australia, in a similar manner to that at an earlier date in the Lombok Ridge region, and Indonesia., through the elimination of some canopy dominants.

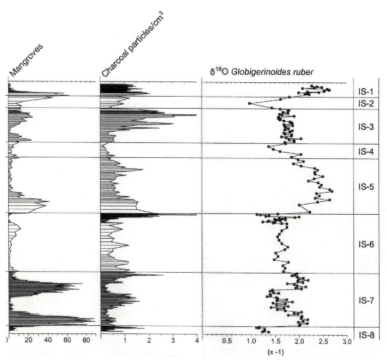

Fig. 7.5 Continuation, right part

ODP Site 820

Cores from this site (16°38'S, 146°18'E) were collected by the Ocean Drilling Program, Leg 133, from 280m of water on the continental slope, about 8 km outside the Great Barrier Reef and some 30 km east of the present shoreline. The site is adjacent to the largest patch of rainforest in Australia existing under rainfall levels from about 1500 mm to in excess of 3000 mm along the coast and coastal highlands. Within about 30 km of the coast the rainforest gives way regionally to open sclerophyll woodland with *Eucalyptus* and, less frequently, Casuarinaceae as the canopy dominants. Impeded drainage along coastal plains supports swamp woodland and grassland while extensive mangroves are found within river estuaries. Small isolated patches of drier rainforest, dominated by the austral conifer, *Araucaria,* are occasionally encountered at the margins of the complex rainforest-dominated area and on offshore islands. Due to the dominance of the south-easterly trades, it is likely that most of the pollen deposited at the site has been transported the from adjacent communities along the major rivers and through a major offshore passage on the continental shelf. The south-easterlies also bring much of the summer-dominated rainfall, and this is supplemented by rain from the north-west monsoon and from cyclones.

Peter Kershaw, Bruno David, Nigel Tapper, Dan Penny and Jonathan Brown (Eds): Bridging Wallace´s Line

The top 65 m of the 400 m core has been analysed to provide a detailed record through the last 250,000 years. Although different age models have been proposed from the isotope record (Peerdeman *et al.* 1993), the pollen data (Figure 7.5), particularly the distribution of mangrove pollen, and a comparison with the record from Lynch's Crater have helped resolve the dating uncertainties beyond the limit of radiocarbon dating (Moss and Kershaw 2000; Moss 1999). As in the other marine records, mangrove pollen shows most pronounced peaks during major marine transgressions from glacials to interglacials. Other peaks, with the notable exception of that within the later part of IS-3, occur where there is a transgression within a glacial or interglacial, but only one, within the 'double-headed' IS-7 interglacial, achieves glacial/interglacial proportions. The distinction between stages is not as clear as in the other marine records but the predominance of rainforest angiosperms which signify complex rainforest and low values for grasses at the beginning of IS-7, in isotope substage 5e and in IS-1 indicate that highest rainfall levels were achieved within interglacials.

The terrestrial pollen record is characterised by a number of stepwise changes, superimposed on the glacial cycles, which appear to have changed the nature of the vegetation of the region. The decline of *Araucaria,* the dominant of drier rainforest, is dramatic and suggests that this vegetation type has contracted from being the most widespread community during IS-8 to IS-6 (and through most of the Quaternary period, Kershaw *et al.* 1993) to almost total elimination from the region at the present day. Major sustained declines are noted around 135,000 and 36,000 years BP. There appear to have been similar though less dramatic declines in other austral conifers, *Agathis, Dacrydium* and perhaps also *Podocarpus,* with *Dacrydium* showing a final occurrence around 22,000 years BP. The latter genus is no longer present on the Australian mainland. The major beneficiary of the decline in araucarian forest was *Eucalyptus* whose increases generally coincide with the declines in *Araucaria.* This suggests the replacement of araucarian forest by eucalypt woodland.

Charcoal is present throughout the record and fluctuating values indicate a great deal of variation in fire activity. The two largest peaks in charcoal relate closely with the proposed times of sustained vegetation changes and are consistent with the replacement of a fire-sensitive by a fire-promoting vegetation. However, the fact that the height of the earlier charcoal peak occurs after the initiation of the decline in *Araucaria* suggests that fire may have been responding to the 135,000 years BP vegetation change rather than initiating the change. Poaceae, which forms the understorey of eucalypt woodlands, does not show similar responses to the eucalypts but does show a sustained increase at the earlier date of about 175,000 years BP. It is possible that Poaceae is reflecting changed conditions on the coastal plain. There is also a general increase in Cunoniaceae, the best represented complex rainforest taxon, which is indicative of wetter and cooler forests. The increased percentages of this taxon may indicate that complex rainforest also expanded, to some degree, into areas occupied by araucarian forest under a long term increase in rainfall. Alternatively, its increase may simply reflect a reduction in total pollen influx with the demise of the high pollen-producing araucarians.

The association of mangroves with the later charcoal peak and expansion of

Eucalyptus relative to *Araucaria* is interesting in that, as previously mentioned, this is the only mangrove expansion that does not relate to a marine trans-gres-sion. In fact, it occurs within what appears to be the most stable period of the isotope record. It is also the only time when mangroves and charcoal peak together. This unusual combination might best be explained by substantial erosion of soil with the destruction of rainforest vegetation and its deposition in marine environments forming a suitable substrate for mangrove expansion.

Lynch's Crater

Comparison with ODP Site 820 has provided a more certain chronology for this longstanding record which occurs within the presumed pollen catchment area of the ODP core (Moss and Kershaw 2000). Lynch's Crater (17°37'S, 145°70'E; altitude 760 m) was formed by volcanic activity and is essentially a closed depression. It occurs on the Atherton Tableland about 30 km from the coast and is surrounded, prior to clearance, by complex lowland to lower montane rainforest supported by a seasonally distributed rainfall of about 2600 mm. It is considered that the 60m core provides a continuous record through the last 215,000 years (Figure 7.6).

There is a clear distinction between 'interglacials' where high percentages of rainforest angiosperms, particularly Cunoniaceae and *Elaeocarpus,* indicate a dominance of complex rainforest around the site, and glacials which are characterised by araucarian forest and sclerophyll vegetation with Casuarinaceae and *Eucalyptus* as co-dominants above an understorey composed largely of grasses. Highest rainfall levels were probably achieved in isotope substage 5e where both conifers and sclerophyll taxa are virtually excluded, with other rainfall peaks at the base of the diagram in IS-7, in substage 5a and in the Holocene (IS-1). Temperatures have been difficult to determine but generally track the trends in rainfall.

A major sustained change in the record occurs between about 38,000 and 26,000 years BP when araucarian forest was progressively replaced by sclerophyll vegetation and virtually eliminated from the area. At the end of this phase *Dacrydium* has its last occurrence. The earliest major peak in charcoal accompanies the beginning of this change and, as in the ODP core around this time, was probably the main responsible agent. Charcoal levels are reduced during the Last Glacial Maximum when the vegetation may have been too open to carry intense or frequent fires but then increase around the IS-1-2 boundary. This charcoal pattern is also evident in the ODP core where, unlike previous glacial/interglacial transitions, there is a delayed response of the vegetation to increased precipitation. It is likely in this region that burning, associated with the established fire-promoting sclerophyll vegetation, effectively delayed the expansion of rainforest from retreats occupied during the last glacial period. Charcoal disappears from the record during the middle part of the Holocene when complex rainforest existing under high rainfall probably excluded fires. The re-occurrence of charcoal towards present has been attributed to Aboriginal and subsequently European settlement burning predominantly on the swamp surface (Kershaw 1983).

Lynch's Crater

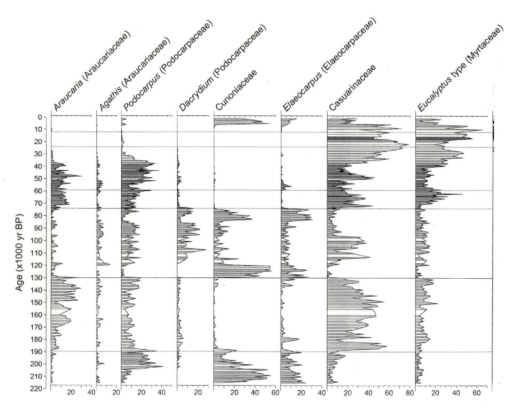

Fig. 7.6 Selected features of the pollen record from Lynch's Crater. All predominantly dry land arboreal taxa composed the pollen sum of each sample upon which all taxon percentages are based. Araucaria, Agathis, Podocarpus and Dacrydium compose the rainforest gymnosperms, Cunoniaceae and Elaeocarpus are the major rainforest angiosperms while Casuarinaceae and Eucalyptus are the dominant sclerophyll tree and shrub taxa. Data from Kershaw (1986), Moss and Kershaw (1999).

Regional patterns of environmental change

Despite the geographical spread and differences in climatic controls and environments of deposition of the sites, all records show a similar general vegetation and climatic response to glacial cyclicity with interglacials being substantially wetter than glacials. This may be predictable as the whole region is strongly affected by changes in the extent of land as a result of sea level varia-tions controlled by changes in global ice volume. In addition to the reduction in marine environments, associated lower sea surface temperatures would also bring about a reduction in the amount of water vapour available for precipitation

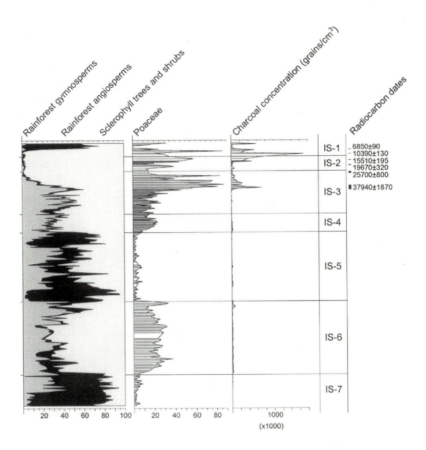

Fig. 7.6 Continuation, right part

and also most likely result in a reduction of monsoon activity within the region. However, there is some notable variation within individual isotope stages, particularly IS-5, where sea level fluctuations and associated changes do not appear to have had an over-riding control and where examination of other influences may prove profitable. There is also some variability but no major inconsistency in the temperature signal from records. Those records within the pollen catchment area of high mountains support the view that tropical temperatures were depressed substantially, and possibly in the order of $6^{0}C$ during glacial periods (see Peterson *et al.* this volume).

From the evidence of biomass burning, it appears that fire has been a constantly operating factor through the recorded periods. From the records of Lombok Ridge and Banda Sea it is suggested that fire activity was greater during drier glacials and interglacials and that much of this burning probably occurred in sclerophyll vegetation of northern Australia. Support for this view is provided by the carbon isotope data from the Banda Sea record which indicates that charcoal was derived predominantly from C4 grasses. The Bandung Basin record suggests that rainforest-type vegetation was also capable of carrying fire although some burning may have derived from herbaceous swamp vegetation. The very low values for charcoal through much of the Lynch's Crater record though indicate that, even during drier periods, fire seldom entered the forest here.

It is possible that disturbances in the volcanically and tectonically active Indonesian region, perhaps including human impact, were responsible for more frequent rainforest burns in this region, but the charcoal data from Bandung need to be refined before more detailed record comparisons can be made. Many portions of charcoal curves also display a great deal of variability with clear peak times of fire activity. Many of these peaks occur in close association with times of climate change suggesting that climate and vegetation instability promotes burning. Other peaks may be related to human impact.

A feature of all records is that they show some sustained changes in vegetation and or burning which are superimposed on the cyclical patterns of change (see Table 7.1). In all cases the change is towards more open and/or sclerophyllous vegetation with increased burning.

Table 7.1. Times of sustained change in vegetation and/or burning recorded within the pollen diagrams.

Date	Records
c. 40 kyr	Banda Sea, Lynch's Crater, ODP 820, Lombok Ridge, Bandung
c. 65 kyr	Bandung, Lombok Ridge
c. 135 kyr	ODP 820
c. 195-175 kyr	Lombok Ridge, ODP 820

One explanation for this directional environmental change is human impact. It has long been postulated that changes in the Lynch's Crater diagram around 38,000 years BP were most likely the result of increased burning with the arrival of Aboriginal people and subsequent archaeological dates for the first presence of people have tended to firm up this explanation (Flood 1995). The ODP 820 record event around the same time indicates that this was likely to have been a regional feature. Furthermore, the apparently synchronous decline of dipterocarps and reduction in eucalypts relative to grasses in association with increased burning in the Banda Sea record suggest that occupation of and impact within Indonesia and Australia by *Homo sapiens* may have been a rapid and dramatic event. Combined with recent archaeological evidence from the Indonesian region (Flood 1995; O'Connor *et al.* this volume), the more northerly of postulated human entry routes into Australia (Birdsell 1977) might be postulated (see Figure 7.7).

The Bandung Basin record and possibly also the record from Lombok Ridge suggest that there may have been a slightly earlier increase in charcoal, around 65,000 years BP, which is not related to a major change in climate and consequently could possibly be attributed to human impact. However, the records are either not firmly dated or well documented around this time. A southern entry into Australia through Java and the Lesser Sunda Islands at this time though could explain the old dates of around 60,000 years BP for human occupation in the Northern Territory (Roberts *et al.* 1993) (Figure 7.7).

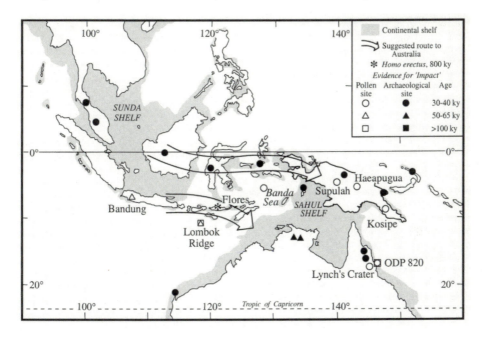

Fig. 7.7 Possible entry routes of people into Australia based on earliest evidence of Homo sapiens from archaeological sites and evidence for impact on the vegetation in palaeoecological sites from the SE Asian-northern Australian region.

Older sustained vegetation changes, around 195-175,000 and 135,000 years BP, are even more difficult to attribute to human impact especially as recent dating of the archaeological site of Jinmium in northern Australia, originally dated to between 120,000 and 176,000 years BP (Fullagar *et al.* 1996), now suggests a date of less than 10,000 years BP (Roberts *et al.* 1998). However, there is evidence for the presence of *Homo erectus* in the Lesser Sunda Islands by 800,000 years BP (Morwood *et al.* 1998) and it would have required only one or two more sea crossings to reach Australia. In any case, an impact cause for vegetation changes around 135,000 and 195,000 to 175,000 years BP is not convincing as there is insufficient association between vegetation change and charcoal peaks to fully support a causal relationship.

A second possible explanation for the sustained vegetation changes is that, in Australia, the vegetation has been undergoing a gradual change from rainforest to sclerophyll vegetation since the climate became drier and more variable in the late Tertiary with fire being the driving force (Kershaw *et al.* 1994). However,

from limited available evidence, it appears that much of the Quaternary supported a much more stable vegetation than is evident in these records and it might
be considered a coincidence that the effect of ecological drift to a more sclerophyllous vegetation is so apparent over a large area in the recent geological past.
Such a process is also unlikely to have influenced the dipterocarps of Asia and
so this hypothesis would exclude the 38,000 year 'event' which can most readily
be attributable to human impact.

A third possibility is that the climate of this region, in contrast to other parts
of the world, has been changing over the last few hundred thousand years. This
is a possibility in light of the tectonic and volcanic activity in SE Asia and the
slow but continuing movement northwards of Australia into the Indonesian
region which could have altered both atmospheric and oceanic circulation
patterns. It has been suggested from the extended oxygen isotope record of ODP
Site 820 (Peerdeman *et al.* 1993). and also from another core record from the
Coral Sea (Isern *et al.* 1996), that there is likely to have been a 4°C increase in
sea surface temperatures between Stages 11 to 8 in the Coral Sea area and that
this increase resulted in the formation of reef systems including the Great Barrier
Reef within the region. Such an increase may have related to the formation of
the West Pacific Warm Pool (Isern *et al.* 1996). This warm pool would have
increased the sea surface temperature gradient across the Pacific Ocean, an
important prerequisite for El Nino-Southern Oscillation activity. A resultant
increase in climatic variability could then explain the trends towards increased
burning and the development of more open vegetation. The combination of
higher precipitation resulting from increased sea surface temperatures and
climatic variability might also have allowed the expansion of both complex
rainforest and sclerophyll vegetation suggested from ODP Site 820. Further
research aimed at extending existing palaeoecological records and understanding
how the warm pool may have influenced regional climates could be of major
benefit, considering that this region may have already experienced a temperature
increase equivalent to that proposed with Greenhouse warming and awaited by
the rest of the world.

Acknowledgments

Much of this research was undertaken as part of a programme on the
Environmental and Cultural History and Dynamics of the Maritime Continent
supported by grants from the Australian Research Council and Monash
University and by a Logan Fellowship to Sander van der Kaars and postgraduate
scholarships to Partick Moss and Sue Wang, and being undertaken in collabora-
tion with Rien Dam of the Netherlands Geological Survey, François Guichard of
LSCE, CNRS-CEA, Gif-sur-Yvette, France, Fred Jansen of the Netherlands
Institute of Sea Research, Patrick De Deckker and Michael Bird of the
Australian National University, and Henk Heijnis of the Australian Nuclear
Science and Technology Organisation. We thank the Ocean Drilling Program
for access to sample material and Gary Swinton for drafting some of the text
figures.

References

Birdsell, J.B. 1977. The Recalibration of a Paradigm for the First Peopling of Greater Australia. In: J. Allen, J. Golson and R. Jones (Editors) *Sunda and Sahul: Prehistoric Studies in Southeast Asia, Melanesia and Australia.* Academic Press, London: 113-67.

Flood, J. 1995. *Archaeology of the Dreamtime (Revised Edition).* Angus and Robertson, Sydney: 328 pp.

Fullagar, R.L.K, Price, D.M. and Head, L.M. 1996. Early human occupation of northern Australia: archaeology and thermoluminescence dating of Jinmium rock-shelter, Northern Territory. *Antiquity,* 70: 751-73.

Garstang, M., Tyler, P.D., Cachier, H. and Radke, L. 1997. Atmospheric Transports of Particulate and Gaseous Products by Fires. In: J.S. Clark, H. Cachier, J.G. Goldammer and B. Stocks (Editors) *Sediment Records of Biomass Burning and Global Change.* Springer, Berlin: 207-50.

Harding, R.M., Fullerton, S.M., Griffiths, R.C., Bond, J., Cox, M.J., Scneider, J.A., Moulin, D.S. and Clegg, J.B. 1997. Archaic African and Asian lineages in the genetic ancestry of modern humans. *American Journal of Human Genetics,* 60: 772-89.

Isern, A.R., McKenzie, J.A. and Feary, D.A. 1996. The role of sea-surface temperature as a control on carbonate platform development in the western Coral Sea. *Palaeogeography, Palaeoclimatology, Palaeoecology,* 124: 247-72.

Kershaw, A.P. 1983. A Holocene pollen diagram from Lynch's Crater, northeastern Queensland, Australia. *New Phytologist,* 94: 669-82.

Kershaw, A.P. 1986. The last two glacial-interglacial cycles from northeastern Australia: implications for climate change and Aboriginal burning. *Nature,* 322: 47-9.

Kershaw, A.P., Martin, H.A. and McEwen, M.J. 1994. The Neogene - a Period of Transition. In: R. Hill (Editor) *Australian Vegetation History. Cretaceous to Present.* Cambridge University Press, Cambridge: 435-62.

Kershaw, A.P., McKenzie, G.M. and McMinn, A. 1993. A Quaternary vegetation history of northeastern Australia from pollen analysis of ODP Site 820. *Proceedings of the Ocean Drilling Program Scientific Results,* 133: 107-14.

Martinson, D.G., Pisias, N.G., Hays, J.D., Imbrie, J., Moore, T.C. and Shackleton, N.J. 1987. Age dating and orbital theory of the Ice Ages: development of a high resolution 0 to 300,000-year chronology. *Quaternary Research,* 27: 1-29.

Morwood, M.J., O'Sullivan, P.B., Aziz, F. and Razza, A. 1998. Fission track ages of stone tools and fossils in central Flores, Indonesia. *Nature,* 392: 173-6.

Moss, P.T. 1999. *Late Quaternary Environments of the Humid Tropics of Northeastern Australia.* Unpublished PhD thesis, School of Geography and Environmental Science, Monash University,

Moss, P.T. and Kershaw, A.P. 2000. The last glacial cycle from the humid tropics of Australia: comparison of a terrestrial and a marine record. *Palaeogeography, Palaeoclimatology, Palaeoecology,* 155: 155-76.

Peerdeman, F,M., Davies, P.J. and Chivas, A.R. 1993. The stable oxygen isotope signal in shallow-water, upper-slope sediments off the Great Barrier Reef (Hole 820A). *Proceedings of the Ocean Drilling Program Scientific Results,* 133: 163-73.

Roberts, R.G., Jones, R. and Smith, M.A. 1993. Optical dating at Deaf Adder Gorge, Northern Territory, indicates human occupation between 53,000 and 60,000 years ago. *Australian Archaeology,* 37: 58-9.

Roberts, R.G., Bird, M., Olley, J., Galbraith, R., Lawson, E., Laslett, G., Yoshida, H., Jones, R., Fullagar, R., Jacobsen, G. and Hua, Q. 1998. Optical and radiocarbon dating at Jinmium rock shelter in northern Australia. *Nature,* 393: 358-62.

Shackleton, N.J., Hall, M.A. and Pate, D. 1995. Pliocene stable isotope stratigraphy of Site 846. *Proceedings of the Ocean Drilling Program Scientific Results*, 138: 337-355.

Sturman, A.P. and Tapper, N.J. 1996. *The Weather and Climate of Australia and New Zealand*. Oxford University Press, Melbourne: 476 pp.

van der Kaars, W.A. 1991. Palynology of eastern Indonesian marine piston-cores: A Late Quaternary vegetational and climatic record for Australasia. *Palaeogeography, Palaeoclimatology, Palaeoecology*, 85: 239-302.

van der Kaars, W.A. 1998. Marine and terrestrial pollen records of the last glacial cycle from the Indonesian region: Bandung basin and Banda Sea. *Palaeoclimates: Data and Modelling*, 3: 209-19.

van der Kaars, W.A. and Dam, M.A.C. 1995. A 135,000-year record of vegetational and climatic change from the Bandung area, West-Java, Indonesia. *Palaeogeography, Palaeoclimatology, Palaeoecology*, 117: 55-72.

van der Kaars, W.A., Wang, X, Kershaw, A.P., Guichard, F. and Arifin Setiabudi, D. 2000. A late Quaternary palaeoecological record from the Banda Sea, Indonesia: patterns of vegetation, climate and biomass burning in Indonesia and northern Australia. *Palaeogeography, Palaeoclimatology, Palaeoecology*, 155: 135-53.

Wang, X., van der Kaars, S., Kershaw, A.P., Bird, M. and Jansen, F. 1999. A record of fire, vegetation and climate through the last three glacial cycles from Lombok Ridge core G6-4, eastern Indian Ocean, Indonesia. *Palaeogeography, Palaeoclimatology, Palaeoecology*, 147: 241-56.

Late Quaternary Mangrove Pollen Records from Continental Shelf and Ocean Cores in the North Australian-Indonesian Region

John Grindrod, Patrick Moss and Sander van der Kaars

Introduction

Mangroves are trees and shrubs that habitually grow in soils subject to tidal inundation. Greatest mangrove diversity lies in the tropical western Pacific, from mainland Malaysia through the Indonesian archipelago to New Guinea and northern Australia. This region straddles the zone of interaction between two major tectonic plates. It encompasses the biogeographic boundary known as the Wallace Line which, from a traditional perspective, marks the separation of the contrasting biological realms of SE Asia and Greater Australia. While in many respects a dramatic biological divide, it is no less true that the Wallace Line and other nearby divisionary markers (e.g. Huxley's Line) are selective in their opposition to biological interaction between the two realms. Predictably, it is amongst the terrestrial biological components, whose species do not readily achieve ocean crossings, that regional demarcation is most readily observed. Many plant and animal species, not so restrained by modest ocean barriers, show scant regard to the Wallace Line as an impediment to geographic dispersal.

It is not surprising then, given the dispersal abilities in sea water of most mangrove species, that mangrove floristic composition across the region is highly uniform, with most between-site variation reflecting local climatic and physiographic conditions. The West Pacific region contains at least 40 common mangrove tree and shrub species (Bunt *et al.* 1982; Tomlinson 1986), with relatively few locally endemic taxa even at species level. For instance, the tropical shores of mainland SE Asia and Sumatra contain 31 core mangrove tree and shrub species, from 15 genera and 12 families. The count for northern Australia is 31 species, 15 genera and 13 families. Fourteen genera are held in common between the two locales, with *Kandelia* (Rhizophoraceae) absent from Australia and *Pemphis* (Lythraceae) absent from the Asian mainland and Sumatra. At least 27 species are common across the entire region. Prominent mangrove families are Rhizopohoraceae (including 4 mangrove genera, and at least 13 species), Avicenniaceae (1 mangrove genus, 4 species), Sonneratiaceae (1 mangrove genus, 5 species), Meliaceae (1 mangrove genus, 2 species) and Combretaceae (1 mangrove genus, 2 species).

ISBN 3-923381-47-6

At the local scale, mangrove luxuriance and diversity is strongly influenced by climate and coastal physiography. Sheltered tidal settings and high rainfall provide optimum mangrove habitat. Species diversity attenuates as coasts become more arid.

This paper investigates the late Quaternary history of mangroves in the north Australian and Indonesian region, based on the pollen analysis of sediment cores from continental shelf and ocean locations. The continental shelf records derive from former mangrove shorelines below present sea level. They provide details of the dynamics of mangrove successional development on drowning coasts during the early Holocene phase of the Post Glacial Marine Transgression (PGMT). The results are used in the interpretation of three deep ocean records, which provide a regional picture of mangrove development and decline through the last full glacial cycle. The potential for climatic reconstruction from mangrove pollen studies is also assessed.

Regional Context

The study region includes the tropical north coast of Australia and the nearby archipelago of SE Indonesia (Figure 8.1). Core sites lie on the continental shelf and slope in NE Queensland, Australia, and in the Banda Sea and at Lombok Ridge (Timor Sea), Indonesia. The region lies within the tropics between 4° and 18° S, and is close to the zone of contact of the Asian and Indo-Australian continental plates. The SE Asian continental shelf incorporates the major Indonesian islands of Sumatra, Java and Borneo. Indonesian islands further east exist in a tectonically active zone in deep ocean, and do not feature broad submarine shelves. New Guinea and Australia lie on the Indo-Australian Plate. The mid-plate location and low relief of the Australian continent ensure tectonically stable coasts and an extensive continental shelf.

Climate within the region varies according to latitude and local physiographic factors. Where rainfall is seasonal it is a summer phenomenon. In the Cairns-Townsville region of NE Queensland, most rain is delivered by southeast trade winds from the Coral Sea. In the vicinity of Cairns, where mountains stand near to the coast, humid tropical conditions prevail with only moderate seasonality of rainfall. In contrast, the low lying coast around Townsville is drier, with pronounced seasonality of rainfall. Strong seasonality is a feature also of rain in NW Australia and SE Indonesia, where summer rains mainly derive from westerly monsoons, and winters are reliably dry.

In keeping with its dual continental affinities, range of climates and physiography, the region contains high biological diversity. Australia's Gondwanan origins, and permanent deep ocean barriers at plate margins, ensure a high degree of biological separation of Greater Australia and continental SE Asia, although the degree to which such barriers have dictated the regional evolution of some components of the terrestrial vegetation has been questioned (Truswell *et al.* 1987; Morley this volume).

Tropical rainforests are a feature of the most humid localities. Clear taxonomic affinities exist between the rainforest floras in SE Asia and northern Australia. Non-rainforest components of the terrestrial vegetation are more clearly separated between the two regions. Australia's widespread sclerophyll vegetation has developed in some degree of isolation during the late Cainozoic drying of the Continent, long since its final Gondwanan separation. Scleorphyll vegetation in northern Australia now dominates sub-humid to semi-arid environments. It exists as forests, woodlands and savannahs with shrubby or grassy understories. The common trees include species of *Eucalyptus*, *Melaleuca*, *Casuarina* and *Callitris*.

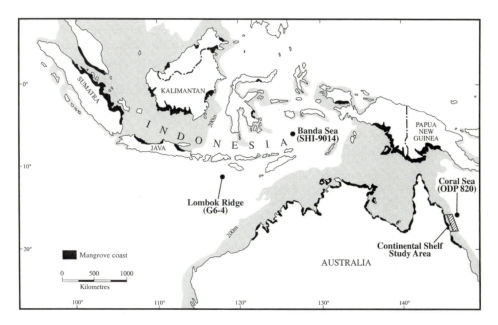

Fig. 8.1 The north Australian-Indonesian region showing ocean core locations and the continental shelf study area. Water depths at the ocean core sites are: SHI-9014, 3136 m; G6-4, 3510 m; ODP 820, 280 m. Distributions of mangrove coast follow Chapman (1977) and Hutchings and Saenger (1987).

As detailed above, mangroves occur throughout the area. The most common elements include *Avicennia*, and the Rhizophoraceae taxa *Rhizophora*, *Bruguiera* and *Ceriops*. These genera are represented at most tropical Australian mangrove localities, with the Rhizophoraceae frequently providing a dominant core of species. However, marked differences occur in the structure and floristic composition of mangrove and associated landward vegetation communities according to regional climates. Species richness is highest at humid locations, where mangroves typically also include a number of species with modest salinity tolerances, such as *Heritiera littoralis*, *Lumnitzera littorea*, *Sonneratia caseolaris*, and *Xylocarpus granatum*. Under drier conditions mangroves generally retain the common Rhizophoraceae species and *Avicennia*. Less conspicuous elements may include highly salt tolerant species such as *Aegialitis annulata*, *Excoecaria agallocha* and *Lumnitzera racemosa*.

Peter Kershaw, Bruno David, Nigel Tapper, Dan Penny and Jonathan Brown (Eds): Bridging Wallace's Line

Mangrove communities that contain at least a few species generally display clear species zonations arranged more or less parallel to the coast. These are characterised by a regionally predicatable sorting of species and species groups across the tidal gradient. It is often suggested that zonations reflect underlying successional processes as coastlines variously migrate seaward or landward, or as estuaries infill with sediment. In NE Queensland a common sequence of dominant elements in a zonation is *Avicennia marina* and *Rhizophora apiculata* in the seaward zone, *R. stylosa* and *Bruguiera gymnorrhiza* in the middle zone, and *Ceriops tagal* and *A. marina* in the most landward zone. While this genera-lised sequence commands a relatively broad climatic range within the study area, various other species may also occur according to local environmental conditions. In humid locations, mangrove communities grade landward to freshwater swamps, woodland or rainforest. Under more arid climates chenopod-dominated saltmarsh communities commonly fringe the landward mangrove boundary.

The palynology of mangroves and saltmarshes

Mangrove and saltmarsh communities lend themselves to palynological study by virtue of their relatively clear taxonomic distinctiveness from other vegetation types, the prodigious pollen output of some key mangrove taxa and identifiable pollen grain morphologies. In addition mangroves commonly occupy annoxic muddy substrates suited to the preservation of fossil pollen. The constraints to interpretation of fossil mangrove and saltmarsh pollen assemblages include factors that are common to all palaeoecological studies involving pollen analysis. Interpretations are especially challenging where the attempt is made to reconstruct former climate from fossil pollen evidence.

As mentioned above, the link between humid climate and optimal mangrove development in modern settings is a strong one. While it may be tempting to interpret palynological evidence for former mangrove abundance directly in terms of climate, it is salutary to examine some potential shortcomings in this regard. Problems arise through the inequitable representation of plant species by their pollen, and the incomplete match between plant species distributions and climate. Examples of very strong representation of Rhizophoraceae pollen in mangrove pollen assemblages are common (Crowley *et al.* 1994; Grindrod 1988; Muller 1964). While some species may have relatively precise climatic require-ments, the Rhizophoraceae as a whole has wide geographic distribution. Diffi-culties arise in identifying pollen types to species level and in some cases to a single genus. Modern pollen studies also indicate that pollen representation is poor for some tropical mangrove species which show the clearest preferences for humid climates. In particular species within *Xylocarpus, Sonneratia, Nypa*, and *Heritiera* are implicated in this regard (Crowley *et al.* 1994; Grindrod 1988).

To further complicate climatic reconstruction from mangrove pollen assem-blages, examination of existing mangrove distributions in the Pacific (Wood-roffe and Grindrod 1991) reveals distributional patterns that appear strongly influenced by factors other than the direct link with climate. In particular Quater-nary sea level adjustments may play a significant role in aberrant or incomplete mangrove distributions. For example, Wells (1982) reports a low species count for mangroves at Limmen Bight in the southwestern Gulf of Carpentaria, which he attributes to the influence of a locally dry climate. However the mangrove flora at Limmen Bight does not clearly reflect a distributional limit imposed by aridity alone. Species present include some that are suited to humid environments (for example *Acanthus ilicifolius*, *Xylocarpus australasica*) but do not include others more suited to aridity (for example *Avicennia officinalis*, *Bruguiera parviflora*, *Ceriops tagal var.tagal*). Consequently it is likely that a broader range of mangrove species could successfully occupy this location under present climatic conditions (Grindrod 1995). It is possible that the low species count is a function of slow species dispersal along modern shorelines since rapid shoreline relocations during the PGMT. Limmen Bight, tucked into the western corner of the Gulf of Carpentaria lies approximately 1000 km from the closest Glacial Maximum shoreline. Distribution anomalies such as this suggest that shoreline migrations, of the scale dictated by Quaternary sea level change, may have a significant influence on the present day biogeography of mangroves (Woodroffe and Grindrod 1991).

Regardless of the potential shortcomings for climate reconstruction, a number of recent mangrove palynological studies have contributed to knowledge in this area as well as to an understanding of modern pollen deposition, plant community successions, and relative sea level reconstruction. Each of these areas of interest is relevant to the present study. The selected studies outlined below provide a background for interpretation and discussion of palynological results of our continental shelf and ocean core records.

Palynology of mangrove succession in northern Australia

Several pollen studies describe mangrove successional events on open coasts and in infilling estuaries since the end of the PGMT around 6000 years BP. These studies confirm the link between mangrove species zonations in existing vegetation and predictable sequences of succession through time. For example a pollen sequence from a littoral zone core from a prograding coast at Princess Charlotte Bay in north Queensland (Grindrod 1985) traces the development from a seaward *Avicennia* zone, through progressively landward associations of *Rhizophora*, *Bruguiera* and *Ceriops* communities, followed by the eventual establishment of a chenopod saltmarsh community at the core site. The sequence is entirely compatible with the ecology of mangrove species presently occupying the site. It provides good evidence of an uninterrupted successional sequence during a period of relative sea level stability, and continued vertical and lateral accretion of sediments as the shoreline advanced steadily seaward. A high degree of stratigraphic resolution is indicated by the pollen sequence, despite the potential for sediment mixing through bioturbation, tidal action and even the wholescale redistribution of material during storms.

Similar detail of local vegetation change in mangrove and associated vegetation types has been recorded in pollen core studies from infilled estuaries in the Northern Territory (Woodroffe *et al.* 1985; Clark and Guppy 1988) and NE Queensland (Crowley *et al.* 1990). These studies trace the developmental

Peter Kershaw, Bruno David, Nigel Tapper, Dan Penny and Jonathan Brown (Eds): Bridging Wallace´s Line

stages of mangrove community successions as riverine sedimentation eventually
leads to the development of freshwater floodplain environments. In each case
the estuarine systems are responding to the cessation of relative sea level rise
and the readjustment of sediment deposition by tidal and fluvial processes. In
the South Alligator River estuary of the Northern Territory (Woodroffe *et al.*
1985), the process of mangrove replacement followed a sequence of Rhizophoraceae and *Sonneratia* mangrove, *Avicennia* mangrove and the eventual
development of freshwater swamp communities, since relative sea level
stabilisation around 6000 years BP.

Modern pollen distributions

Crowley *et al.* (1994) report pollen assemblages from coastal sediments at
contrasting humid and arid locations in the Cairns to Townsville region of NE
Queensland. The data include analyses of mangrove sediments at two high rainfall locations (Mutchero Inlet and Hinchinbrook Channel), and one seasonally
dry location (Cleveland Bay). At each site mangrove pollen types dominate the
pollen assemblages. High values for *Rhizophora* and *Brugiera/ Ceriops* are
ubiquitous in all mangrove samples, while *Avicennia* is a consistent component
of the Cleveland Bay assemblages only. The results demonstrate a clear climatic
signal in the non-mangrove pollen component. The Mutchero Inlet and Hinchinbrook Channel assemblages consistently feature rainforest and freshwater
swamp pollen types, and lack the coastal saltmarsh component Chenopodiaceae.
In contrast, the Cleveland Bay assemblages consistently include Chenopodiaceae
and lack rainforest pollen types. Studies of this kind provide a basis for the
interpretation of fossil mangrove sequences. They demonstrate also the potential
to derive a climatic signal, at least in terms of relative humidity, from mangrove
pollen records, particularly as they reveal associated coastal vegetational
environments.

Climate reconstruction from mangrove pollen assemblages

A few long pollen records from deep ocean cores provide regional mangrove
pollen histories. Climatic reconstructions drawn from these records are generally
based on the terrestrial pollen components rather than mangroves. However, van
Campo (1986) provides an example of late-Quaternary climate reconstruction
directly from pollen evidence for former mangrove development. This study
describes two marine pollen sequences from cores collected in the Arabian Sea,
approximately 150 km from India's west coast. Both sequences cover the period
from the Last Glacial Maximum of around 18,000 years BP to present. A high
level of consistency between the two records exists in the relative representation
of some pollen groups, particularly the mangroves group – *Avicennia, Rhizophora* and *Sonneratia*. The author establishes three major late Quaternary
climatic phases for the region based on mangrove pollen frequencies. The interpretation rests on the assumption that high mangrove pollen content in the cores
reflects extensive mangrove development, as a consequence of a more humid

than present climate and higher than present discharge from coastal rivers. The climatic interpretation can be summarised as follows. Low mangrove pollen counts around 18,000 years BP reflect poorly developed mangrove communities and arid conditions in western India. High mangrove pollen counts around 11,000 years BP reflect extensive mangroves and humid conditions (a peak in summer monsoon). A steady decline in mangrove pollen since 11,000 years BP reflects increasingly dry conditions to the present day (a weakening monsoonal system). These interpretations are discussed below in the light of our own ocean core pollen records from northern Australia and Indonesia.

Methods

The Continental Shelf Cores

Core Collection

Thirty sediment cores from mid to outer continental shelf locations were collected by vibrocorer aboard the MV James Kirby during February 1995. Core locations lay between 28.5 and 64.5 m water depth. At each core site the aim was to recover organic mangrove sediments suitable for pollen analysis and carbon dating. Upper core sections consisting of unconsolidated marine carbonates were mostly discarded on board ship, and only the lower sections retained for ease of stowage. Core site locations were identified using a ship-board Precision Depth Recorder (PDR) which provides good stratigraphic detail in unconsolidated sediments. Detailed mapping of seismic data in the region (Johnston *et al.* 1982) reveals an extensive pre-Holocene system of palaeo-channels likely to provide estuarine habitat during the Last Glacial Marine Transgression (LGMT). Core locations were preferentially selected at palaeo-channel margins where PDR profiles indicate an organic horizon of likely mangrove origin (visible as a dark band on PDR profiles), within the 4.0 m depth range of the corer. Water depth was recorded by PDR, while core site locations were recorded by the ships satellite assisted Global Positioning System (GPS).

Core descriptions and sampling

All continental shelf cores were logged and described in the Centre for Palaeo-ecology and Palynology, Monash University, prior to sampling for radiocarbon and pollen analyses. Twelve out of the set of 30 cores have been analysed for this study. Cores that included a basal soil horizon and/or likely organic mangrove sediments were preferentially selected for pollen analysis. The pollen extraction technique from raw sediment samples followed the procedure described by van der Kaars *et al.* (2000). It includes sediment disaggregation in tetra sodium pyrophosphate, digestion of carbonates in hydrochloric acid, removal of extraneous organic material by acetolysis, and the isolation of the inorganics through heavy liquid separation in sodium polytungstate.

Peter Kershaw, Bruno David, Nigel Tapper, Dan Penny and Jonathan Brown (Eds): Bridging Wallace´s Line

Radiocarbon Analyses

Core sequence VC1 was dated by the conventional radiocarbon technique at Beta Analytical Laboratories, Florida. This determination was done on a mangrove rootlet. The remaining continental shelf cores were dated by AMS radiocarbon analysis at the laboratories of the Australian Nuclear Science and Technology Organisation, Lucas Heights, Sydney. These analyses were done on pollen preparations selected from the point of maximum mangrove pollen content in each core sequence, as detected in preliminary pollen counts. The pollen extraction method for AMS samples followed the preparation method described above, except that extraneous organics were removed by a nitric acid cold Schultz method instead of acetolysis. This avoids the use of acetic anhydride which may introduce carbon to the sample. In each case the aim of dating the continental shelf cores was to determine the age of mangrove occupation at each core site.

Ocean Cores

Core collection and sampling

The ODP 820 core was collected during Leg 133 of the Ocean Drilling Program, in a joint exercise with the Australian Geological Survey. ODP site 820 (16° 38'S, 146° 18'E) is situated in 280 m of water on the Australian continental slope, 60 km offshore from Cairns in NE Queensland. Two cores were collected at this site. The core analysed in this study (820A) is 146 m in length. The analyses presented here cover the upper 67 m of core 820A, providing a record back to oxygen isotope Stage 8 (Moss 1999; Kershaw *et al.* this volume). 251 samples have been analysed for pollen, with a coarse sampling interval of 40 cm and a finer interval of 20 cm at selected locations.

The Banda Sea piston core (SHI-9014) was collected in 1990 as part of a French-Indonesian marine geological program. The core site (5° 46'S, 126° 58'E) lies at 3163 m water depth in the central Banda Basin. Core length is 7.94 m, providing a record from Isotope stage 6 to present (van der Kaars *et al.* 2000; Kershaw *et al.* this volume). A total of 96 samples have been analysed for pollen. The sampling interval for pollen analysis is 10 cm, except for an interval of 4 cm across the Stage 4/5 boundary.

The Lombok Ridge core (G6-4) was collected by piston corer in 1985 during an Indonesian-Dutch marine scientific expedition (the Snellius II Expedition) (van der Kaars 1991; Wang *et al.* 1999; Kershaw *et al.* this volume). Water depth at the core site (10° 47'S, 118° 04'E) is 3510 m. The site lies approximately 160 km southwest of the island of Sumba, and at least 800 km from Australia's northwestern coast. The core length is 9.0 m; 127 samples have been analysed at sampling intervals of 5 and 10 cm.

The pollen preparation technique for the deep sea cores broadly follows the technique developed specifically for ocean cores by van der Kaars (1991). In this procedure sediment samples are disaggregated in 10% sodium-tetra pyrophosphate. The coarse sand and clay fractions are removed by sieving through 200 micron and then 5 micron mesh. Heavy liquid separation procedure using sodium polytungstate (2.0 specific gravity) is then applied for removal of residual inorganic material. The carbonate fraction is digested in concentrated hydrochloric acid. Prepared pollen samples were placed in silicon oil (ODP 820) or glycerol jelly (Banda Sea and Lombok Ridge) for analysis under the microscope.

Pollen Counting

All pollen identification and counting was done on Zeiss or Olympus light microscopes, with identifications aided by comparison with reference material held at the Centre for Palaeoecology and Palynology at Monash University. Pollen counts for continental shelf samples continued until at least 100 identified pollen grains were recorded, while counts on ocean core samples were continued until at least 100 dry land pollen taxa were recorded. Results are displayed on pollen diagrams, which provide percentage data based on a pollen sum including all identified pollen grains, and excluding pteridophyte spores.

Isotope records and dating the deep sea cores

The oxygen isotope analyses for ODP 820 (Peerdeman and Davies 1993) and Lombok Ridge (Wang *et al.* 1999) were done at the Research School of Earth Sciences, Australian National University, Canberra. Isotope analyses for Banda Sea (Ahmad *et al.* 1995) were done at the Laboratoire des Sciences du Climat et de l'Environnement, CNRS Gif-sur-Yvette, France. In each case the analyses were conducted on fossil remains of the surface planktonic foraminifer *Globigerinoides ruber*. The isotope curves provide the basis for correlation with oxygen isotope stages reported by Chappell and Shackleton (1986).

Results

The Continental Shelf cores

Stratigraphy

Figure 8.2 gives a representative stratigraphy of the continental shelf cores, based on the VC28 sediment sequence. The basal palaeosol corresponds to the Reflector A Horizon of Johnson *et al.* (1982) which is readily detected throughout the region. This is overlain by light coloured clay with sparse organic fines, and dark organic mud with coarse organic remains. The latter grade upwardly to uniform grey mud, which in turn is overlain by carbonate sand and mud. This sequence is common on the Queensland continental shelf. It is generally interpreted as indicating former dryland conditions (palaeosol) during the late Pleistocene sea level low stand, followed by littoral zone migration (organic clay and mud), low-tide or sub-tidal environments (uniform mud), and the eventual establishment of full marine conditions (carbonate sand and mud). Our PDR surveys suggest that the whole sequence is common to all the continental shelf core locations in this study. However, of the 12 continental shelf cores analysed for pollen, only three (VC25, VC27, and VC28) extend into the palaeosol. All others extend at least to the overlying organic mud.

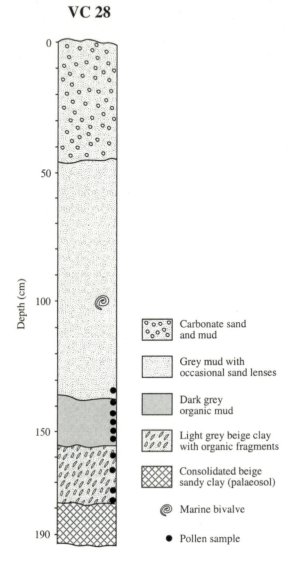

Fig. 8.2. Representative stratigraphy and interpretation for
the continental shelf cores,
based on core VC28. Basal unit
(palaeosol) represents former
subaerial Pleistocene soil.
Organic clay
and mud (188-136 cm) include
the transgressive mangrove unit.
Occasional macroscopic organic fragments in the organic
mud resemble mangrove roots.
Uniform grey mud (136-46 cm)
is a low-tide to sub-tidal facies
type. Carbonate sand and mud
(46-0 cm) represents post-
transgression marine deposition. PDR surveys and coring
indicate that this sequence is
widespread at varying
thicknesses in the continental
shelf study area. All continental
shelf cores analysed for pollen
extend at least to the organic
mud unit.

Core locations and radiocarbon ages

Table 8.1 provides details of core location, depth of water at core site and radiocarbon ages for mangrove organic sequences in the continental shelf cores. The
results indicate organic sedimentation on the mid to outer shelf below water
depths from 28.5 to 64.5 m, for the period approximately 10,000 to 8500 years
BP.

Table 8.1. Core locations, water depth and radiocarbon dates for mangrove muds.

Core #	Location	Water depth (m)	Radiocarbon Laboratory Code	Radiocarbon date (BP)
VC1	16°44'14"S 146°11'86"E	60.0	Beta 88448	8720±60
VC3	17°03'64"S 146°22'86"E	64.5	OZC 407	10,050±190
VC8	17°08'96"S 146°17'92"E	48.0	OZC 408	8570±110
VC12	17°02'61"S 146°17'50"E	52.5	OZC 409	9560±340
VC13	17°02'25"S 146°08'21"E	37.0	OZC 410	9640±340
VC17	17°13'17"S 146°08'88"E	28.5	OZC 417	8990±110
VC20	17°12'10"S 146°11'09"E	33.0	OZC 412	9140±240
VC25	17°32'83"S 146°27'03"E	49.5	OZD 484	9820±100
VC27	17°58'51"S 146°25'02"E	32.0	OZC 415	9060±230
VC28	17°56'85"S 146°24'88"E	34.0	OZC 416	9750±130
VC29	17°56'34"S 146°24'75"E	34.0	OZC 418	8770±210
VC30	17°55'88"S 146°24'61"E	34.0	OZC 419	9420±110

Continental shelf pollen records

Palynological results for the 12 continental shelf cores are shown in Figure 8.3. Mangrove pollen types consistently dominate. In each sequence Rhizophora-ceae is very strongly represented by the *Rhizophora* and *Bruguiera/ Ceriops* pollen types. *Avicennia* is the only other commonly recorded mangrove pollen type. It attains strong representation in cores VC25, VC27 and VC28 and is a minor component elsewhere. Other mangrove taxa recorded are *Aegiceras, Excoecaria, Liumnitzera, Sonneratia* and *Xylocarpus*. These attain minor and sporadic representation only. The saltmarsh component Chenopodiaceae is present in all cores except VC1 and VC13. While it is usually a minor component of the pollen sum, it attains strong representation in cores VC27 and VC28.

The former terrestrial vegetation for the region is represented by a variety of pollen types. Poaceae is the most strongly represented of the non-mangrove/saltmarsh components. It is recorded in all pollen sequences, with strongest representation in cores VC1, VC25, VC27 and VC28. Cyperaceae attains moderate representation in a few cores only (VC1, VC8, VC25 and VC27). Myrtaceae is consistently present, best represented by *Melaleuca* and *Eucalyptus* pollen types in cores VC1 and VC25. Many other terrestrial pollen types, common as regional components in the present vegetation of the study area, appear as minor elements in the continental shelf pollen sequences.

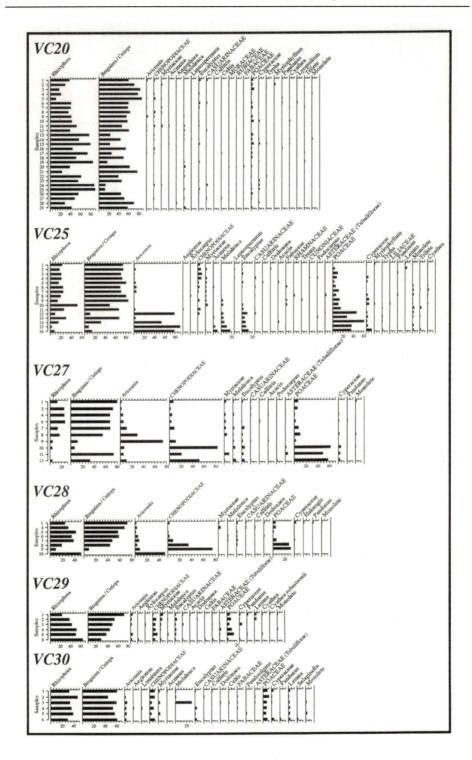

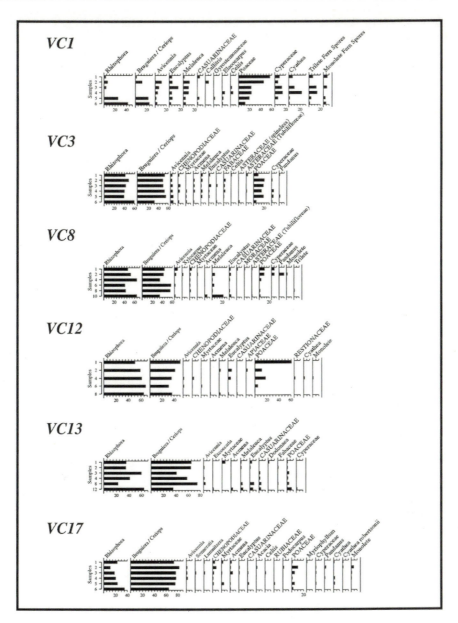

Fig. 8.3 Percentage pollen data for continental shelf pollen cores.

While the continental shelf pollen sequences are strongly dominated by mangrove pollen types, three of the records contain evidence of change from terrestrial to littoral zone vegetation communities.

VC25 contains high values of non-mangrove taxa *Eucalyptus*, *Acmena*, *Melaleuca*, Cyperaceae and Poaceae in lower levels of the core. The mangrove taxon *Avicennia* is also strongly represented in lower levels. In contrast the upper section of the core is dominated by the Rhizophoraceae mangrove types — *Rhizophora* and *Bruguiera/Ceriops*. The saltmarsh component Chenopodiaceae is relatively well represented in the top few levels. Other pollen types and fern spores are recorded at minor frequencies only.

VC27 has high values of Poaceae in the lowest three levels of the sequence. These coincide with relatively strong values for Chenopodiaceae, *Eucalyptus*, *Melaleuca* and Myrtaceae (undif). Very strong values for *Avicennia* are recorded midway through the sequence. These attenuate steadily through upper levels. The upper section of the sequence is dominated by *Rhizophora* and *Bruguiera/ Ceriops*, with minor representation of other pollen types.

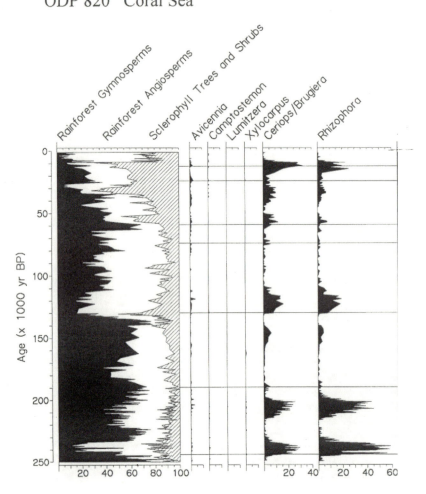

Fig. 8.4 Percentage pollen data for selected taxa, Coral Sea core ODP 820.

VC28 has very high Poaceae and Chenopodiaceae values towards the base of the sequence. *Avicennia* and *Rhizophora* are strongly represented in the most basal sample. Upper levels are dominated by *Rhizophora* and *Bruguiera/Ceriops*.

All other continental shelf pollen records are dominated by *Rhizophora* and/or *Bruguiera/Ceriops* through all levels. Poaceae and Chenopodiaceae are generally the best represented non-mangrove elements. Other pollen and spore types have low or sporadic representation only.

Pollen in the ocean cores

Figures 8.4 to 8.6 show detailed mangrove pollen and summarised terrestrial pollen components of the three ocean records (see Kershaw *et al.* this volume for more detailed accounts of the terrestrial pollen components for these records).

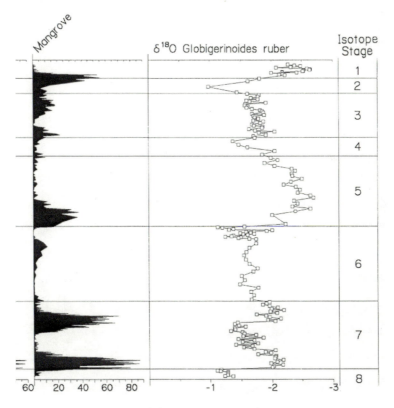

Fig. 8.4 Continuation, right part

Peter Kershaw, Bruno David, Nigel Tapper, Dan Penny and Jonathan Brown (Eds): Bridging Wallace´s Line

ODP 820 (Coral Sea)

The ODP 820 record (Figure 8.4) covers the last 2 glacial cycles (isotope Stages 1 to 8). *Rhizophora* and *Ceriops/Bruguiera* are the predominant mangrove pollen types recorded. Increases in the relative representation of these pollen types correspond closely with deccreases in oxygen isotope values. Other mangrove pollen types – *Avicennia, Camptostemon, Lumnitzera* and *Xylocarpus* – have relatively minor representation only, and do not display clear correspondence with isotope values.

   The main terrestrial pollen groups provide a regional record of the relative abundance of rainforest and sclerophyll elements for NE Queensland adjacent to the ODP 820 core site. The record suggests a long term increase in sclerophyll vegetation and reduction of rainforest gymnosperms. Fluctuations in the relative abundance of rainforest angiosperm and sclerophyll components are considered to reflect regional climatic variations, particularly in regard to contrasting humid and arid phases within the glacial cycle (Moss 1999).

G6-4 (Lombok Ridge)

The G6-4 record (Figure 8.5) also covers two full glacial cycles (isotope Stages 1 to 8). Mangrove pollen is best represented by the Rhizophoraceae, with minor representation of *Sonneratia* and the mangrove associate *Brownlowia*. Values for combined mangrove pollen are highly variable, showing strong correspondence with isotope values.

   The non-mangrove pollen component of the G6-4 record includes a diverse array of regional rainforest, woodland and grassland components which derive from nearby Indonesian shores and from further afield. Consistently high values for *Eucalyptus*, Poaceae and Cyperaceae indicate consistently strong representation of northwestern Australian elements throughout the period represented (Wang *et al.* 1999), even though the core site lies around 850 km from the present Australian coastline.

SHI-9014 (Banda Sea)

The SHI-9014 record (Figure 8.6) covers the last full glacial cycle and into isotope Stage 6. *Rhizophora* type is the predominant mangrove pollen type. Other mangroves recorded at minor frequencies include *Brownlowia, Nypa* and *Sonneratia*. Major increases in combined mangrove values occur near the isotopic Stage 6 to Stage 5 transition, and early in Stage 1. In keeping with the other ocean records increases in mangrove pollen abundance correspond to decreases in ^{18}O isotope values.

   The terrestrial pollen record of SHI-9014 represents a diverse assemblage of components from surrounding landscapes. The range of lowland to upper montane elements reflects the physiographic diversity of the region, and the proximity of the core site to high mountains. Grassland and woodland elements include significant contributions from Australian and New Guinean sources (van der Kaars 1998). Pronounced peaks in the mangrove pollen data are not reflected in the terrestrial pollen components.

Lombok Ridge, core G6-4

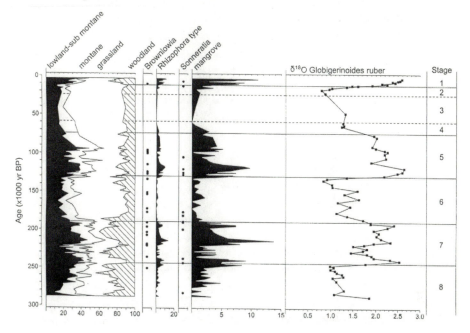

Fig. 8.5 Percentage pollen data for selected taxa, Lombok Ridge core G6-4.

Banda Sea, core SHI-9014

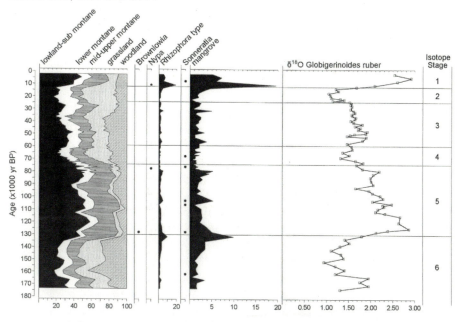

Fig. 8.6 Percentage pollen data for selected taxa, Banda Sea core SHI-9014.

Discussion

Mangrove pollen on the continental shelf

The precision depth recorder and palynological surveys of continental shelf sediments provide a convincing record of mangrove and saltmarsh occupation on transient shorelines of the Post Glacial Marine Transgression. Our pollen evidence derives from former estuaries only, as identified by PDR. Whilst it is likely that mangrove communities also occupied open coasts, such communities are as yet palynologically unrecorded. Indeed it is uncertain if intact littoral zone sediment sequences from former open coasts of the PGMT, exist on the Queensland continental shelf.

The stratigraphic section in Figure 8.2 is representative of the main elements of the sedimentary sequence at each continental shelf core site. It records the development of littoral zone systems over terrestrial environments, through to the eventual establishment of full marine conditions. Pollen preservation is variable in the sediment sequences recovered, according to facies type and the environment of deposition represented. None of the Pleistocene soil samples recovered in this study adequately preserve pollen for analysis, in terms of both quantity and quality of preservation. This result is not unexpected as pollen preservation is inherently poor in biologically active soils. Saltmarsh pollen assemblages are restricted to the often thin sediment transition between soil horizon and mangrove mud. The potential for dry season deflation of sediment from coastal saltmarsh flats in the high-tide to supra-tidal range in northern Australia, has been demonstrated in modern coastal studies elsewhere (Rhodes 1980; Grindrod 1988). Continuous sedimentation cannot be assumed within this facies type. While strong chenopod pollen values are sometimes recorded in modern saltmarsh sediments (Grindrod and Rhodes 1984; Grindrod 1985; Crowley *et al.* 1994), the quality of preservation is variable and often poor. This is an outcome of strong pollen production by the saltmarsh chenopods, and marginal conditions for preservation on saltmarsh flats. The latter is a reflection of low sedimentation rates and enhanced potential for oxidation of organic material as surface sediments undergo occasional tidal flooding and longer periods of subaerial desiccation. The strong representation of Poaceae and *Avicennia* pollen, in conjunction with saltmarsh assemblages in cores VC28 and VC29, indicates the close proximity of terrestrial communities to the most landward mangroves. Similar pollen signatures are recorded at mangrove-saltmarsh transitions elsewhere in northern Australia (Grindrod 1985).

Pollen is richly preserved in the organic clay and mud units which overly soil and saltmarsh sediments. The high quality of preservation and characteristic low pollen diversity, are consistent with pollen signatures for modern mangrove sediments recorded in various other studies (Crowley *et al.* 1994) and in core sequences interpreted as mangrove (Grindrod and Rhodes 1984). Such assemblages are invariably dominated by the Rhizophoraceae, and contain few other mangrove pollen types. Regional pollen components are also present, but often only marginally represented.

Low pollen concentrations in marine carbonate deposits, which overly mangrove muds in all continental shelf cores, reflect distance from the pollen source once full marine conditions are established.

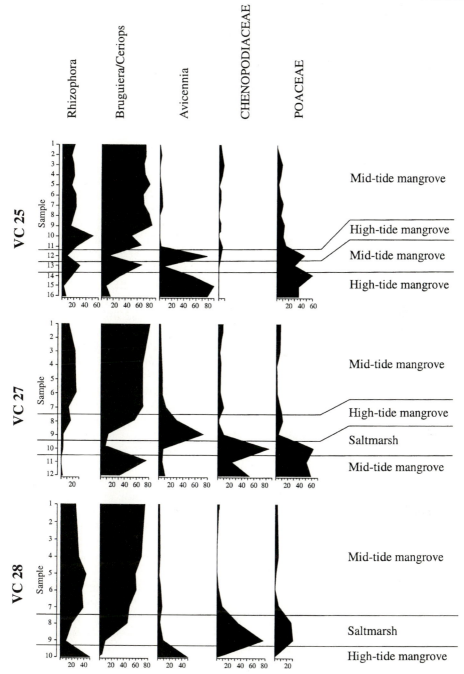

Fig. 8.7 Summarised pollen data and reconstructed successional sequences for three continental shelf cores. In each case reconstructed successions differ from predicted littoral sequence (saltmarsh ⇒ rear mangrove ⇒ mid-tide mangrove) for smoothly rising sea level.

Mangrove successions during the PGMT

Figure 8.7 shows summaries of the interpreted plant successional sequences for the pollen records from cores VC25, VC27 and VC28. These are the first pollen records of mangrove successional processes on the continental shelf during the PGMT. Undisturbed successional sequences on transgressive shores would be expected to show the reverse order of plant community change as that described earlier for prograding shores, assuming a similar suite of species for each and a smooth relative sea level rise.

At the most general level the continental shelf sequences indicate former local successions from grassy terrestrial communities through saltmarsh to mangrove. The mangroves subsequently drown as sea level continues to rise. This essentially is the reverse sequence described earlier for the late Holocene prograding shore at Princess Charlotte Bay, where pioneering mangroves give way to more landward mangroves and then saltmarsh communities.

However none of the pollen records for VC25, VC27 and VC28 provide evidence for uninterrupted successional sequences, as would be predicted under a steadily rising relative sea level. The expected sequence from landward to seaward would be: terrestrial communities (here dominated by Poaceae) ⇒ supra-tidal saltmarsh (Chenopodiaceae) ⇒ high-tide back mangrove (*Avicennia, Ceriops*) ⇒ mid-tide mangrove (*Rhizophora, Bruguiera*). In light of the taxa most strongly represented by pollen, the predicted pollen sequence through time (upwards from base of core) would be: Poaceae (and/or other terrestrial pollen types) ⇒ Chenopodiaceae ⇒ *Avicennia* ⇒ *Bruguiera/Ceriops* ⇒ *Rhizophora*. None of the pollen sequences clearly show this sequence. For instance VC28 has strong *Avicennia* and *Rhizophora* values in the lowest sample followed by short-lived increases in Chenopodiaceae and Poaceae, and eventual establishment of high *Rhizophora* and *Bruguiera/Ceriops* values up the sequence. The sequence for VC27 begins with high *Bruguiera/Ceriops*, Chenopodiaceae and Poaceae, followed by high *Avicennia* with the eventual return to *Bruguiera/Ceriops* and *Rhizophora*. VC25 provides the nearest pollen sequence to that predicted for an uninterupted succession, but shows early fluctuation of *Avicennia* and *Bruguiera/Ceriops* values and a later minor increase in Chenopodiaceae.

The difference between predicted and observed successional sequences may be a consequence of vagaries of sedimentation at each core site, real plant successions that do not reflect smoothly rising relative sea level, or a combination of these factors.

Vagaries of sedimentation may include tidal reworking of sediments and loss of material through scouring. Neither of these processes adequately explains the recorded sequences. The expected result of tidal reworking, which would presumably operate to a daily, even diurnal, schedule would be a highly homogenised pollen record. The well defined transition in the pollen records of each of the three pollen sequences in Figure 8.7 suggest that tidal reworking has not had a significant influence in this regard. This is in keeping with other published mangrove and saltmarsh pollen sequences which commonly retain very clear pollen zone boundaries. Loss of material through scouring, resulting in a discontinuous sedimentary record, cannot be discounted. As mentioned above, the VC25 pollen sequence provides the nearest to predicted successional sequence, except that it does not contain an initial saltmarsh transitional zone as would be expected in a transgressive sequence. The loss of this section in the sediment column is possible, through deflation or scouring of saltmarsh sediments. However, it is important to bear in mind that erosional loss of sediment does not explain reversals of plant successional events as are evident in each of the three pollen records.

From the forgoing it is likely that the sedimentary sequences in cores VC25, VC27 and VC28 are essentially intact. In this event the reconstructed vegetation successions are real, and apparently reflect a non-constant rate of relative sea level rise at each core site. Relative sea level change at each location is potentially a balance between the rate of eustatic sea level change and rates of sediment accumulation. While pulses in sedimentation at the core sites cannot be ruled out, especially as they are located at channel margins, there is no indication of sedimentary changes in the cores that coincide with the successional reversals in the pollen records.

For these reasons it is concluded that the sediment sequences in cores VC25, VC27 and VC28 are substantially intact, and that the sequences of plant succession in the pollen records provide tentative evidence for a variable rate of eustatic sea level rise. This result is in keeping with claims by Carter *et al.* (1986) and Larcombe *et al.* (1995), based on PDR and sedimentological data, for variable rates of sea level rise on the Queensland continental shelf during the Post Glacial Marine Transgression.

Climate reconstructions and the continental shelf cores

While inherent problems of disproportionate pollen representation and unfilled climatic ranges may reduce the detail of climatic reconstructions from mangrove (and other) pollen assemblages, evidence for former climate on the continental shelf can be derived from the shelf cores. A consistent feature of coastal vegetation in modern north Queensland settings is the occurrence of chenopod saltmarshes adjacent to mangroves in low rainfall environments. The study of modern pollen deposition by Crowley *et al.* (1994) demonstrates that the mangrove pollen components are indistinguishable between assemblages representing the climatic range of modern settings in the region. This is largely the result of ubiquitous high Rhizophoraceae values at all mangrove sites combined with very poor representation of key mangrove species of either humid or arid locations. However the non-mangrove components of pollen assemblages provide a convincing indication of local climate. In general the pollen assemblages from the high rainfall locations Mutchero Inlet and Hinchinbrook Channel include strong values for rainforest pollen types and fern spores, and lack saltmarsh (chenopod) pollen. In contrast pollen assemblages from the low rainfall location at Cleveland Bay are low in rainforest and fern taxa, have a relatively strong sclerophyll component (particularly Myrtaceae), and invariably contain a saltmarsh component (Crowley *et al.* 1994). None of the continental shelf pollen sequences provide evidence of humid coastal environments of the kind described by Crowley *et al.* (1994). Where pollen sequences extend to the former Pleistocene surface (VC25, VC27 and VC28) saltmarsh successions are indicted. These pollen records compare most closely with modern pollen assemblages described for Cleveland Bay, suggesting relatively dry climatic conditions, at least for the three core locations involved, at the early Holocene coast.

Given the general correlation in modern settings between luxuriant mangrove growth and high humidity, it is tempting to draw a cruder level of interpretation that a high pollen count for fossil mangroves indicates former humid climate. While this interpretation may make intuitive sense, it pays little regard to the dynamics of mangrove development in relation to sea level change and the physiographic evolution of coasts. Evidence from north Australia suggests that mangrove abundance is greatly influenced by environmental imperatives linked to the physical evolution of coasts, beyond any direct link with climatic humidity. Important influences are relative sea level change and stages in the geomorphological development of coasts (Chappell and Grindrod 1984; Woodroffe *et al.* 1985).

In relation to the present study, a combination of evidence indicates that high mangrove abundance, at least as it is reflected in the ocean core pollen records, is a poor indicator of former humid climate. Figure 8.8 compares the timing of mangrove representation in ODP 820 from the late Pleistocene to present, with reconstructions for humidity (Lynch's Crater precipitation curve, Kershaw 1994) and aridity (Australian Arid Period: Kershaw and Nanson 1993). Included as well is a summarised curve for rainforest angiosperm pollen from ODP 820 (Moss 1999), and evidence from the present study of former saltmarsh shorelines in continental shelf cores. The data show that regionally strong mangrove development, reflected in the ODP mangrove curve, is initiated around 15,000 years BP. This corresponds with the period of late-Pleistocene dryness (Australian Arid Period). Furthermore, peak ODP mangrove values between 10,000 and 9000 years BP coincide with pollen evidence for saltmarsh shorelines on the adjacent continental shelf. In contrast mangrove values decline with the onset of humid Holocene conditions, reflected in strong ODP rainforest pollen values, and the Lynch's Crater precipitation curve. Strongest precipitation values in the Lynch's Crater record clearly postdate the mangrove phase.

From these comparisons it is difficult to conclude that the periodically strong mangrove representation in our ocean pollen records is an indication former episodes of humid climate. This finding contrasts with published claims for former humid climate based on fossil evidence for extensive mangrove growth (van Campo 1986; Jennings 1975), and highlights the need for careful regard of other environmental factors that influence mangrove abundance.

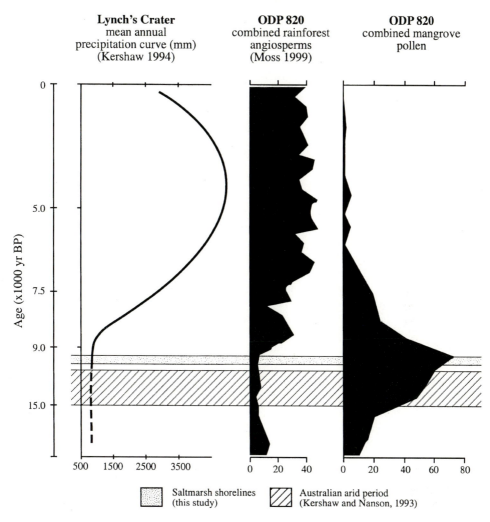

Fig. 8.8 Comparison of selected pollen data and climatic reconstructions for the late Pleistocene and Holocene in northern Australia. The Lynch's Crater precipitation curve is based on rainforest angiosperm pollen. Rainfall is low prior to 9000 years BP, and at its highest around 5000 years BP. This trend is supported by the ODP 820 record for rainforest angiosperms. The timing of evidence for saltmarsh shorelines on the continental shelf (this study), and the Australian arid period (Kershaw and Nanson 1993) also shown. The comparisons indicate that mangrove pollen representation increases during Pleistocene aridity and declines with the onset of moister conditions during the Holocene.

Mangrove development on transgressive and regressive shorelines

An important outcome of the ocean records is the clear relationship, during the late Quaternary, that exists between oxygen isotope fluctuations and mangrove abundance. It is generally accepted that isotopic fluctuations reflect ocean volume change (Chappell and Shackelton 1986), and so provide a convenient record of eustatic sea level change. For the purposes of this discussion the isotope record may also be used as a surrogate trace of relative sea level, particularly for the tectonically benign Australian coastline. The record shows that mangroves occur in abundance during transgressive phases, and have restricted occurrence during regressions and in rarer periods of relative sea level stability.

In light of our continental shelf PDR surveys and pollen cores, it is evident that in the north Queensland example at least, strong mangrove development occurs in drowning tidal channels and estuaries. The model presented here to account for the observed differences in mangrove development on transgressive and regressive shores invokes a contrast in the physiographic responses of coasts to sea level change. Channel incision occurs on the exposed continental shelf during periods of low relative sea level. While there is some debate as to whether incision in the study area is predominantly a product of fluvial or tidal processes (Wolf *et al.* 1998), a system of palaeochannels is readily identified in PDR surveys, and has been described in detail for some regions (Johnson *et al.* 1982). During periods of sea level rise the coastal reaches of these channels transform to marine estuaries and embayments. Thus a coast of intricate plan form develops, affording maximum mangrove habitat in sheltered settings. Low gradients on Australian shores ensure extensive estuarine development through this process. This is particularly so in macrotidal settings of the north and northwest, where today tidal influence in estuaries extends more than 100 km inland from the coast.

The ocean core pollen records also indicate subdued mangrove development during phases of relative sea level stability and lowering. Woodroffe *et al.* (1985) demonstrate that mangroves that flourished in the South Alligator River towards the end of the PGMT were rapidly replaced as sea level stabilised, as ongoing fluvial processes ensured the continued vertical accretion of sediments beyond tidal influence. Hence widespread mangrove habitat decline proceeds purely as an outcome of predictable processes of sedimentation. During periods of relative sea level fall, mangrove habitat is also restricted due to a lack of suitable river estuaries on the newly exposed continental shelf. Under these conditions coasts assume a simple plan form characterised by fewer estuaries, and a predominance of open shorelines less suited to mangrove colonisation.

Hence, each phase of the cycle of mangrove development and decline identified in the ocean core pollen records is consistent with hypothetical and reconstructed physiography and sedimentological process on north Australian coasts during defined phases of relative sea level adjustment. While the broader environmental changes involved are climate-linked, the cycle of mangrove development is more intimately connected to physical processes that dictate habitat availability. This finding is in general agreement with other published models for shoreline evolution based on geomorphological responses to sea level adjustment (Posamentier and Vail 1988; Posamentier *et al.* 1988; Morley 1995; Woodroffe *et al.* 1985).

Conclusions

The results of this study indicate that continental shelf sediments in northern Australia include stratigraphically intact mangrove sequences from the Post Glacial Marine Transgression. The PDR surveys indicate that this sediment sequence is widespread in former estuaries and tidal channels. It is uncertain whether similarly intact sequences representing open coasts of the transgressive period also exist. The former mangrove communities on the continental shelf included species of *Rhizophora*, *Bruguiera*, *Ceriops* and *Avicennia* as major floristic components, and are palynologically indistinguishable from common mangrove communities on the north Queensland coast today. Reconstructed mangrove and saltmarsh successions do not conform to the sequence predicted for an uninterrupted vegetation sequence on a transgressive shore, suggesting a variable rate of relative sea level rise on the north Queensland coast during the early Holocene period. The strong representation of chenopod saltmarsh communities in continental shelf cores that penetrate to the Pleistocene soil surface suggest relatively dry climatic conditions at the early Holocene coast.

Mangrove pollen is also well represented in ocean cores at widespread locations across the study region. The reconstructed late-Quaternary history of mangroves is one of fluctuating development and decline on an extensive regional scale. Strong mangrove development coincides with rising sea levels and consequent marine transgressions across continental shelves. Conversely, modest mangrove development coincides with falling sea levels and regressive shores. Comparisons with a range of available palaeoclimatic data indicate that the large-scale mangrove changes cannot be directly linked to climate change. It is concluded that regional mangrove histories are strongly influenced by physiographic changes to coasts, as they variously respond to Quaternary sea level adjustments.

Acknowledgements

Fieldwork and radiocarbon dating were funded by grants from the Australian Research Council and the Australian Institute for Nuclear Science and Engineering. We thank Bob Carter and the Department of Geology, James Cook University, for access to the RV James Kirby; Jane Blevin and Kevin Hooper for core site location and core collection on the continental shelf. We thank also Fred Jansen of the Netherlands Instute for Sea Research for access to the Lombok Ridge core, and Francois Guichard of the Laboratoire de Sciences du Climat et de l'Environnement, France, for access to the Banda Sea core. This paper is a contribution to IGCP Project 437 'Coastal Environmental Change During Sea-Level Highstands: A Global Synthesis with Implications for Management of Future Coastal Change.

References

Ahmad, S.M., Guichard, F., Hardjawidjaksana, K., Adisaputra, M.K. and Labeyrie, L.D. 1995. Late Quaternary paleoceanography of the Banda Sea. *Marine Geology*, 122: 385-9.

Bunt, J.S., Williams, W.T. and Duke, N.C. 1982. Mangrove distributions in north-east Australia. *Journal of Biogeography*, 9: 111-20.

Carter, R.M., Carter, L. and Johnson, D.P. 1986. Submergent shorelines in the southwest Pacific: evidence for an episodic post-glacial transgression. *Sedimentology*, 33: 629-49.

Chapman, V. J. 1977. Wet coastal formations of Indo-Malesia and Papua-New Guinea. In: V.J. Chapman (Editor) *Ecosystems of the World: Wet Coastal Ecosystems.* Elsevier, Amsterdam: 261-70.

Chappell, J. and Grindrod, J. 1984. Chenier plain formation in northern Australia. In: B.G. Thom (Editor) *Coastal Geomorphology in Australia.* Academic Press, London: 197-229.

Chappell, J. and Shackelton, N.J. 1986. Oxygen isotopes and sea level. *Nature,* 324(13): 137-40.

Clark, R.L. and Guppy, J.C. 1988. A transition from mangrove forest to freshwater wetland in the monsoon tropics of Australia. *Journal of Biogeography*, 15: 665-84.

Crowley, G.M., Anderson, P., Kershaw, A.P. and Grindrod, J. 1990. Palynological reconstruction of mangrove succession during the Holocene marine transgression in humid, tropical north Queensland. *Australian Journal of Ecology.* 15: 231-20.

Crowley, G.M., Grindrod, J. and Kershaw, A.P. 1994 Modern pollen deposition in the Tropical lowlands of northeast Queensland, Australia. *Review of Palaeobotany and Palynol*ogy, 83: 299-327.

Grindrod, J. 1985. The palynology of mangroves on a prograded shore, Princess Charlotte Bay, North Queensland, Australia. *Journal of Biogeography,* 12: 323-48.

Grindrod, J. 1988. The palynology of Holocene mangrove and saltmarsh sediments, particularly in northern Australia. *Review of Palaeobotany and Palynology,* 55: 229-45.

Grindrod, J. 1995. Mangrove vegetation responses to late-Quaternary environmental changes. In: G. Dixon and D. Aitken (Editors) *Institute of Australian Geographers Conference Proceedings (1993).* Monash Publications in Geography No. 45. Monash University, Melbourne: 103-10

Grindrod, J. and Rhodes, S. 1984. Holocene sea level history of a tropical estuary: Missionary Bay, north Queensland. In: B.G. Thom (Editor) *Coastal Geomorphology in Australia.* Academic Press, London: 151-76.

Hutchings, P. and Saenger, P. 1987. *Ecology of Mangroves.* University of Queensland Press, Brisbane: 388 pp.

Jennings, J.N. 1975. Desert dunes and estuarine fill in the Fitzroy Estuary (north-western Australia). *Catena,* 2: 215-62.

Johnson, D.P., Searle, D.E. and Hopley, D. 1982. Positive relief over buried post-glacial channels, Great Barrier Reef province, Australia. *Marine Geology*, 46: 149-59.

Kershaw, A.P. 1994. Pleistocene vegetation of the humid tropics of northeastern Queensland, Australia. *Palaeogeography, Palaeoclimatology, Palaeoecology* 109: 399-412.

Kershaw, A.P. and Nanson, G.C. 1993. The last full glacial cycle in the Australian region. *Global and Planetary Change*, 7: 1-9.

Larcombe, P., Carter, R.M., Dye, J., Gagan, M.K. and Johnson, D.P. 1995. New evidence for episodic post-glacial sea-level rise, central Great Barrier Reef, Australia. *Marine Geology,* 127: 1-44.

Morely, R.J. 1995. Biostratigraphic characterisation of systems tracts in Tertiary sedimentary basins. *Proceedings of the Indonesian Petroleum Association Symposium on Sequence Stratigraphy in SE Asia.* Indonesian Petroleum Association, Jakarta: 49-71.

Moss, P. 1999. *Late Quaternary Environments of the Humid Tropics of Northern Australia.* Unpublished PhD Thesis, Monash University, Clayton.

Muller, J. 1964. A palynological contribution to the history of the mangrove vegetation of Borneo. In: L.M. Cranwell (Editor) *Ancient Pacific Floras.* University of Hawaii Press, Honolulu: 33-42.

Peerdeman, F.M. and Davies, P.J. 1993. Sedimentological response of an outer-shelf, upper-slope sequence to rapid changes in Pleistocene eustatic sea level: Hole 820A, northeastern Australian margin. *Proceedings of the Ocean Drilling Program Scientific Results,* 133: 303-13.

Posamentier, H.W., Jervey, M.T. and Vail, P.R. 1988. Eustatic controls on clastic deposition. 1 - conceptual framework. In: C.K. Wilgus, B.S. Hastings, H. Posamentier, J. van Wagoner, C.A. Ross and C.C.StC. Kandall (Editors) *Sea-level Changes - An Integrated Approach.* Society of Economic Palaeontologists and Mineralologists Special Publication No. 42: 109-24.

Posamentier, H.W. and Vail, P.R. 1988. Eustatic controls on clastic deposition. 2 - sequence and systems tract models. In: C.K. Wilgus, B.S. Hastings, H. Posamentier, J. van Wagoner, C.A. Ross and C.C.StC. Kandall (Editors) *Sea-level Changes - An Integrated Approach.* Society of Economic Palaeontologists and Mineralologists Special Publication No. 42: 126-54.

Rhodes, E.G. 1980. *Modes of Holocene Coastal Progradation, Gulf of Carpentaria.* Unpublished Ph.D. Thesis, Australian National University, Canberra.

Tomlinson, P.B. 1986. *The Botany of Mangroves.* Cambridge University Press, Cambridge: 413 pp.

Truswell, E.M., Kershaw, A.P. and Sluiter, I.R. 1987. The Australian/Malayasian connection: Evidence from the palaeobotanical record. In: T.D. Whitmore (Editor) *Biogeographic Evolution of the Malay Archipelego.* Oxford University Press, Oxford: 32-49.

van Campo, E. 1986. Monsoon fluctuations in two 20,000 yr BP oxygen isotope pollen records off southwestern India. *Quaternary Research, 26:* 376-88.

van der Kaars, W.A. 1991. Palynology of eastern Indonesian marine piston-cores: A Late Quaternary vegetational and climatic record for Australasia. *Palaeogeography, Palaeoclimatology, Palaeoecology,* 85: 239-302.

van der Kaars, S. 1998. Marine and terrestrial pollen records of the last glacial cycle from the Indonesian region: Bandung Basin and Banda Sea. *Palaeoclimates,* 3: 209-19.

van der Kaars, S., Wang, X., Kershaw, A.P., Guichard, F. and Setiabudi, D.A. 2000. A late Quaternary palaeoecological record from the Banda Sea, Indonesia: Patterns of vegetetion, climate and biomass burning in Indonesia and northern Australia. *Palaeogeography, Palaeoclimatology, Palaeoecology,* 155: 135-53.

Wang, X., van der Kaars, S., Kershaw, A.P. Bird, M. and Jansen, F. 1999. A record of fire, vegetation and climate through the last three glacial cycles from Lombok Ridge core G6-4, eastern Indian Ocean, Indonesia. *Palaeogeography, Palaeoclimatology, Palaeoecology,* 147: 241-56.

Wells, A.G. 1982. Mangrove vegetation of Northern Australia. In: B.F. Clough (Editor) *Mangrove Ecosystems in Australia. Structure, Function and Management.* ANU Press, Canberra: 57-78.

Woodroffe, C.D. and Grindrod, J. 1991. Mangrove biogeography: the role of Quaternary environmental and sea level change. *Journal of Biogeography,* 18: 479-92.

Woodroffe,C.D., Thom, B.G. and Chappell, J. 1985. Development of widespread mangrove swamps in mid-Holocene time in northern Australia. *Nature,* 317: 711-7.
Wolfe, K.J., Larcombe, P., Naish, T. and Purdon, R.G. 1998. Lowstand rivers need not incise the shelf: an example from the Great Barrier Reef, Australia, with implications for sequence stratigraphic models. *Geology,* 26: 75-8.

Palaeoceanography of the
Western Pacific Warm Pool
During the Last Glacial Maximum:
Long-term Climatic Monitoring of the Maritime Continent

J. Ignacio Martinez, Patrick De Deckker and Timothy T. Barrows

Introduction

Until recently, palaeoceanographic changes in the North Atlantic, through the Atlantic Conveyor Belt, were put forward as a plausible mechanism to explain Pleistocene global oceanic circulation and climate changes (e.g. Broecker and Denton 1989). This view has been challenged by the realisation that oceano-graphic and climatic phenomena in the tropics may also significantly contribute to global change (e.g. Hirst and Godfrey 1993). The Western Pacific Warm Pool (WPWP), the warmest region of the global ocean (>28°C, Yan *et al.* 1992), is part of the El Niño-Southern Oscillation phenomenon (ENSO) and, as such, variations in its extent and dynamics through time play an important role in global climate (e.g. Godfrey 1996). Furthermore, the Indonesian passageway, as the connection at the surface between the Pacific and Indian Oceans, constitutes a major component of the renewal of the North Atlantic Deep Water (NADW; e.g. Gordon 1986). However, this view has been questioned by a number of researchers (e.g. Pariwono *et al.* 1986; Rintoul 1991; Schmitz and McCartney 1993; Schmitz 1995; Godfrey 1996; Wijffels *et al.* 1996; Macdonald and Wunsch 1996; Macdonald 1998; Ganachaud and Wunsch 2000) who consider the Indonesian throughflow as a minor constituent for the advection of NADW, but probably a major contributor to the exchange of upper water in the Southern Hemisphere. Furthermore, it has been suggested that the wind field around Australia exerts a major influence on the magnitude of the throughflow (e.g. Pariwono *et al.* 1986; Godfrey 1996; McBride 1998), and in particular, the strength of the wind field in the Southern Ocean as well as the topographic barrier imposed by New Zealand (Godfrey 1996).

If this last scenario is valid, then we have to look for palaeoceanographic and palaeoclimatic evidences all around Australia when trying to reconstruct the WPWP and the dynamics of the Indonesian throughflow during the last glacial maximum (LGM), a time when wind strength and, in concordance, oceanic currents in the Southern Ocean may have been stronger (Petit *et al.* 1981; de Angelis *et al.* 1987); this would have caused a more intense throughflow between the Pacific and Indian Oceans.

ISBN 3-923381-47-6
© 2002 by CATENA VERLAG, 35447 Reiskirchen

Recent interest in reconstructing the WPWP during the LGM include the works of Thunell *et al.* (1994) and Martinez *et al.* (1997, 1999), who compiled previous palaeoceanographic studies in the western Pacific (e.g. CLIMAP 1981; Thompson and Shackleton 1980; Wells and Wells 1994; Martinez 1994b; Miao *et al.* 1994).

The present paper is a review of published data up to 1997 when this paper was accepted for publication in this special issue [some updated references have been added herein] and an examination of the palaeoceanographic and palaeo-climatic conditions of the Indonesian throughflow under a LGM scenario when 1) sea level was lowered by 125 ± 4 m (e.g. Yokoyama *et al.* 2000, 2001), 2) the Indonesian passageway was reduced in extent, 3) precipitation in the region was reduced by ~30% (e.g. Flenley 1979; van der Kaars and Dam 1995), 4) sea-surface temperatures were lower by ~2°C (e.g. CLIMAP 1981; Thunell *et al.* 1994), and 5) the monsoons were less intense in August and more intense in February (e.g. Duplessy 1982; Sirocko *et al.* 1996). This chapter largely relies on published information, in addition to new palaeoceanographic data from the eastern Indian Ocean. As with oceanographic studies for the present, attention will be given to the sources of the throughflow water, its physical parameters (salinity and temperature), its paths, and a discussion of its possible effects on global climate.

Oceanography of the Western Pacific Warm Pool

The Indonesian Throughflow

The Indonesian region is characterised by at least eight deep basins (and seas with water depths generally reaching more than 4500 m) connected through relatively shallow sills (Tomczak and Godfrey 1994).

The high atmospheric and ocean temperatures in the region induce evaporation of oceanic water, formation of low pressure cells and precipitation; the overall balance is a gain of freshwater from precipitation. Consequently, SSTs are high and SSSs relatively low (Wyrtki 1961). As indicated above, waters with SSTs exceeding 28°C are referred to as the WPWP and occur in response to the west-flowing equatorial currents (e.g. Tomczak and Godfrey 1994; Gordon and Fine 1996) (Figure 9.1).

During the Southern Hemisphere winter (August), trade winds blow over Australia's northwest region from the southeast; during the Southern Hemisphere summer (February), winds take the opposite direction. Maximum throughflow occurs in August as a result of a lowering of sea level south of Indonesia in response to the strong SE trade winds. Consequently, water moves from the Pacific Ocean to the Indian Ocean as a response to a higher steric height in the former region (e.g. Godfrey and Golding 1981; Godfrey and Ridgway 1985). Estimates of the volume of water transported from the Pacific to the Indian Ocean varies between 2 Sv and 24 Sv (1 Sverdrup, Sv = 10^6 m^3/s) (Tomczak and Godfrey 1994) or more conservatively, 7 Sv to 18.6 Sv (Murray and Arief 1988; Godfrey 1996). Recent estimates of ~19 Sv in August and ~3 Sv in February evidence the wide range of monsoonal variation of the throughflow and the difficulty of the problem of measuring it (Godfrey 1996; but see also Wiffels *et al.* 1996 and McBride 1998).

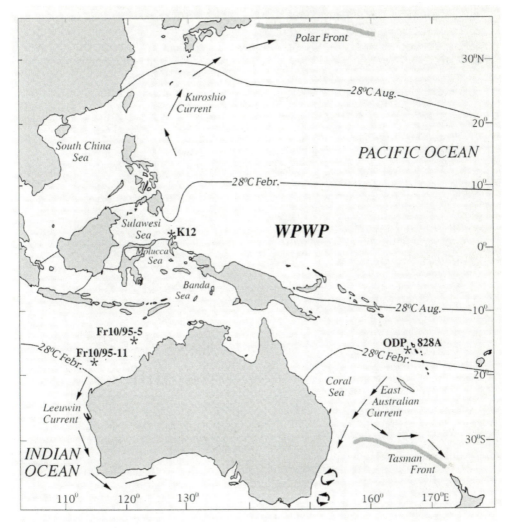

Fig. 9.1 Map of the western Pacific showing the location of cores (stars) and the limits of the Western Pacific Warm Pool (WPWP; the 28°C isotherm) in August and February (from Levitus et al. 1994). The Kuroshio, Leeuwin and East Australian Currents are indicated as outflows of the WPWP.

The WPWP and the atmospheric pressure system are coupled phenomena that operate as a west-east oscillator causing El Niño events to occur every 2 to 5 years (e.g. Enfield 1989; Clarke 1991; Yan *et al.* 1992). Water that passes the Indonesian throughflow eventually spreads into the Indian Ocean as the west-flowing South Java and the South Equatorial Currents, and as the south-flowing Leeuwin Current (e.g. Godfrey and Ridgway 1985; Wijffels *et al.* 1996).

As early as 1961, Wyrtki noticed the sharp contrast between the WPWP surface mixed-layer characterised by SST >28°C, SSS <34‰ and density st <22.0, and by intermediate water masses being colder, saltier and with a st =27.7. Its boundary is a strong discontinuity layer located between 120 and 160 m water depth, as indicated by a sharp density variation; temperature (i.e. the 25°C isothermal) is a good proxy of the density gradient because salinity changes little in the sub-surface (Wyrtki 1961). Localised seasonal upwelling areas in the WPWP are the only places where the surface layer is disrupted (Wyrtki 1961).

The uppermost 1000 m of the Indonesian throughflow are occupied by a modified Pacific Central Water mass whose vertical salinity profile progressively becomes uniform prior to reaching the Indian Ocean. Subsequently, the latter mixes with Indian Central Water to restore this gradient (Tomczak and Godfrey 1994). One possibility is that, in the Indonesian throughflow, turbulent mixing occurs at sill depths, thus causing a vertical homogenisation of salinity, but not of water temperature which retains its gradient (Tomczak and Godfrey 1994). Another explanation would be that most of the flow passing through the upper thermocline of the Lombok Strait consists of low-salinity, well oxygenated water of North Pacific origin (Gordon and Fine 1996). In contrast, only a small amount of high salinity water may pass through the lower thermocline of the east Indonesian throughflow as export flux from the South Pacific (Gordon and Fine 1996). Most of the Indonesian throughflow is assumed to contain water of low salinity, North Pacific origin (the North Equatorial Current) and moving from Mindanao through Makassar Strait and then towards the Lombok, Savu and Timor Straits into the Indian Ocean (Gordon and Fine 1996). Conversely, the more saline South Pacific water (the South Equatorial Current) is mostly recycled in the Halmahera Eddy and returns to the Pacific (e.g. Gordon 1986; Gordon and Fine 1996). No claim for a turbulent mixing mechanism is made in the latter interpretation to explain uniform salinity in sub-thermocline waters. It should be noted that the WPWP displays strong seasonal SSS variations that can reach 2‰ in the Banda and Arafura Seas (Wyrtki 1961).

Upwelling occurs in the Banda and Arafura Seas during August, when the SE monsoon forces the Monsoon Current through the Java and Flores Seas. As the volume of water entering the Indonesian region from the Pacific Ocean through the Halmahera Sea is insufficient to compensate the volume of water transported to the west (and to the south through the Timor Sea), upwelling occurs thus bringing cold and salty sub-surface water to the sea-surface (the Subtropical Lower Water located at 125-150 m; Wyrtki 1958). In February, the opposite mechanism occurs; downwelling is induced by the NW monsoon that brings an excess of water to the Banda and Arafura Seas. At ~1000 m depth, the excess of water in the Banda Sea finds its way into the Indian Ocean (Wyrtki 1958).

Upwelling south of Java occurs during August when the throughflow is at its maximum. It is expected that SSTs would be low (Wyrtki 1962). However, the reduced steric heights in the region then cause the throughflow to increase and warm water to flood into the eastern Indian Ocean (Godfrey 1996).

Outflows of the Western Pacific Warm Pool (WPWP)

As indicated above, difficulties still exist to precisely calculate the Indonesian throughflow partly due to the fact that part of the WPWP is recycled into the North Pacific through the north-flowing Kuroshio Current and into the South Pacific through the south-flowing East Australian Current. Both the Kuroshio and the East Australian Currents are fed by the North and South Equatorial Currents and, presumably, would be stronger if the Indonesian throughflow were closed (refer to modelling results by Godfrey and Golding 1981; Hirst and Godfrey 1993). A third important outflow of the WPWP comprises the South Equatorial and Leeuwin Currents in the Indian Ocean. Consequently, under a LGM scenario when the Indonesian throughflow could have been highly restricted due to a sea-level drop of 125 ± 4 m (e.g. Yokoyama *et al.* 2000, 2001), all possible outflows of the WPWP (and their associated seas) should be considered; namely, the NW Pacific and the South China Sea, the SW Pacific and the Coral Sea, and the eastern Indian Ocean.

The South China Sea has a maximum depth of ~4300 m and connects the northwest extreme of the Indonesian throughflow with the North Pacific. Ocean currents in this region respond to the monsoon forcing with a general circulation that is clockwise during the SW monsoon in August, and anti-clockwise during the NE monsoon in February when some water is supplied to the Kuroshio Current. Salinities in the South China Sea vary seasonally in response to monsoonal rains, i.e. lower SSS in the eastern South China Sea occur during the SW monsoon (Tomczak and Godfrey 1994).

The Coral Sea is an open sea receiving the influx of the South Pacific Equatorial Current that is affected by seasonal variations and gets stronger during the Southern Hemisphere winter (when Trade Winds are stronger). At ~18°S, the South Equatorial Current diverges: part of it travels to the north into the Solomon Sea to ultimately feed the North Equatorial Counter Current or the Kuroshio Current – and part of it travels southward to feed the East Australian Current that eventually deviates to the east at the Tasman Front (at ~30°S), or ultimately disappears as warm-water eddies in the southern portion of the Tasman Sea (Tomczak and Godfrey 1994).

In the eastern Indian Ocean, the Leeuwin Current is a narrow eastern boundary current that runs along the Western Australian margin to eventually reach the Great Australian Bight in the Southern Ocean transporting up to ~5 Sv of water (e.g. Tomczak and Godfrey 1994). The action of the Trade Winds on the westward flowing South Equatorial Current (in the Indian Ocean) triggers a southward Ekman transport whose cooling at high latitudes results in a lower steric height in the eastern Indian Ocean and causes the Leeuwin Current to flow southward (e.g. Cresswell and Golding 1980; Godfrey 1996). In a LGM scenario, a reduced throughflow, and presumably, a less intense Leeuwin Current would imply that the Ekman transport was less efficient in lowering the steric height of the eastern Indian Ocean. Recent work on modern-day clay distribution on the sea floor in the eastern Indian Ocean by Gingele *et al.* (2001a), in contrast with LGM clay distribution in the same area (Gingele *et al.* 2001b), confirm that the Leeuwin Current, if at all existent at the LGM, would have taken the direction of the central Indian Ocean.

Peter Kershaw, Bruno David, Nigel Tapper, Dan Penny and Jonathan Brown (Eds): Bridging Wallace's Line

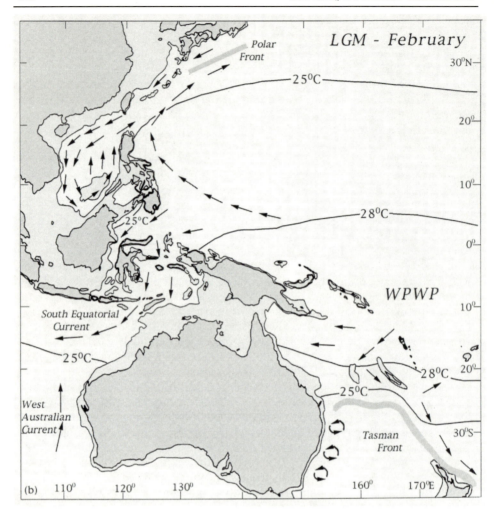

Fig. 9.2 Palaeoceanographical representation of the WPWP during the LGM for the austral a) summer (February), and b) winter (August). The 28°C and 25°C SST isotherms are from CLIMAP (1981), Anderson et al. (1989), Miao et al. (1994), Wang et al. (1995) and Barrows et al. (1996). Reconstruction of the South China Sea relies on information obtained in Wang et al., (1995), the Polar Front from Thompson and Shackleton (1980) and Oda and Takemoto (1992), the Tasman Front from Martinez (1994b), and the Leeuwin Current from Wells and Wells (1994).

Table 9.1. Core locations. Selected sites in the WPWP.

Core	Latitude	Longitude	Water depth (m)
K12	02°41'20"N	127°44'10"E	3510
ODP 828A	15°47.34'S	166°17.04'E	3086.7
Fr10/95-5	14°00.55'S	121°01.58'E	2472
Fr10/95-11	17°38.57'S	114°59.93'E	2458

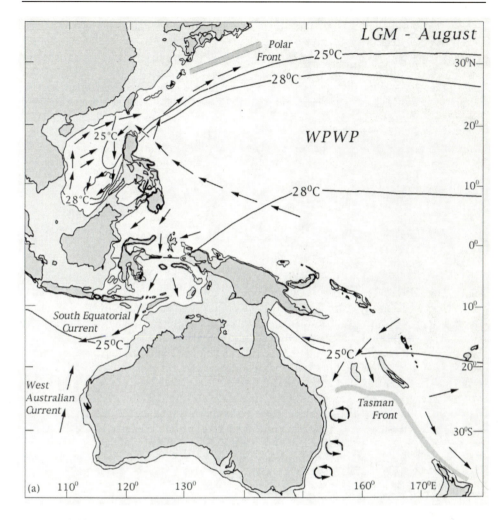

Fig. 9.2b

Materials and methods

As indicated above, much of this presentation relies on published information as well as on four sediment cores carefully selected as to represent the pattern of fluctuations in the WPWP. Planktonic foraminifera analyses from cores K12 from the Molucca Sea (Barmawidjaja *et al.* 1993), ODP 828A from the eastern Coral Sea (Martinez 1994a, 1994b), Fr10/95-11 and Fr10/95-5 from the eastern Indian Ocean (see Table 9.1 and Figure 9.1 for core locations) are used to derive sea-surface temperatures (SST) for the last ~30,000 years. The modern analogue technique (MAT) is applied herein to reconstruct SSTs and is regarded as more reliable than the transfer function method used in previous studies (e.g. Prell 1985; Thunell *et al.* 1994; Ortiz and Mix 1997). The MAT world data base (Prell

Peter Kershaw, Bruno David, Nigel Tapper, Dan Penny and Jonathan Brown (Eds): Bridging Wallace's Line

1985) was modified by, 1) adding a new foraminiferal data set of core-tops recently collected in the eastern Indian Ocean (Martinez *et al.* 1998) and from around Australia (Barrows and Hunt 1996) and, 2) removing data points from the North Pacific and North Atlantic Oceans as they are not regarded as analogous to conditions in the Southern Hemisphere. This modified MAT world database provides better modern analogues for the Indonesian region. We also selected *N. dutertrei*, a characteristic intermediate to deep-dwelling species (Bé 1977; Hemleben *et al.* 1989), in order to determine the influence of intermediate waters (and the depth of the deep chlorophyll maximum, DCM) as well as deciphering the possible bias in estimating SSTs when using the MAT.

Results

Oceanographic parameters in the WPWP during the LGM: A brief review

Sea-surface temperature patterns

The reduction in extent of the WPWP during the LGM was already implied by the CLIMAP (1981) results that show a 28°C SST isotherm restricted to the western Pacific in August (Southern Hemisphere winter) and split in two areas in February (Southern Hemisphere summer) (Figure 9.2). The CLIMAP (Climate/Long-Range Investigations, Mapping and Prediction 1981) results also suggested: 1) a reduced change in SST in the WPWP area, 2) a stronger seasonality in the WPWP area, 3) the absence of the WPWP in the South China Sea, the eastern Indian Ocean, and most of the Coral Sea, 4) a warm-water tongue in the Coral Sea apparently representing a more dynamic East Australian Current, 5) the equatorward deflection of isotherms in the NW Pacific Ocean, and 6) the equatorward deflection of isotherms in the eastern Indian Ocean. Anderson *et al.* (1989) confirmed some of these results and showed no evidence of a more dynamic East Australian Current, but the northern displacement of the 18° to 22°C isotherms (see Figure 9.2). The northward displacement of cool subtropical waters, i.e. at the Tasman Front, was confirmed by factor analyses on assemblages of planktonic foraminifera (Martinez 1994b), benthic foraminifera (Kawagata 2001) and recent SST reconstructions in the Tasman Sea (Barrows and Hunt 1996).

Similarly, Miao *et al.* (1994) and Wang *et al.* (1995) showed that the 28°C isotherm was restricted to the South China Sea in August (see Figure 9.2). The equatorward displacement of isotherms in the NW Pacific Ocean, implying an eastward deflection of the Kuroshio Current by the southern deflection of the Oyashiro Current (the southward migration of the Polar Front), was also suggested (e.g. Thompson and Shackleton 1980; Ujiié *et al.* 1991; Oda and Takemoto 1992).

Finally, the results of Wells and Wells (1994) and Barrows *et al.* (1996) seem to agree with the CLIMAP SST reconstructions for the eastern Indian Ocean, thus supporting the idea of van Andel *et al.* (1967) that the West Australian Current was stronger during the LGM (see Figure 9.2). Van Andel *et al.* (1967) also suggested that SSTs at the Timor Sea region were considerably cooler due to the blockage of warm currents through Torres Strait, although recent work by

Spooner (2001) clearly demonstrates that there was little SST change in the Banda Sea east of Timor during the LGM. The reduced SSTs in the eastern Indian Ocean and the lowering of sea-level in northern Australia caused the disappearance of the Gulf of Carpenteria as a major cyclogenic area (Webster and Streten 1978). Because tropical cyclones today originate in oceanic areas where SST exceeds 28°C (i.e. within the WPWP), then important climatic consequences are expected from the exclusion of cyclones in the eastern Indian Ocean that would therefore face a reduced rainfall (Webster and Streten 1978).

Beside microfossil analyses for the reconstruction of sea-surface conditions in the western Pacific, Sr/Ca in corals (e.g. Beck *et al.* 1997), the U_{37}^k technique (Lyle *et al.* 1992), and oxygen-isotope analyses of planktonic foraminifera (Broecker 1986) have been used. Unfortunately, corals studied so far have been collected at the margins or outside the WPWP and Sr/Ca ratios have not been measured for the LGM (Beck *et al.* 1997). The oxygen-isotope records of planktonic foraminifera for the western Pacific appeared to support CLIMAP results (Broecker 1986). However, those records came from low-sedimentation rate areas where the glacial-interglacial $\delta^{18}O$ difference is probably blurred by bioturbation (Broecker 1986; Martinez *et al.* 1997). U_{37}^k analyses for the central Pacific and the Indian Ocean tend to confirm the CLIMAP results (Lyle *et al.* 1992; Rostek *et al.* 1993; Sonzogni *et al.* 1998).

Overall, the WPWP (between 20°S and 20°N) did register a reduced drop in SST (<1.5°C) as confirmed by Thunell *et al.* (1994) who used the modern analogue technique (MAT) on planktonic foraminifera assemblages. The reliability of the MAT and the transfer function method to infer past SSTs will be discussed below with new data available from the WPWP region.

Sea-surface salinity changes

Sea-surface salinities in the Bonaparte Depression (NW Australia) during the LGM were interpreted to be similar to today's based on palaeontological and sedimentological inferences (van Andel *et al.* 1967). However, van Andel *et al.* (1967) also noted that in order to keep SSS close to normal in the Bonaparte Depression, river runoff had to be substantially lower; similarly, the occurrence of caliche nodules in the Sahul Rise was interpreted as an indicator of more arid conditions in Australia.

On the basis of a higher relative abundance of *N. dutertrei* during the LGM, it was suggested that SSSs were reduced in the Sulu Sea (Linsley *et al.* 1985), and increased in the Molucca Sea (Barmawidjaja *et al.* 1993). The latter researchers argued that the presence of *N. dutertrei* in both region should be regarded as indicative of a 'deep' chlorophyll maximum closer to the sea-surface, the presence of upwelling, and possibly saltier conditions (by as much as 1.9‰) all over the western Pacific. Barmawidjaja *et al.* (1993) also found an increase in the content of *Neogloboquadrina pachyderma* right-coiling form (=labelled *N. pachyderma* R herein) in the Molucca Sea that they postulate to indicate lower SSTs.

From a recent compilation done on deep-sea cores from the WPWP, large planktonic $\delta^{18}O$ differences were interpreted to represent a general increase of

SSS during the LGM (Martinez *et al.* 1997). This interpretation relies on the assumption that SST reconstructions, by applying the MAT, are reliable. Sea-surface salinity was found to increase by ~1‰ at the centre of the WPWP and probably to a higher value south of ~12°S and north of ~8°N (Figure 9.3), thus implying that evaporation minus precipitation was higher in the region for the LGM. To interpret this phenomenon, two approaches were then followed: (1) the first one consists of using *a residual approach* where the glacial-interglacial change in SSS resulted from the difference of the LGM and Holocene $\delta^{18}O$ signals in planktonic foraminifera, after a correction for the ice-volume effect (~1.2‰; Fairbanks 1989) and SST variations (Thunell *et al.* 1994), and (2) the second one being *a mathematical approach*, where the SSS difference is calcula-ted as:

$$\Delta SSS = SSS_{LGM} - SSS_{Recent} = (\Delta\delta^{18}O - a - b\Delta SST)/c$$

where SSS_{LGM} is the local salinity during the LGM, and SSS_{Recent} represent today's local salinity. $\Delta\delta^{18}O$ is the difference between the LGM and the modern values, a is the ice-volume effect, ΔT is the SST difference between the LGM and the Recent, b is the slope of the $\delta^{18}O$ versus temperature relationship, and c is the slope of the $\delta^{18}O$ versus salinity relationship (Broecker 1989; Rostek *et al.* 1993). Sea-surface salinity during the LGM (SSS_{LGM}) was derived by assuming an ice-volume effect of 1.2‰, the slope of the $\delta^{18}O$ versus temperature relationship being -0.23‰/°C for *Globigerinoides sacculifer* and -0.2‰/°C for *Globigerinoides ruber* (Duplessy *et al.* 1981), and the slope of the $\delta^{18}O$ versus salinity relationship being $0.5\delta^{18}O$‰/S‰ (Broecker 1989) (Figure 9.3 and Martinez *et al.* 1997). The SST difference between the LGM and the Recent along a latitudinal transect in the region was taken from MAT results on planktonic foraminiferal assemblages (Thunell *et al.* 1994).

As mentioned before, both approaches seem to indicate that SSS during the LGM was higher by ~1‰ at the centre of the WPWP, implying that evaporation minus precipitation was higher with important consequences for the coupled ocean-atmosphere system, the fresh-water balance and global circulation.

Sea-surface currents and the wind field

As indicated above, physical oceanographers have found that the Indonesian throughflow responds to the dynamics of currents and wind fields around Australia and most particularly in the Southern Ocean. CLIMAP's (1981) results indicate that, on a global scale, the wind field and oceanic currents were more dynamic during the LGM than today, whereas the overall albedo was larger due to drier conditions on land and increased polar ice cap. In fact, the wind strength in the Southern Ocean during the LGM has been suggested to have been stronger, based on the presence of aeolian dust (Petit *et al.* 1981) and chloride concentrations in ice cores (de Angelis *et al.* 1987). This view has been supported by others (e.g. Sarnthein *et al.* 1981; Rea 1994) who argued that the increase upwelling in equatorial regions in the Pacific and Atlantic Oceans was due to a stronger Southern Ocean circulation induced by a stronger wind field.

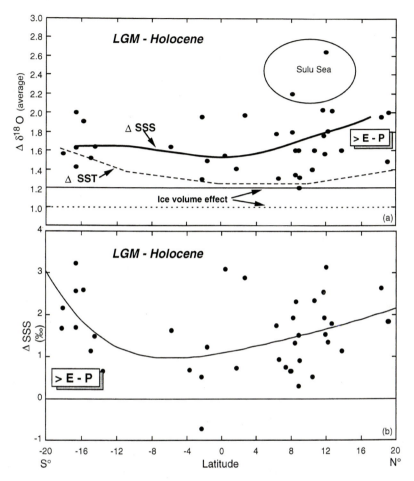

Fig. 9.3 SSS changes in the WPWP. Residual approach a), and mathematical approach b). In a), glacial-interglacial Dδ¹⁸O values are plotted against latitude; residuals attributable to salinity anomalies (DSSS) are derived after applying a δ¹⁸O 1.2‰ ice-volume effect (Fairbanks 1989), and correcting for SST changes (Thunell et al. 1994). In b), the LGM and Holocene SSS difference (DSSS) is plotted against latitude. Note that SSS and evaporation minus precipitation (>E-P) was higher in all the WPWP during the LGM. Correlation lines were obtained as polynomial fits of order 5. A δ¹⁸O 1.0‰ ice-volume effect as suggested by Schrag et al. (1996) is also indicated by a horizontal dashed line in a). For further explanations see text and Martinez et al. (1997).

It is estimated that the wind increased by a factor of 30% to 70% (Crowley and Parkinson 1988). In contrast to the idea of a stronger wind field in the Southern Ocean, and linked to the northern shift of the Antarctic Polar Front, Bareille *et al.* (1994) showed that aridity and the larger exposure of land would be suffi-cient to explain the dust peaks in ice-cores and deep-sea cores. Furthermore, there is no definitive explanation to the northern shift of the Antarctic Polar

Front, as modelling results show that increasing the strength of the wind by a factor of 0.5 to 2.0 would increase the speed of the water flow without changing the position of the Front. In contrast, by shifting meridionally the present wind field, a change in the strength of the current and the position of the Front (Klinck and Smith 1993) would occur. The dust argument can easily be challenged in contrast with the one favouring chloride input. As pointed out by Crowley and North (1991), the concentration of sea salt in the atmosphere is proportional to the strength of the wind and, consequently, higher values in Antarctic ice-cores can only be explained by a wind increase of 20% to 50% or perhaps even more.

Apart from the study of Wang and Li (1995) in the South China Sea, there is no other attempt to quantitatively model oceanic currents in the WPWP for the LGM. Sea-surface currents in the South China Sea were inferred to follow a similar pattern to today's; i.e. a clockwise circulation in August, and an anti-clockwise circulation in February (see Figure 9.2). However, the southward displacement of the Polar Front in the North Pacific (Thompson and Shackleton 1980; Oda and Takemoto 1992), and the northward displacement of the Tasman Front (Martinez 1994b) were shown to imply a weakening of both the Kuroshio and East Australian Currents respectively during the LGM. As mentioned above, the same reasoning has been applied to the eastern Indian Ocean where the Leeuwin Current would have been weaker (van Andel *et al.* 1967; Wells and Wells 1994; McCorkle *et al.* 1994).

The ocean general circulation model (OGCM) prepared by Hirst and Godfrey (1993) suggests that because the Kuroshio and East Australian Currents are fed by the North and South Equatorial Currents, these would be presumably stronger if the Indonesian throughflow were closed. However, this model was created in an attempt to understand circulation patterns for the present Indonesian through-flow. Consequently, LGM boundary conditions, such as the reduction in SST in high latitudes, were not included in their model. In contrast, an ocean general circulation model (OGCM) for the LGM by Lautenschlager *et al.* (1992) suggests that the strength of the Kuroshio Current and East Australian Current would have decreased during the LGM, thus supporting the Thompson and Shackleton (1980), and Martinez (1994b) results. However, even though the Tasman Front was displaced northward from ~30 to ~26°S during the LGM, tropical planktonic foraminifera species (such as *G. ruber* and *G. sacculifer*) were found east of Tasmania during the LGM (refer to Factor 1 of Martinez 1994b). The presence of the foraminiferal assemblage represented by Factor 1 almost 20° south of their normal habitat, suggests a transport by warm-water eddies such as occurs today. Another possibility is that this fauna may indicate that the East Australian Current was still operating during the LGM and that not all of its volume was diverted towards the central Pacific along the Tasman Front.

SST estimates from sediment core material

Core K12 from the Molucca Sea

Our SST estimates for Core K12, a high-sedimentation rate core from the Molucca Sea (Barmawidjaja *et al.* 1993), are presented in Figure 9.4. The latter authors did not carry out SST reconstructions in this core. Temperatures in August (Northern Hemisphere summer) are higher than 28 °C during most of the Holocene and during two short intervals prior to the LGM. The lowest SSTs were reached during the LGM (~25.6 °C) and during a short interval prior to the LGM. The SST pattern for February (Northern Hemisphere winter) is similar, but the 28 °C value is only exceeded during brief intervals in the Holocene (Figure 9.4).

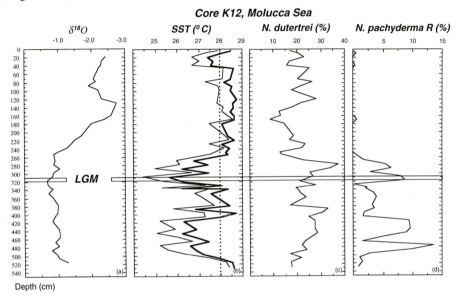

Fig. 9.4 Core K-12, Molucca Sea. a) $\delta^{18}O$ record smoothed to 3-point moving average, from Barmawidjaja et al. (1993); b) February SST (bold line) and August SST (plain line) MAT estimates; c) N. dutertrei percentage curve; d) N. pachyderma R percentage curve. Note the correspondence of the LGM with low SSTs and high percentages of N. dutertrei and N. pachyderma.

The pattern of SSTs in core K12 shows a similarity with the percentage abundance of *N. dutertrei* and *N. pachyderma* R, i.e. higher abundances of these species correspond to low SSTs, and conversely, lower abundances correspond to high SSTs. Noticeably, *N. pachyderma* R almost disappears at the end of the glacial stage (at ~250 cm) whereas *N. dutertrei* displays a declining trend from the LGM towards the Holocene reaching a minimum (~9%) at 170 cm, i.e. at the beginning of the Holocene.

In the MAT, as in the Transfer Function method, it is assumed that the distribution of foraminifera species are mainly controlled by SST. As pointed out by Barmawidjaja *et al.* (1993), the increased content of *N. dutertrei* and *N. pachyderma* R in core K12 may indicate the presence of a deep chlorophyll maximum within the photic zone. The effects of these species on SST estimates will be discussed below.

Core ODP 828A from the eastern Coral Sea

Core ODP 828A is a very-high sedimentation rate core from west of Vanuatu in the eastern Coral Sea (Martinez 1994a, b; Martinez *et al.* 1997; Figure 9.5). As with many Pacific deep-sea cores, carbonate dissolution is severe during inter-glacial intervals, whereas carbonate preservation is good during glacial intervals (Martinez 1994a). Consequently, SST estimates between 4 and 16 m (a preservation interval) in that core are considered herein as highly reliable; at this interval, SSTs in February are invariably above 28 °C and, in many cases, above 29 °C (Figure 9.5). In August, in contrast, SSTs show an average of ~27 °C dropping occasionally to 25 °C and rising to >28 °C. Noticeably, SSTs in August show higher amplitude fluctuations than SSTs in February (Figure 9.5).

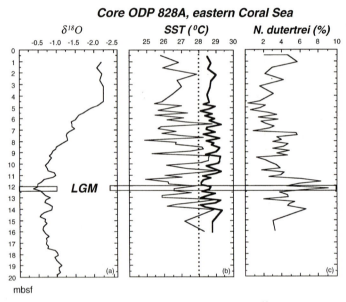

Fig. 9.5 Core ODP 828A, Vanuatu (eastern Coral Sea). a) $\delta^{18}O$ record smoothed to 3-point moving average, from Martinez et al. (1997); b) February SST (bold line) and August SST (plain line) MAT estimates; c) N. dutertrei percentage curve. Note the correspondence of the LGM with high percentages of N. dutertrei, but the lack of any conspicuous drop in SST.

The percentage abundance of *N. dutertrei* shows maximum values during the LGM, a declining trend towards the Holocene and a minimum value at the beginning of the Holocene (Figure 9.4). In contrast to core K12 from the Molucca Sea, the percentage abundance of *N. dutertrei* is low in core ODP 828A; in the latter, it is the percentage of other deep-dwelling species (Martinez 1994b) which engender a minimum bias in the estimation of SSTs using the MAT.

Cores Fr10/95-5 and Fr10/95-11 from the eastern Indian Ocean

February SST reconstructions on core Fr10/95-5 display a steadily declining trend, from temperatures >28°C in the late Holocene to ~26°C during the glaciation phase (Figure 9.6). There is no significant drop in SST during the LGM as it is observed between 30 and 45 cm depth interval that corresponds to a trough in the $\delta^{18}O$ record (heavy values). The 28°C value is exceeded only during the Holocene, particularly above the 20 cm depth interval; this interval also corresponds to maximum values (>1) for the *G. sacculifer/N. dutertrei* ratio which Martinez *et al.* (in press) suggested to be a good proxy of the WPWP boundary in the eastern Indian Ocean. August SSTs display a declining trend from ~25°C in the late Holocene to ~23°C during the glaciation phase, thus paralleling February SSTs (Figure 9.6).

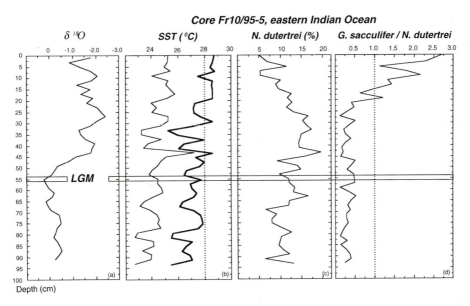

Fig. 9.6 Core Fr10/95-5, eastern Indian Ocean. a) $\delta^{18}O$ record smoothed to 3-point moving average, from a work in preparation; b) February SST (bold line) and August SST (plain line) MAT estimates; c) N. dutertrei percentage curve; d) G. sacculifer/N. dutertrei ratio. Note the correspondence of the LGM with intermediate percentages of N. dutertrei, but the lack of a conspicuous drop in SST. Note also that the 28°C value is exceeded during the late Holocene in correspondence to a >1 G. sacculifer/N. dutertrei ratio.

In contrast to the previous two cores (K12 and ODP 828A), maximum percentages of *N. dutertrei* do not coincide with the LGM, but to the beginning of the Holocene instead. A declining trend from a maximum of ~20% during the early Holocene to a minimum of ~5% during the late Holocene is also observed (Figure 9.6). During the LGM, values vary between ~10 and ~15%.

August and February SST reconstructions on core Fr10/95-11 parallel each other and display a maximum reduction of ~3°C during the LGM (Figure 9.7).

The 28°C value is exceeded all throughout the core, except during the glacial phase. As with core Fr10/95-5, the *G. sacculifer/N. dutertrei* ratio shows a good correspondence with warm intervals, i.e. values exceeding 1.0 which correspond to SSTs in excess of 28°C, whereas values <1.0 correspond to the glacial interval (see Figure 9.7).

As with cores K12 and ODP 828A, maximum percentages of *N. dutertrei* show a maximum value close to the LGM, and a declining trend towards the beginning of the Holocene (Figure 9.7). Percentages of *N. dutertrei* in this core are comparable to those in core ODP 828A.

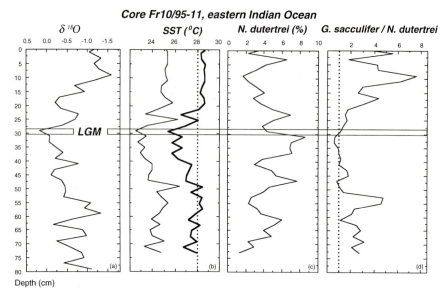

Fig. 9.7 Core Fr10/95-11, eastern Indian Ocean. a) $\delta^{18}O$ record smoothed to 3-point moving average, from a work in preparation; b) February SST (bold line) and August SST (plain line) MAT estimates; c) N. dutertrei percentage curve; d) G. sacculifer/N. dutertrei ratio. Note the correspondence of the LGM with high percentages of N. dutertrei,

Discussion

SST reconstructions and the percentage abundance of N. dutertrei

By comparing SST reconstructions for the LGM in cores K12, ODP 828A, Fr10/95-5 and Fr10/95-11 with maps for the same time-slice (Figures 9.2(a) and 9.2(b); reconstructed herein using a composite of previously published data for the LGM), a number of discrepancies appear. These are: 1) the 28°C isotherm in August was located north of the Molucca Sea, 2) the 25°C isotherm was located south of the Molucca Sea, and 3) the 25°C isotherm was located slightly north-ward in the eastern Indian Ocean in August. Our new results may imply that the WPWP was further reduced in extent in comparison to the   reconstructions presented in Figure 9.2. Similarly, the almost constant SSTs in the eastern Coral Sea (as represented by the ODP 828A record) imply that the region was enclosed within the WPWP for the last ~30,000 years in February and outside of it in August. This finding strongly contrasts with Beck *et al.*'s (1997) Sr/Ca SST reconstructions for a Vanuatu coral which suggest a 6.5°C reduction in temperature 10,000 years ago. Seasonality in the region, however, gives similar results using the two palaeothermometers, i.e. ~3°C.

Before discussing the implications of these results, we should examine the statistical methods, based on plankton assemblages, used to reconstruct SSTs. Reconstruction of sea-surface environmental parameters using the transfer function method and the MAT are based on similar assumptions, namely, that the ecological requirements of foraminiferal species have remained unchanged during the Pleistocene period and that the yearly average SST and SSS are directly related to ecological parameters of surface waters (Imbrie and Kipp 1971). The reliability of both methods has been tested by several authors (e.g. Molfino *et al.* 1982; Prell 1985; Le 1992; Ortiz and Mix 1997). Concordance in estimating SST for the past using different microfossil groups, i.e. foraminifera, coccolithophorids and radiolarians, was pointed out as a strength of the transfer function method (Molfino *et al.* 1982). However, Molfino *et al.* (1982) also showed that, despite the ability to discern water masses using groups of microfossils with diverse ecological requirements, in low latitudes major dissimilarities are found that may be due to other ecological factors more important than SST, such as nutrients and thermocline depth. The implied assumption in the transfer function and the MAT is that foraminiferal species (as well as radiolarians and coccolithophorids) record sea-surface conditions despite their transit through different water depths during their life-cycle or the almost exclusive intermediate to deep-water habitat of some species. It may be that, following our recent investigations on 54 core-tops from the eastern Indian Ocean (Martinez *et al.* in press), the distribution of planktonic foraminifera in that area is directly linked with SST and nutrients. Then, a change in the thermocline in the past may require a dropping in optimum SST requirements for some species. If that was the case, past SST reconstructions would have to be modified accordingly and this would explain the discrepancies obtained between SSTs and land temperature reconstructions using pollen assemblages. As indicated above, we wish to explore the influence of intermediate waters and its possible bias when estimating past SSTs, by selecting the percentage abundance of *N. dutertrei*, a characteristic intermediate to deep-dwelling species (Bé 1977; Hemleben *et al.* 1989).

Neogloboquadrina dutertrei is mainly a tropical-subtropical, herbivorous species considered to preferentially occur in upwelling regions (Bé 1977; Thiede 1975, 1983; Thunell *et al.* 1983; Hemleben *et al.* 1989). Based on glacial-interglacial $\delta^{18}O$ difference in *N. dutertrei*, Shackleton (1977) suggested that this species may have preferred warmer waters and 'followed' a particular density level in a saltier glacial ocean. However, this view has been changed by the observation that the depth of the chlorophyll maximum, rather than density, is the major controlling factor on the distribution of planktonic foraminifera (Fairbanks and Wiebe 1980; Ravelo *et al.* 1990). *N. dutertrei* has been found in

Peter Kershaw, Bruno David, Nigel Tapper, Dan Penny and Jonathan Brown (Eds): Bridging Wallace's Line

abundance during the upwelling season (e.g. in the Panama Basin) when the thermocline is shallow and primary production is high (Thunell and Reynolds 1984), or during a post-upwelling stage (e.g. in the San Pedro basin, NE Pacific Ocean) when the upper water column becomes warmer (between 12° to 20°C) and stabilises, and with the chlorophyll maximum being strong, coinciding with abundant numbers of diatom species (Sautter and Thunell 1991). In his global distribution study, Hilbrecht (1996, 1997) shows that *N. dutertrei* exhibits narrower tolerances during summer, i.e. SST above 21°-22°C with an optimum at ~27°C, a temperature of ~15°C at 200 m, SSS of ~35‰ (but it can be found in waters with SSS between ~34‰ and ~37‰) and a surface density of ~22.5. Hilbrecht (1996) also points out that *N. dutertrei* increases when *Globigerina bulloides* decreases. The latter species is also regarded as typical of upwelling systems and shows preference for productive cold waters (e.g. Bé 1977; Hemleben *et al.* 1989). Thus, the exclusion of one of the two species in eutrophic regions may reflect a different availability of phytoplankton preys and the advantageous competence of one species over the other.

Our results corroborate the suggestion by Barmawidjaja *et al.* (1993) that the relative increase of *N. dutertrei* (and *N. pachyderma* R when present) during the LGM, accounts for most of the drop in SSTs in the WPWP (see Figures 4 and 7). In fact, high percentage abundances (>30%) of *N. dutertrei* are only recorded in core K12 from the Molucca Sea where SST dropped by ~3°C during the LGM; in other cores, the abundance of the species is low coinciding with a reduction in SST during the LGM. We also agree with the interpretation that the deep chlorophyll maximum was probably within the photic zone and nutrients were more abundant and close to the surface; seasonal upwelling was also more frequent and intense. However, we do emphasise that it may well be that the drop in SST is apparent and most probably assemblages of planktonic foraminifera reflect a shallower thermocline where SST remained almost the same as today. Similarly, we also consider the alternative possibility that even though the MAT accurately compares down-core assemblages with modern analogues, and those similar assemblages presumably can be translated in SST estimates, it may be that those tropical species whose optimum is close to the maximum SST of the present ocean, did live in slightly colder waters (perhaps by ~1° to 3°C) where nutrients were abundant and close to the sea-surface.

A shallower nutricline/thermocline (and stronger oceanic upwelling) during the LGM implies an increase in the strength of oceanic divergence along the equatorial Pacific. This can be easily explained by evidence of stronger wind fields around Australia during the LGM. However, the intensification of trade winds also implies a more efficient accumulation of warm water in the western Pacific and a deepening of the thermocline as happens today during non-El Niño years (e.g. Enfield 1989). Consequently, to solve this apparent contradiction, the El Niño events may have been presumably more frequent, thus favouring upwelling by the equatorial divergence of currents during the LGM.

SSS reconstructions and the pattern of sea-surface currents

In today's Indonesian throughflow, most of the surface water is of low-salinity North Pacific origin rather than of a high-salinity South Pacific origin (e.g.

Wyrtki 1961; Gordon and Fine 1996; Godfrey 1996). Under a saltier LGM scenario (SSS higher by ~1‰ at the centre of the WPWP), we suggest that the source of Indonesian throughflow water would mainly have been of South Pacific origin. Today, a salinity maximum (35.2‰ to 35.5‰) is found at ~100 m to 150 m water depth and is referred to as the Subtropical Lower Water originating at mid latitudes (Wyrtki 1961). Also this water mass becomes diluted by the abundant fresh water precipitating in the Indonesian region (Gordon and Fine 1996). Under an increased evaporation minus precipitation balance in the region (Martinez *et al.* 1997), salty South Pacific water would not be significantly diluted and could occur closer to the sea-surface, therefore explaining not only the regional increase in SSS, but also the shallow deep chlorophyll maximum as suggested by Barmawidjaja *et al.* (1993). It may also be that the mixed layer of the ocean during the LGM was further enriched in nutrients supplied by river runoff from the extensively exposed continental shelves of the region. Iron – a biolimiting element in many regions of the ocean (e.g. Martin 1990) – may have been more efficiently supplied to the sea-surface by the intense wind field promoting primary productivity and increasing export production of organic matter out of the euphotic zone (e.g. Herguera 1994; Kawahata and Suzuki 1994, 1995). Already, recent work by Hesse (1997) implies a greater supply of aeolian dust during the LGM at least for the Exmouth Plateau. However, more recently Hesse and McTainsh (1999) postulated that mid-latitude transport of dust during the LGM was no more prominent than during the Holocene • at least in the Tasman Sea • but that wind strength was more predominant in high latitudes. Further, Gingele *et al.* (2001b) confirm that winds at low latitudes north of Australia were quite different from today´s arrangement. Nevertheless, the relative decrease in the percentage abundance of *N. dutertrei* in core Fr10/95-5 (Figure 9.6) may indicate that upwelling conditions close to the Australian continent were significantly reduced as compared to other regions of the WPWP. Another interpretation would be that the nutricline was closer to the surface and the thermocline much steeper compared to today. In a saltier (and cooler?) glacial ocean, we would expect denser surface waters in the Indonesian throughflow moving, and perhaps, sinking at intermediate depths in the eastern Indian Ocean; this would be the glacial Leeuwin Current. It could be that the Leeuwin Current did not circumnavigate the West Australian coast but instead migrated westward to the centre of the Indian Ocean between latitude 10° and 20°S (as the South Equatorial Current?). Gingele *et al.* (2001b) agree to this proposed scenario.

Conclusions

A review of the palaeoceanography of the western Pacific Warm Pool (WPWP), focusing on sea-surface temperature (SST) and salinity (SSS) reconstructions and the pattern of currents, was presented. New SST reconstructions, by using the modern analogue technique (MAT), are shown for cores K12 (Molucca Sea), ODP 828A (Vanuatu, eastern Coral Sea), and Fr10/95-5 and Fr10/95-11 (eastern Indian Ocean). The WPWP, a seasonally and inter-annual changing feature of the ocean, was saltier (by ~1‰) and highly reduced in extent during the Last

Glacial Maximum (LGM). However, the Indonesian throughflow may have been
stronger in response to stronger wind fields in the Southern Ocean and around,
as well as north, of Australia. The reduction in extent of the WPWP was accompanied by the equatorward migration of the Tasman Front in the South Pacific,
and the Polar Front in the North Pacific. The Kuroshio and East Australian
Currents may still have been strong at that time. The Leeuwin Current may have
been saltier and colder and still in operation during the LGM, but its overall path
different. Reduced SSTs during the LGM may be a bias of the increase abundance of the intermediate to deep-dwelling planktonic foraminiferal species
Neogloboquadrina dutertrei and *N. pachyderma* right-coiling. We support a
previously proposed idea that during the LGM, the deep chlorophyll maximum
was within the photic zone, the thermocline gradient steeper and primary productivity was higher in the WPWP. Under a LGM scenario, when wind fields
were stronger, oceanic divergence may have been more efficient.

Tropical planktonic foraminifera may have lived in a slightly colder WPWP
thus biasing SST reconstructions (by ~1°to 3°C ?).

Epilogue: Significance of palaeoceanographic work to modern conditions in the Maritime Continent north of Australia

We have demonstrated through the palaeoceanographic investigations of some
selected cores in the WPWP that this area is subject to some long-term, very
dynamic changes. The evaporation-precipitation balance did change, and so did
the nutricline/thermocline depth. The position of the latter two may have
resulted from a different wind strength as postulated for the LGM. The glacial
ocean, even in the WPWP, would have been different even though high SSTs
were maintained close to 28°C.

The WPWP is not only a dynamic region, but it is also climatically a very
significant region of the globe. It is where high values of heat and moisture
transfer occur between the ocean and atmosphere. This area is also directly
linked with the eastern Pacific through ENSO signals.

Our preliminary study of deep-sea cores provides a basis for future work on
other aspects linked to climate change in the WPWP. These include: 1) vegetational changes and ecological successions (to be detected through pollen
examination in the cores; see van der Kaars and De Deckker submitted); 2) fire
intensity and frequency (through the analysis of charcoal particles in the cores);
3) mineralogical changes (through compositional changes of sediment; see
Gingele *et al.* 2001a, 2001b) - some of which may be related to erosional
changes on land (some caused by human activity?); and 4) aeolian transport and
volcanic dust. The evidence of all these can be obtained using the same cores,
and therefore would provide a link between oceanic, continental as well as atmospheric phenomena.

Future study of the physicochemical phenomena linked with the palaeoceanography of the WPWP promises to be very exciting indeed, and ought to
help better understanding the interplay of many different components, including
human-related ones, in what is best called the Maritime Continent north of
Australia.

Acknowledgments

We thank the officers and the crew of the *RV Franklin* for their unconditional help in obtaining the cores for this study. Fr10/95 and Fr2/96 cruises were financed by 3 ARC grants to P. De Deckker. Sonia Bremstaller, Margareth Chorley, and D. Ryan enthusiastically participated on the cruises and also helped with sample preparation. Judith Shelley and Timothy Munson kindly prepared the samples for stable-isotope analyses, and Joe Cali performed the analyses at the Research School of Earth Sciences. Drs N. Tapper and G. Cresswell provided useful reviews of the original manuscript and we sincerely thank them for their efforts.

References

Anderson, D.M., Prell, W.L. and Barratt, N.J. 1989. Estimates of sea surface temperature in the Coral Sea at the last glacial maximum. *Paleoceanography*, 4: 615-27.

Bareille, G., Grousset, F.E. and Labracherie, M. 1994. Origin of detrital fluxes in the southeast Indian Ocean during the last climatic cycles. *Paleoceanography*, 9(6): 799-819.

Barmawidjaja, B.M., Rohling, E.J., van der Kaars, W. A., Vergnaud-Grazzini, C. and Zachariasse, W. J. 1993. Glacial conditions in the northern Molucca Sea (Indonesia). *Palaeogeography, Palaeoclimatology, Palaeoecology*, 101: 147-67

Barrows, T.T. and Hunt, G.R. 1996. A reconstruction of Last Glacial Maximum sea-surface temperatures in the Australasian region. *Quaternary Australasia*, 14(1): 27-31.

Bé, A.W.H. 1977. An Ecological, Zoogeographical and Taxonomic Review of Recent Planktonic Foraminifera. In: T.S. Ramsay (Editor) *Oceanic Micropaleontology*. Academic Press, London: 1-100.

Beck, J.W., Récy, J., Taylor, F., Edwards, R.L. and Cabloch, G. 1997. Abrupt changes in early Holocene tropical sea surface temperature derived from coral records. *Nature*, 385: 705-7.

Broecker, W.S. 1986. Oxygen isotope constraints on surface ocean temperatures. *Quaternary Research*, 26: 121-34.

Broecker, W. S. 1989. The salinity contrast between the Atlantic and Pacific Oceans during glacial time. *Paleoceanography*, 4(2): 207-12.

Broecker, W. S. and Denton, G. H. 1989. The role of ocean-atmosphere reorganizations in glacial cycles. *Geochimica et Cosmochimica Acta*, 53: 2465-501.

Clarke, A.J. 1991. On the reflection and transmission of low-frequency energy at the irregular western Pacific Ocean boundary. *Journal of Geophysical Research*, 96: 3289-305.

CLIMAP Project Members. 1981. *Seasonal Reconstructions of the Earth's Surface at the Last Glacial Maximum*. Geological Society of America Map and Chart Series No. MC-36. Geological Society of America, Boulder: 18 maps.

Cresswell, G.R. and Golding, T.J. 1980. Observations of a south-flowing current in the southeastern Indian Ocean. *Deep Sea Research*, (27A): 449-66.

Crowley, T.J. and Parkinson, C.I. 1988. Late Pleistocene variations in Antarctic sea ice II: effects of interhemispheric deep-ocean heat exchange. *Climate Dynamics*, 3: 93-103.

Crowley, T.J. and North, G.R. 1991. *Paleoclimatology*. Oxford University Press, Oxford: 339 pp.

de Angelis, M., Barkov, N.I. and Petrov, V.N. 1987. Aerosol concentrations over the last climatic cycle (160 kyr) from an Antarctic ice core. *Nature,* 325: 318-21.

Duplessy, J. C. 1982. Glacial to interglacial contrasts in the northern Indian Ocean. *Nature,* 295: 494-98.

Duplessy, J.C., Be, A. and Blanc, P.L. 1981. Oxygen and carbon isotope composition and biogeographic distribution of planktonic foraminifera on the Indian Ocean. *Palaeogeography, Palaeoclimatology, Palaeoecology,* 33: 9-46.

Enfield, D.B. 1989. El Niño, past and present. *Review of Geophysics,* 27(1): 159-87.

Fairbanks, R.G. 1989. A 17,000-year glacio-eustatic sea level record: influence of glacial melting rates on the Younger Dryas event and deep-ocean circulation. *Nature,* 342: 637-42.

Fairbanks, R.G. and Wiebe, P.H. 1980. Foraminifera and chlorophyll maximum; vertical distribution, seasonal succession and paleoceanographic significance. *Science,* 209: 1524-6.

Flenley, J.R. 1979. *The Equatorial Rain Forest: A Geological History.* Butterworths, London: 162 pp.

Ganachaud, A. and Wunsch, C. 2000. Improved estimates of global ocean circulation, heat transport and mixing from hydrographic data. *Nature* 408: 453-7.

Gingele, F.X., De Deckker, P. and Hillenbrandt, C.-D. 2001a. Clay mineral distribution in surface sediments between Indonesia and NW Australia - source and transport by ocean currents. *Marine Geology* 179: 135-46.

Gingele, F.X., De Deckker, P. and Hillenbrand, C.-D. 2001b. Late Quaternary fluctuations of the Leeuwin Current and palaeoclimates on the adjacent land masses - clay mineral evidence. *Australian Journal of Earth Sciences* 48.

Godfrey, J.S. 1996. The effect of the Indonesian throughflow on ocean circulation and heat exchange with the atmosphere: a review. *Journal of Geophysical Research,* 101(C5): 12217-37.

Godfrey, J.S. and Golding, T.J. 1981. The Sverdrup relation in the Indian Ocean, and the effect of the Pacific-Indian Ocean throughflow on the Indian Ocean circulation and on the East Australian Current. *Journal of Physical Oceanography,* 11: 771-9.

Godfrey, J.S. and Ridgway, K.R. 1985. The large-scale environment of the poleward-flowing Leeuwin Current, Western Australia: Longshore steric height gradients, wind stresses and geostrophic flow. *Journal of Physical Oceanography,* 15: 481-95.

Gordon, A.L. 1986. Interocean exchange of thermocline water. *Journal of Geophysical Research,* 91(C4): 5037-46.

Gordon, A.L. and Fine, R.A. 1996. Pathways of water between the Pacific and Indian oceans in the Indonesian seas. *Nature,* 379: 146-9.

Hemleben, G., Spindler, M. and Anderson, O.R. 1989. *Modern Planktonic Foraminifera.* Springer, New York: 363 pp.

Herguera, J.C. 1994. Nutrient, Mixing and Export Indices: a 250 kyr Paleoproductivity Record from the Western Equatorial Pacific. In: R. Zhan, T.F. Pedersen, M.A. Kaminski and L. Labeyrie (Editors) *Carbon Cycling in the Glacial Ocean: Constraints on the Ocean's Role in Global Change.* NATO ASI Series No. 117: 482-519.

Hesse, P. 1997. Desert dust from northwestern Australia in Indian Ocean sediments: the balance between aridity and the Australian monsoon. In: D. Price and G. Nanson (Editors) *Quaternary Deserts and Climatic Change Conference.* School of Geoscience, University of Wollongong, Australia.

Hesse, P.P. and McTainsh, G.H. 1999. Last glacial maximum to early Holocene wind strength in the mid-latitudes of the Southern Hemisphere from aeolian dust in the Tasman Sea. *Quaternary Research* 52: 343-9.

Hilbrecht, H. 1996. *Extant Planktic Foraminifera and the Physical Environment in the Atlantic and Indian Oceans* - Mitteilungen aus dem Geologischen Institut der Eidgen.

Technischen Hochschule und der Universität Zürich, Neue Folge. No. 300: 93 pp.

Hilbrecht, H. 1997. Morphologic gradation and ecology in *Neogloboquadrina pachyderma* and *N. dutertrei* (planktic foraminifera) from core top sediments. *Marine Micropaleontology*, 31: 31-43.

Hirst, A.C. and Godfrey, J.S. 1993. The role of Indonesian throughflow in a global ocean GCM. *Journal of Physical Oceanography*, 23: 1057-85.

Imbrie, J. and Kipp, N.G. 1971. A New Micropaleontological Method for Paleoclimatology: Application to a Late Pleistocene Caribbean Core. In: K.K. Turekian (Editor) *The Late Cenozoic Glacial Ages.* Yale University Press, New Haven: 71-179.

Kawagata, S. 2001. Tasman Front shifts and associated paleoceanographic changes during the last 250,000 years: foraminiferal evidence from the Lord Howe Rise. *Marine Micropaleontology*, 41: 167-191.

Kawahata, H. and Suzuki, A. 1994. The fluctuation of primary productivity during the last 300 kyr in the West Caroline Basin. *The Journal of the Geological Society of Japan,* 100(10): 762-70. (in Japanese)

Kawahata, H. and Suzuki, A. 1995. The Record of Late Pleistocene Biogenic Sedimentation in the Western Pacific. In: Y. Kharaka and O.V. Chudaev (Editors) *Water-Rock Interaction.* Proceedings of the 8th InternationalSymposium on Water-Rock Interaction, Vladivostok, Russia. Balkema, Rotterdam: 255-8.

Kinkade, C., Marra, J., Langdon, C., Knudson, C. and Ilahude, A.G. 1997. Monsoonal differences in phytoplankton biomass and production in the Indonesian Sea: tracing vertical mixing using temperature. *Deep-Sea Research I* 44: 581-592.

Klinck, J. M. and Smith, D. A. 1993. Effect of wind changes during the last glacial maximum on the circulation of the Southern Ocean. *Paleoceanography,* 8(4): 427-33.

Lautenschlager, M., Mikolajewicz, U., Maier-Reimer, E. and Heinze, C. 1992. Application of the ocean models for the interpretation of atmospheric general circulation model experiments on the climate of the last glacial maximum. *Paleoceanography,* 7(6): 769-82.

Le, J. 1992. Palaeotemperature estimation methods: sensitivity test on two western equatorial Pacific cores. *Quaternary Science Reviews,* 11: 801-20.

Levitus, S., Burgett, R. and Boyer, P. 1994. *World Ocean Atlas, Volume 4: Temperature.* NOAA Atlas Series, U.S. Department of Commerce, Washington D.C.: 117pp. and CD.

Linsley, B.K., Thunell, R.C., Morgan, C. and Williams, D.F. 1985. Oxygen minimum expansion in the Sulu Sea, western equatorial Pacific, during the last glacial low stand of sea level. *Marine Micropaleontology,* 9: 395-418.

Lyle, M., Prahl, F. and Sparrow, M. 1992. Upwelling and productivity changes inferred from a temperature record in the central equatorial Pacific. *Nature,* 355: 812-5.

Macdonald, A.M. 1998. The global ocean circulation: a hydrographic estimate and regional analysis. *Progress in Oceanography* 41: 281-382.

Macdonald, A.M. and Wunsch, C. 1996. An estimate of global ocean circulation and heat fluxes. *Nature* 382: 436-39.

Martin, J.H. 1990. Glacial-interglacial CO2 change: the iron hypothesis. *Paleoceanography,* 5: 1-13.

Martinez, J.I. 1994(a). Late Pleistocene dissolution cycles in the Vanuatu region, western Pacific Ocean. *Proceedings of the Ocean Drilling Program Scientific Results,* 134: 293-308.

Martinez, J.I. 1994(b). Late Pleistocene palaeoceanography of the Tasman Sea: Implications for the dynamics of the warm pool in the western Pacific. *Palaeogeography, Palaeoclimatology, Palaeoecology,* 112: 19-62.

Martinez, J.I., De Deckker, P. and Chivas, A. 1997. New estimates for salinity changes in the western Pacific Warm Pool during the Last Glacial Maximum: oxygen isotope evidence. *Marine Micropaleontology,* 32: 311-40.

Martinez, J.I., Taylor, L.C., De Deckker, P. and Barrows, T.T. 1998. Planktonic foraminifera from the eastern Indian Ocean: distribution and ecology in relation to the Western Pacific Warm pool. *Marine Micropaleontology* 34: 121-51.

Martinez, J.I., De Deckker, P. and Barrows, T.T. 1999. Palaeoceanography of the Last Glacial Maximum in the eastern Indian Ocean: Planktonic foraminiferal evidence. *Palaeogeography, Palaeoclimatology, Palaeoecology* 147: 73-99.

McBride, J. 1998. Indonesia, Papua New Guinea, and Tropical Australia: The Southern Hemisphere Monsoon. *Meteorological Monographs* 49: 89-99.

McCorkle, D.C., Veeh, H.H. and Heggie, D.T. 1994. Glacial-Holocene Paleoproductivity off Western Australia: a Comparison of Proxy Records. In: R. Zahn, T.F. Pedersen, M.A. Kaminski and L. Labeyrie (Editors) *Carbon Cycling in the Glacial Ocean: Constraints on the Ocean's Role in Global Change.* NATO ASI Series No. 117: 443-79.

Miao, Q., Thunell, R.C. and Anderson, D.M. 1994. Glacial-Holocene carbonate dissolution and sea surface temperatures in the South China and Sulu seas. *Paleoceanography,* 9(2): 269-90.

Molfino, B., Kipp, N.G. and Morley, J.J. 1982. Comparison of Foraminiferal, Coccolithophorid, and Radiolarian paleotemperature equations: assemblage coherency and estimated concordancy. *Quaternary Research,* 17: 279-313.

Murray, S.P. and Arief, D. 1988. Throughflow into the Indian Ocean through the Lombok Strait, January 1985 - January 1986. *Nature,* 333: 444-7.

Oda, M. and Takemoto, A., 1992. Planktonic foraminifera and paleoceanography in the domain of the Kuroshio Current around Japan during the last 20,000 years. *The Quaternary Research (Japan)* 31(5): 341-57. (in Japanese with English abstract)

Ortiz, J. and Mix, A.C. 1997. Comparison of Imbrie-Kipp transfer function and modern analog temperature estimates using sediment trap and core top foraminiferal faunas. *Paleoceanography,* 12(2): 175-90.

Pariwono J.I., Bye, J.A.T. and Lennon, G.W. 1986. Long-period variations of sea-level in Australasia. *Geophysical Journal Research Astronomical Society,* 87(1): 43-54.

Petit, J., Briat, M. and Royer, A. 1981. Ice age aerosol content from east Antarctic ice core samples and past wind strength. *Nature,* 293: 391-4.

Prell, W.L. 1985. *The Stability of Low-Latitude Sea Surface Temperatures: An Evaluation of the CLIMAP Reconstruction with Emphasis on the Positive SST Anomalies.* Technical Report TRO25, Department of Energy, Washington D.C.: 66 pp.

Ravelo, A.C., Fairbanks, R.G. and Phillander, S.G.H. 1990. Reconstructing tropical Atlantic hydrography using planktonic foraminifera and an ocean model. *Paleoceanography,* 5(3): 409-31.

Rea, D.K. 1994. The paleoclimatic record provided by eolian deposition in the deep sea: The geologic history of wind. *Review of Geophysics,* 32: 159-95.

Rintoul. S.T. 1991. South Atlantic interbasin exchange. *Journal of Geophysical Research,* 96: 2675-92.

Rostek, F., Ruhland, G., Bassinot, F.C., Muller, P.J., Laberyrie, L.D., Lancelot, Y. and Bard, E. 1993. Reconstructing sea surface temperature and salinity using $\delta^{18}O$ and alkenone records. *Nature,* 364: 319-21.

Sarnthein, M., Tetzlaff, G., Koopmann, B., Wolter, K. and Pflaumann, U. 1981. Glacial and interglacial wind regimes over eastern subtropical Atlantic and north-west Africa. *Nature,* 293: 193-4.

Sautter, L.R. and Thunell, R.C. 1991. Seasonal variability in the $\delta^{18}O$ and $\delta^{13}C$ of planktonic foraminifera from an upwelling environment: sediment trap results from the San Pedro basin, Southern California Bight. *Paleoceanography,* 6(3): 307-34.

Shackleton, N.J. 1977. Carbon-13 in *Uvigerina*: Tropical Rain Forest History and the Equatorial Pacific Carbonate Dissolution Cycles. In: N. Anderson and A. Malahoff (Editors) *The Fate of Fossil Fuel CO_2 in the Oceans*. Plenum, New York: 412-8.

Schmitz, W.J. and McCartney, M.S. 1993. On the North Atlantic circulation. *Review of Geophysics,* 31: 29-49.

Schmitz, W.J. 1995. On the interbasin-scale thermohaline circulation. *Review of Geophysics,* 33(2): 151-73.

Schrag, D.P., Hampt, G. and Murray, D.W. 1996. Pore fluid constraints on the temperature and oxygen isotopic composition of the glacial ocean. *Science,* 272: 1930-2.

Sirocko, F., Garbe-Schonberg, D., McIntyre, A. and Molfino, B. 1996. Teleconnections between the subtropical monsoons and high-latitude climates during the last deglaciation. *Science,* 272: 526-9.

Sonzogni, C., Bard, E. and Rostek, F. 1998. Tropical sea-surface temperatures during the last glacial period: a view based on alkenones in Indian Ocean sediments. *Quaternary Science Reviews* 17: 1185-1201.

Spooner, M.I. 2001. The late Quaternary palaeoceanography of the Banda Sea east of Timor with implications for past monsoonal climates. B.Sc. (Hons) thesis, Australian National University, Canberra (unpubl.).

Thiede, J. 1975. Distribution of foraminifera in surface waters of a coastal upwelling area. *Nature,* 253: 712-714.

Thiede, J. 1983. Skeletal Plankton and Nekton in Upwelling Water Masses off Northwestern South America and Northwestern Africa. In: E. Suess and J. Thiede (Editors) *Coastal Upwelling: Its Sedimentary Record, Part A.* Plenum Press, New York: 183-208.

Thompson, P.R. and Shackleton, N.J. 1980. North Pacific palaeoceanography: late Quaternary coiling variations of planktonic foraminifer. *Neogloboquadrina pachyderma. Nature,* 287: 829-33.

Thunell, R.C., Curry, W.B. and Honjo, S. 1983. Seasonal variation in the flux of planktonic foraminifera: time series sediment trap results from the Panama Basin. *Earth and Planetary Science Letters,* 64: 44-55.

Thunell, R.C. and Reynolds, L.A. 1984. Sedimentation of planktonic foraminifera: seasonal changes in species flux in the Panama Basin. *Micropaleontology,* 30(3): 243-62.

Thunell, R., Anderson, D., Gellar, D. and Miao, Q. 1994. Sea-surface temperature estimates for the tropical Western Pacific during the Last Glaciation and their implications for the Pacific warm pool. *Quaternary Research,* 41: 255-64.

Tomczak, M. and Godfrey, J, S. 1994. *Regional Oceanography: An Introduction.* Pergamon, New York: 422 pp.

Ujiié, H., Tanaka, Y. and Ono, T. 1991. Late Quaternary paleoceanographic record from the middle Ryukyu Trench slope, Northwest Pacific. *Marine Micropaleontology,* 18: 115-28.

van Andel, T.H., Heath, G.R., Moore, T.C. and McGeary, D.F.R. 1967. Late Quaternary history, climate and oceanography of the Timor Sea northwestern Australia. *American Journal of Science,* 265: 737-58.

van der Kaars, W.A. and Dam, M.A.C. 1995. A 135,000-year record of vegetational and climatic change from the Bandung area, West-Java, Indonesia. *Palaeogeography, Palaeoclimatology, Palaeoecology,* 117: 55-72.

van der Kaars, S. and De Deckker, P. submitted. A late Quaternary pollen record from deep-sea core Fr10/95-GC17 offshore Cape Range Peninsula, northwestern Western Australia. *Reviews in Palaeobotany and Palynology.*

Peter Kershaw, Bruno David, Nigel Tapper, Dan Penny and Jonathan Brown (Eds): Bridging Wallace's Line

Wang, P. and Li, R. 1995. Numerical simulation of surface circulation of South China Sea during the last glaciation and its verification. *Chinese Science Bulletin,* 40(21): 1813-7.

Wang, P., Wang L., Bian, Y. and Zhimi, J. 1995. Late Quaternary paleoceanography of the South China Sea: surface circulation and carbonate cycles. *Marine Geology,* 127: 145-65.

Webster, P.J. and Streten, N.A. 1978. Late Quaternary Ice Age climates of tropical Australasia: interpretations and reconstructions. *Quaternary Research,* 10: 279-309.

Wells, P.E. and Wells, G.M. 1994. Large-scale reorganization of ocean currents offshore Western Australia during the Late Quaternary. *Marine Micropaleontology,* 24: 157-86.

Wijffels, S.E., Bray, N., Hautala, S., Meyers, G. and Morawitz, W.M.L. 1996. The WOCE Indonesian Throughflow Repeat Hydrography Sections: I10 and IR6. *International WOCE Newsletter* 24: 25-8.

Wyrtki, K. 1958. The water exchange between the Pacific and the Indian Oceans in relation to upwelling processes. *Proceedings 9th Pacific Science Congress,* 16: 61-5.

Wyrtki, K. 1961. *Scientific Results of Marine Investigations of the South China Sea and the Gulf of Thailand, 1959-1961, Volume 2: Physical Oceanography of Southeast Asian Waters.* NAGA Report No. 2, Scripps Institute of Oceanography, University of California, San Diego: 195 pp.

Wyrtki, K. 1962. The upwelling in the region between Java and Australia during the south-east monsoon. *Australian Journal of Marine and Freshwater Research,* 13: 217-25.

Yan, X-H., Ho, C-R., Zheng, Q. and Klemas, V. 1992. Temperature and size variabilities of the Western Pacific warm pool. *Science,* 258: 1643-5.

Yokoyama, Y., Lambeck, K., De Deckker, P., Johnston, P. and Fifield, L.K. 2000. Timing of the Last Glacial Maximum from observed sea-level minima. *Nature* 406: 713-16.

Yokoyama, Y., De Deckker, P., Lambeck, K., Johnston, P. and Fifield, L.K. 2001. Sea-level at the Last Glacial Maximum: evidence from northwestern Australia to constrain ice volumes for oxygen isotope 2. *Palaeogeography, Palaeoclimatology, Palaeoecology* 165: 281-97.

Mountain Environments in New Guinea and the Last Glacial Maximum 'Warm Seas/Cold Mountains' Enigma in the West Pacific Warm Pool Region

Jim Peterson, Geoff Hope, Mike Prentice and Wahyoe Hantoro

Introduction

In palaeoenvironmental reconstructions of the Sunda-Sahul region, the existence (although not without changes of extent) of a western Pacific Warm Pool (WPWP) throughout (at least) the late Quaternary is widely accepted (e.g. see Thunell *et al.* 1994) on the basis of evidence in the ocean cores for relatively little change in proxies for sea surface temperature in the WPWP. These are presumed to reflect stable sea surface temperature (SST) values higher than those for any other large area of the global ocean surface. Thus late Quaternary proximity of terrestrial environments to a form of the WPWP has been assumed, notably for northern New Guinea because of the steepness of the continental shelf there (e.g. see Martinez *et al.* this volume, figure 9.2). The heat and moisture flux associated with the presumed WPWP would have been at least as important in determining global and regional climate during the last glacial maximum (LGM) as it is now, although a somewhat drier LGM atmosphere (as proposed by Barmawidjaya *et al.* 1993) is acknowledged, and greater continentality for regions now near the New Guinea south coast (e.g. see Martin and Peterson 1978, figure 3.6.4) would have resulted in a pattern of climatic gradients different from those of the present day in some places. Additionally, LGM elevations in New Guinea were about 100-120 m greater due to glacial eustasy, but for some areas this is in part cancelled by an uplift rate, which, on the Northern coast, has been defined by Abbott *et al.* (1997) for the Finisterre Mountains (6°S,146°E) as 0.8-2.1 mm yr^{-1} (i.e. there has been uplift of between 14.4 and 37.8 m for the post LGM period). This would account for about 0.5°C of the LGM to present thermal differences.

Results of the analysis of plankton remains and oxygen isotopes from marine cores in the WPWP region (e.g. see Martinez *et al.* this volume, figure 9.1) have suggested that the average SST there during the LGM was no more than about 2°-3°C less than that found today (e.g. Shackleton 1967; CLIMAP 1976, 1981, 1984; Barmawidjaya *et al.* 1993; Thunell *et al.* 1994; Martinez *et al.* this volume). In contrast, from an interpretation of lake bio-stratigraphy at 665 m

ISBN 3-923381-47-6

elevation on nearby Java, van der Kaars and Dam (1995, 1997) argue that the LGM temperature was about 6°C less than found there at present. Such work supports long-known interpretations about the chronology, nature and distribution of glacial landforms in New Guinea (e.g. Gentilli 1961; Hope and Peterson 1975; Hope *et al.* 1976) a result now also known for the tropical Andes (Klein 1999). This, and recent work aimed at deriving palaeo-sea surface temperatures from the Sr/Ca ratios in corals from the Caribbean (e.g. Guilderson *et al.* 1994) and New Guinea (McCulloch *et al.* 1994, 1996, 1999) among other places (e.g. Vanuatu, Beck *et al.* 1997) might give pause to those palaeoenvironmental reconstructionists using the results of palaeoSST determination from the palaeontology of ocean cores.

Thus, to paraphrase Broeker and Denton (1990), and in recognition that incompatibilities have long been recognised (e.g. Webster and Streten 1978; Rhind and Peteet 1985), an 'enigma' of WPWP LGM warm seas and cold mountains still exists for the northern edge of the maritime continent, if not elsewhere in the tropics. Colinvaux *et al.* (2000), suggest that this enigma has been 'solved' (or transferred) in South America given evidence in the lowland Amazon for shifts of 6°C or more. On the northern slopes of New Guinea, the evidence from pollen diagrams at a range of altitudes (including marine sediments) shows that moist communities were present or even enhanced, at LGM times (Hope and Tulip 1994; Hope 1996). This in turn has been taken as indicating that major terrestrial changes to New Guinea environments were probably due to thermal factors, or CO_2 concentration factors (e.g. Flenley 1997) in contrast to drier equatorial mountains in Africa and South America. However steeper lapse rates as for instance, suggested by Walker and Flenley (1979) are not favoured by climatic modellers (e.g. Kutzbach and Guttner 1986, but see Farrera *et al.* 1999). Implied steeper tropical-temperate thermal gradients in New Guinea may have strengthened wind systems, leading to greater orographic precipitation than would otherwise have been the case.

Confirmation of CLIMAP (minimally changed) SST models has been drawn, for instance, from the work of Overpeck *et al.* (1985) and Broeker (1986). Application of LGM (and other) SST data in General Circulation Models (GCMs) for testing against the terrestrial record yields broad acceptance for many regions (e.g. Australia, see Harrison and Dodson 1993) and seems to be supported by the analysis of alkenones in ocean cores (e.g. Sonzogni *et al.* 1998). PalaeoSST estimates from ocean core palaeontology were used for the CLIMAP sea surface temperature Last Glacial Maximum atmospheric general circulation modeling (Kutzbach and Ruddiman 1993). A -4°C shift was used in dynamic ocean model runs by the COHMAP 1995. Regionally-based interpretations of palaeoSSTs are called for because it seems that not all tropical LGM SSTs were depressed the same amount. So far, published accounts of such attempts (e.g. Hostetler and Mix 1999) have included the east, but not the west tropical Pacific.

Clearly, a reappraisal of New Guinean Late Quaternary climate change evidence is called for. This paper reviews the evidence for a large climatic shift (compared with the present and, presumably, other interglacial times, assuming the mountains then were high enough to carry glaciers) at LGM in the New

Guinea mountains. Dateable glacial landforms and deposits provide unequivocal evidence for climatic change. New data are here used to update earlier analyses such as those of Hope and Peterson (1976) based mainly on the work of the Australian Universities' Carstensz Glaciers expeditions in the early 1970s. Since then, data has accumulated from ; a) research carried out by one of us (GSH) over three decades, and b) the results of our joint work on the NSF-funded follow-up expedition to the Kemabu Plateau in 1994. Better topographic elevation data, the availability of satellite images and advances in dating techniques since the 1970s have been of benefit to our later efforts.

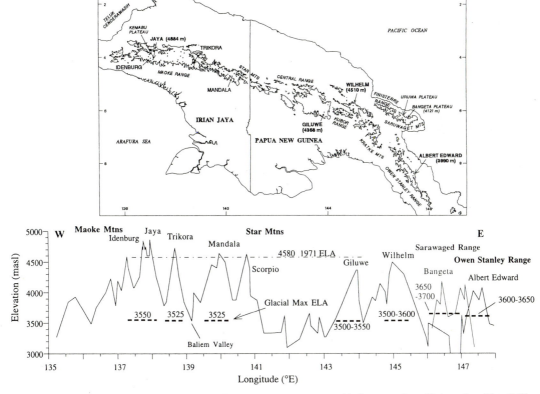

Mountain New Guinea sites of glaciation (a) and ELAs present and past (b). Present and past ELAs projected into E-W oriented topographic profile of New Guinea peaks (ONC chart series at 1:1,000,000) drawn from sources described in text.

Fig. 10.1 New Guinea highlands as defined by the 3000 m contour. Selected peaks (Pk) are also shown. During the height of the neoglacial advances, ice fields were found on Pk Jaya, Pk Trikora, Pk Mandala, and Mt Wilhelm and probably a few other high peaks as well. Today, only shrinking ice fields on Pk Jaya remain. Intact LGM ice tongues reached elevations as low as 3000 m and some reconstituted ice bodies were at even lower altitudes.

New Guinea glaciation

In the whole of the Asian and Australasian tropics, New Guinea provides the best laboratory for estimating the nature and magnitude of glacial change. The extent of glaciation referable to the LGM is estimated to cover about 2000-2200 km^2, all occurring above 3000 m (see Figure 10.1).

Fossil glacial landforms and deposits are found on isolated ranges in Papua New Guinea (total inferred LGM ice cover is about 1100-1400 km^2). At least 800 km^2 of Irian Jaya ranges were glaciated. Here, most ice formed on and above a few relatively extensive high plateaus north of the crestal ridges. In contrast to places where the steep terrain dictated that glaciers would dump ice and detritus over cliffs and precipitous valleys, the existence of these plateaus has led to the preservation of glacial features and stratigraphy. In that most of these plateau areas are dominated by limestone, post-glacial erosion is limited and preservation enhanced. Such is the case for the Kemabu Plateau and adjacent areas, and the slopes north of Pk Trikora, 165 km to the east.

From the mapping of cirques and latero-terminal moraines there, we can set the LGM ELA at about 3500 m from the elevation of clear transitions between erosion and deposition zones, including those referring to glaciers formed on peaks marginal to the plateau and apparently only just high enough to support glaciation. Today, these locations lie on the 8°C average annual temperature isotherm.

Maoke Range

This, the highest mountain block, is located in the west of the island. It includes the Kemabu plateau north of the highest crests: a rolling area 3400-3700 m in altitude onto which a large set of glaciated valleys with differing source areas and altitudes flowed out. During previous work here (Hope and Peterson 1976) we lacked satellite images and control on documentation of elevation. Our current assessment is based on field work carried out in mid-1994 for ground-truthing (including acquisition of GCPs) a 1987 SPOT image (Figure 10.2).

For the first time in New Guinea, cores were collected from a lake dammed behind terminal moraines to obtain the fullest possible post-glacial palaeo-environmental high resolution record. We focussed on the Hogayaku Valley where the nature and distribution of glacial landforms shows that the LGM ice was confined between bedrock ridges. Terminal moraines occur in a 1km-wide zone below Lower Lake Hogayaku, and above glaci-fluvial deposits. Stratigraphy exposed by lake outlet incision displays weathered horizons and buried glaci-fluvials, testifying to three re-advances. These features are well preserved, and so ages referring to the last glacial stage are indicated. From a lake margin core we obtained a date of 14,815 years BP on organic matter immediately overlying rock flour, and infer a greater age for the deepest nearby lake sediments. Thus we regard the moraine that dams the lake as most likely to mark the LGM terminus of the Hogayaku Valley Glacier and the moraines down valley as older but belonging to the last glacial stage (Figure 10.3).

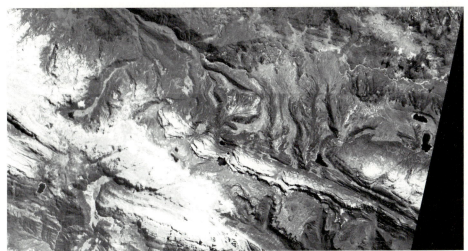

Fig. 10.2 Output from July 1988 SPOT image data depicting the Hogayaku Valley (with its upper and lower lake) and adjacent regions.

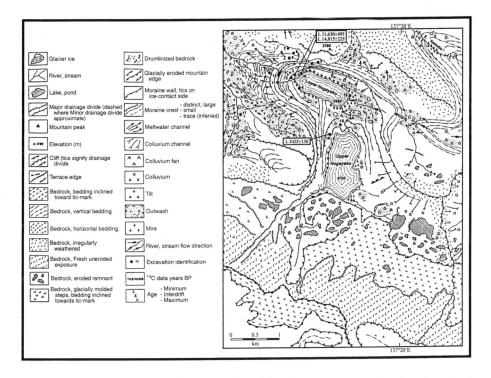

Fig. 10.3 Geomorphic map depicting glacial landforms and core site locations in the Hogayaku Valley, Irian Jaya.

Peter Kershaw, Bruno David, Nigel Tapper, Dan Penny and Jonathan Brown (Eds): Bridging Wallace´s Line

Besides the three sets of terminal moraines and glaci-fluvials downstream from them, retreat moraines in the upper Hogayaku Valley are depicted. The brief retreat-phase equilibria implied are not yet dated. From the Ijomba Basin, an adjacent valley at 3630 m, a lower resolution 14,000 year post-glacial record provides comparative results (Hope and Peterson 1976).

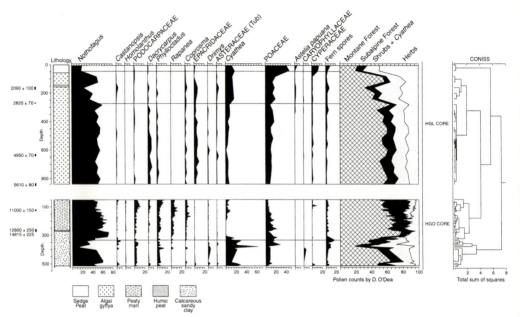

Fig. 10.4 Summary pollen diagram from the basin of Lower Lake Hogayaku. The earliest record from the core is post-LGM, but from a time when the basin was still receiving meltwater/rockflour from retreating ice up valley.

Lithology and plant micro-fossil stratigraphy from the longer Hogayaku core is depicted in Figure 10.4. The diagram shows that a tundra or open grassland occurred on the slopes through the time when rock flour was filling the lake. Ice was probably at the upper part of the lake and vegetation colonising the fresh moraines. The *Nothofagus* component is a long distance element from montane forest; its contribution diminishes as local scrub is developed around 13,000 years BP, and forest invaded at 11,000 years BP. The Ijomba record also confirms that the subalpine forest trees such as *Dacrycarpus* reach their altitudinal limits about 10,200 years BP.

   Towards the edge of the Kemabu Plateau, surface drainage across glacial and glaci-fluvial deposits and regolith formed on non-limestone rocks has incised the plateau edge. Hence most of these glaciers spilled steeply. This was not the case with the Hogayaku glacier because it encountered the plateau where it was not incised. As seen from Figure 10.2, upstanding bedrock ridges diverted the glacier to the west, so its terminus occurred at higher altitudes than adjacent valley glaciers. This may explain why the Hogayaku Lakes are larger than the other Kemabu Plateau lakes, and why the moraine-dammed lakes there do not have direct counterparts in adjacent valleys. Nevertheless, from the distinct sets of terminal and retreat moraines we have mapped, we can argue that the Hogayaku landform pattern represents events that were widespread on the Kemabu Plateau. Recent field work at Mt Trikora demonstrates that the patterns are repeated in general terms on the northern slopes and plateaus of many other parts of the.range.

Star Mountains

This range, at 5°S, with a maximum altitude of 4214 m at Mt Capella, is approximately in the centre of the island and about 450 km from either coast. The cordillera here is narrow, overlooking no barriers south or north. Extremely wet (annual average above 8000 mm), it is little more or less insular in aspect now than when glacial low sea levels prevailed. Well-developed cirques and associated moraines are well preserved. Good height control is available for Papua New Guinea, so that a mean lowest cirque height of 3460 m can be assigned. At a cirque rock hollow (Yakas Tarn, 3540 m) organic sediments immediately overlying rockflour, date at 13250±395 years BP (GX 3945). They contain pollen indicating scrubby vegetation. No tundra post-glacier phase is evident, and so rapid deglaciation is implied.

Central PNG Highlands (Giluwe, Wilhelm)

Mt Giluwe and Mt Wilhelm are the highest mountains in Papua New Guinea, occurring on the southern and northern cordillera (here about 150 km wide) flanks respectively. Mt Giluwe receives abundant precipitation from the south, whereas Mt Wilhelm has a wetter northern slope. Both mountains have well-developed glacial features all attributed to the LGM and the ELA for these two mountains can be confidently set at 3550 m, there being little obvious difference despite contrasting aspects. The dome-like Mt Giluwe supported an ice cap of about 200 km^2, with a few ice lobes emerging from deeper valleys. Loeffler (1972) described the landforms and later cored some mires with Hope. The deglaciation dates, ranging from 13,100 at 3580 m to 9630 years BP at 4280 m, suggest a steady retreat to the highest cirques through the end of the Pleistocene. These results are in good agreement with those obtained from the Pindaunde Valley of Mt Wilhelm. Here Hope and Peterson (1975) dated four localities which indicated a steady glacial retreat after an estimated 14,000 years BP, with final deglaciation being achieved by 9400 years BP. A lower site, Komanima-buno at 2760 m, has the least tree pollen input at 17-15,000 years BP, corresponding to the coldest or most extreme time in the glacial period.

Peter Kershaw, Bruno David, Nigel Tapper, Dan Penny and Jonathan Brown (Eds): Bridging Wallace´s Line

Owen Stanley Ranges

Mt Albert Edward is a schist mountain reaching 3996 m significantly south (8°S) of the other glaciated peaks, and about 80 km from northern and southern coastlines. There is a tendency to low precipitation and cooler temperatures May-August although the overall precipitation is high. Loeffler (1972) has described the glacial landforms, which include ice plateaux above 3600 m, debouching into valley glaciers. The ELA is 3630 m and a core at Laravita Tarn at about 3780 m provides a basal date above glacially-smoothed bed-rock of 12,850 years BP. The vegetation at this time is an alpine grassland, so some time may have elapsed after deglaciation before this mire commenced to accumulate. The cessation of glacifluvial debris supply is indicated by the drainage of a lake at 2980 m on the Neon Basin which had been dammed by outwash. This lake had accumulated clay with glacially etched quartz grains until 9700 years BP after which peat formation took place.

Summary of LGM ELA

Figure 10.1b depicts a range of deduced ELAs for a traverse along the crestal ridges that takes in all the major glaciated areas. The variation in ELA altitudes can be attributed to differences in latitude, magnitude of moisture supply/ cloudiness, and degree of dependence of positive mass balance upon topographic enhancement of snow accumulation. Today the ELA of the modern Carstensz snowfields occurs at an estimated MAT of +0.5°C, amongst the warmest snowline possible. The LGM snowlines may have been slightly colder than this, but were clearly not like some Andean snowlines which occur in arid conditions at -6°C. Not surprisingly, the lowest ELAs are in the Star Mountains where the highest alpine precipitation totals are found today. Here, insularity (distance-to-sea, notably to the steep northern shelf edge) would have been little different to what prevails today. Relatively high snowfall totals and precipitation/snow percentages, and high cloudiness can be postulated. Drier ranges carried ice at higher elevations. The overall pattern of LGM ELAs in New Guinea can be used to infer LGM climatic gradients. In particular, if the (austral winter) SE trade winds from across the LGM Arafura savanna had been relatively important snow sources for the nearby Maoke Range, the glaciers there would have been small. As we have shown, this was not the case, and so we infer that LGM snow-bearing winds to that range came from the north (ultimately the WPWP) and despite having passed across several high ranges parallel to the Maoke Range, were able to deliver enough snow to sustain some of the largest NG LGM glaciers. The diurnal circulation of afternoon storms would also have been maintained and was important in reducing ablation.

Palaeothermometry from palaeoecology

The discovery that regional New Guinea treelines (limits of closed forest) stood at only 2200 m at LGM compared with about 3800-4000 m today (e.g. see Table 10.1) was made 25 years ago (e.g. Walker and Flenley 1979) but further work has defined other formational boundaries at different altitudes. This suggests that the LGM formation boundary was different from the modern treeline (Hope *et al.* 1988). Hence the modern treeline, which accords fairly well with the 6°C MAT isotherm, is formed by a high altitude tall shrub community. This scrub cannot be identified in pollen diagrams until glacial recession is well underway (after 13,000 years BP). At LGM the treeline forest seems to have been made of montane forest elements, above which a zone of open shrub and grasslands extended as a sub-alpine formation, with frequent tree ferns (*Cyathea* spp.) which do not occur above modern treelines. Other boundaries appear to have had a longer persistence presumably due to the thermal tolerances of some major rainforest dominants, although LGM associations differ in their total floristic make up from modern formations.

Table 10.1. Zonation limits of the northern Maoke and Cyclops Ranges, Irian Jaya. (after Hope 1996).

Zone	Upper Boundary (present)	Upper boundary (18,000 years BP)	Altitude (m)	Biome
Alpine	4600	3450	1150	Grassland/tundra
Sub-alpine	3900	2900	1000	Low forest/Open shrubland
Upper montane	3100	2200	900	Beech or mixed nanophyll forest
Lower montane	1750	850	900	*Castanopsis*-mixed mesophyll forest
Lowland	700	>250	?<450	Complex macrophyll lowland forest

These major floristic and structural shifts (Biome boundaries) represent significant ecophysiological boundaries that reflect climate. The Pleistocene upper limits of lowland vegetation are still not known, but may not have been greatly depressed. The shift in the subalpine zone vegetation from open shrubland-grassland to dense forest may reflect changed competitive abilities as CO_2 concentrations rose (e.g. Street-Perrott *et al.* 1997; Flenley 1996). However thermally limited taxa such as tree ferns (which can withstand only mild frost) are present in both vegetation types within the subalpine zone. Other boundaries, and variations in the types of forests within zones may be locally affected by cloudiness, with associated low light, high humidity and low temperature variability. This probably accounts for the much greater extent of *Nothofagus* dominated forests at LGM compared to the Holocene montane forests.
Assuming that:
a) tropical atmospheric lapse rates were always close to the moist adiabatic rate, even during the (drier) LGM as shown by the persistence of moist commu-

Peter Kershaw, Bruno David, Nigel Tapper, Dan Penny and Jonathan Brown (Eds): Bridging Wallace´s Line

nities at all altitudes (e.g. see Webster and Streten 1978; Kutzbach and Guttner 1986; Hope 1996).

b) the descent of the vegetation zone boundaries was mainly driven by reductions in annual average temperature

c) the indicated shifts in temperature that might account for the average zonal changes are in the range of 8-5°C at high altitudes, 4.5-5.5°C in the montane zone, and 3-5°C in the lowlands.

Scope for interpreting the fossil record has been enhanced as progress is made in ecological studies including biome modelling and the analysis of ecophysiological envelopes (e.g. Read and Hope 1996). However these advances do not always support the broad assumptions listed above so that the fossil record may not be readily interpreted in terms of temperature/climate history. The following matters could also be significant in interpretation of the fossil record, especially that from the upland and mountainous tropics:

a) rather than climatic mean annual value, it may be the frequency of episodes of extreme climate (such as frost and drought) that may control vegetation.

b) the Massenerhebung/Merriam effect (Schroeter 1908; Flenley 1993) changes with increased continentality. This should act to understate the thermal change in the free air column.

c) the effect of LGM low partial pressure of CO_2, which gives competitive advantage to grasses over trees and shrubs and reduces water use efficiency (e.g. see Street-Perrott 1994).

d) the effect (in the high altitude tropics) upon plant metabolism/physiology/speciation of disproportionately-enhanced UV-B radiation there (e.g. see Flenley 1996, 1998)

e) the fact that there is only physiognomic rather than physiological backing for the 'calibration' of vegetation associations with climate, and bioclimates in New Guinea are poorly described.

These effects prevent precision in thermal estimates from palaeoecology which in turn prevent the accurate comparison with other results. This is a pity because palynology is presently the only comprehensive and reasonably well-dated terrestrial data set from a wide range of altitudes. However, as Colinvaux *et al.* (2000) argue, fossil pollen and related remains are more useful in advancing the science of ecology than they are as indicators of past climates, and that the nature and magnitude of climatic forcing that caused the migrations should be established by other means.

Discussion

Several questions about the methods used in palaeoenvironmental reconstruction arise from the work reported here: They encourage further analysis and impose restraints upon synthesis. A solution will come with advancing theory and more accurate thermal and other climatic estimates. In the meantime, interpretation is enhanced with access to new data and improvements in documentation and diffusion not only of the data but also documentation about it (i.e. data lineage).

Allowing for lower sea level, and a WPWP temperature of 24-25°C, and applying a moist adiabatic lapse rate to any of the NG glaciated ranges, it is clear that the 'warm seas and cold mountains' enigma remains. A steepened lapse rate (as proposed by Barmiwidjaja *et al.* 1993) still seems inconsistent with the high atmospheric moisture content implied by the glacial and biogeographical evidence. Resolution might be achieved by testing the applicability of one or more of the following possibilities:

a) a reappraisal of evidence for dominance by 'obligate wet' vegetation during the LGM and post-glacier period such that the dry adiabatic lapse rate can be applied

b) a reappraisal of the applicability of modern lapse rates to LGM times, when, for instance, the partial pressure of one of the most important greenhouse gases (CO_2) was about 88 ppm less in LGM times.

c) a reappraisal of the evidence and methodologies underlying currently accepted LGM SSTs for the WPWP.

d) a reappraisal of the evidence from which LGM WPWP extent is deduced.

As with all Quaternary studies, the scope for cross checking of interpretations is sought here. Regrettably, the outcome of research about the late Quaternary environments of other tropical mountains is not fully relevant because none of it refers to landscapes orientated (sub)parallel to the equator and so close to the warmest SSTs on earth. The tropical Andes display considerably enhanced aridity at LGM, as a result of increased cold water in the Humboldt current. This must have influenced ELA as glaciers advanced to their maxima in Bolivia as warmer conditions prevailed after 14,000 years BP, presumably because precipitation also rose sharply (Clapperton 1993). Despite definite evidence of cold ocean water and aridity, there remains a similar warm oceans/cold mountains puzzle for the tropical East Pacific, derived from not only ocean cores, bogs, and glacial landform studies there but also from ice cores taken from the Quelcaya and other equatorial ice caps (Thompson 1999; Thompson *et al.* 1995). Mount Kenya has also been shown to have been drier during LGM (Street Perrott *et al.* 1997). In Indonesia there appears to be a gradient in LGM results, with drying lakes in Java and Sulawesi (Dam *et al.* 2001; Hope 2001) but substantial open vegetation only noted from Java. (van der Kaars and Dam 1995, 1997). This suggests increasingly dry conditions west and south from New Guinea.

If the recent marine results showing high tropical SST values west of New Guinea (Martinez *et al.* this volume) are accepted these must be associated with moist air production. The moist lapse rate exists in warm humid environments because saturation follows slight cooling, and condensation then produces heat which resists the cooling. Such saturation also reduces diurnal variation and encourages fogs and cloud formation. It is thus difficult to support lapse rates steeper than 6° per km altitude. If slight cooling led to clearer skies and drier air, a rise in ELA would be the result, as reduced precipitation and increased ablation would offset any gains from reduced temperature.

In that the glacial evidence is unequivocal, and the 'steepened lapse rate hypothesis' is, as yet, unproven, the status of the interpretation of data recruited as WPWP SST proxies is worth re-appraisal. Changes in the $^{16/18}O$ ratio in ocean surface water are assumed to reflect regional changes in SST 'instantaneously' in terms of the time frames distinguishable in the fossil records.

Martinez *et al.* (this volume), conclude that 'Tropical planktonic forams may have lived in a slightly colder' LGM WPWP, 'thus biasing SST reconstructions (by ~1° to 3°C ?). Since the WPWP is the warmest area, marine taxa inhabiting it are likely to have the ability to adapt to less warm conditions, as their habitat would disappear if cooler conditions prevailed. We would thus expect thermal tolerance to characterise taxa occupying the most extreme habitats. Martinez *et al.* (this volume) consider this point likely and that the SST estimates might be biased by 1-3°C. Colinvaux *et al.* (2000) make the same point in respect of low altitude rainforests, suggesting that they tolerate lower temperature at glacial time, at which time they are invaded by higher altitude elements. The ocean core data is internally consistent but needs further refinement, and also invites testing against conclusions derived from other evidence of Quaternary change. For instance, as mentioned above, results from SST palaeothermometry using Sr/Ca ratios in fossil coral reefs call into question the results from analysis of fossils in nearby ocean cores.

Conclusions

From an analysis of glacial landforms and deposits referable to wet-based ice under a diurnal temperature regime (i.e. daily melting and nightly freezing) in an apparently cloudy climate, it seems that annual average temperatures during the LGM above 3000 m or so in NG were about 6-8°C lower than found there today. This assessment is supported by results of palaeobotanical analysis. In that the LGM was characterised by a dry Arafura Shelf, it seems that the WPWP was the major source of moisture/snow, even for those ranges closer to the Arafura savanna than the Pacific Ocean.

The enigma of warm seas near cold mountains remains. The ocean core evidence is sufficiently multidisciplinary to survive tests so far made of it. On the other hand, the glacial evidence is unequivocal but apparently incompatible. Possible paths to reconciliation can be identified, but have not yet been explored well enough to yield useful results. Some of the tests may be made anywhere in the high tropics, but most will have to be made in New Guinea after further data and information is collected.

Acknowledgements

LIPI kindly granted permission for the 1994 field trip which was logistically supported by PT Freeport McMoran Indonesia. Funding for the work reported here was provided by ARC (Discovery Grant to G. Hope and J. Peterson) and NSF (Initiation Grant to M. Prentice). We thank Dr Ukat Sukanta for his contribution in the field (1994 expedition). Dominique O´Dea counted the pollen from the Lake Hogayako core. Professor Peter Kershaw provided useful comments on the draft.

References

Abbott, L.D., Silver, E.A., Anderson, R.S., Smith, R., Ingle, J.C., King, S.A., Haig, D., Small, E., Galewsky, J., and Silter, W. 1997. Measurement of tectonic surface uplift rate in a young collisional mountain belt. *Nature,* 385: 501-7.

Barmawidjaja, B.M., Rohling, E.J., van der Kaars, W.A., Vergnaud Grazzmi, C. and Zachariasse, W.J. 1993. Glacial conditions in the North Molluca Sea region (Indonesia). *Palaeogeography, Palaeoclimatology, Palaeoecology,* 10(1): 147-67.

Beck, J.W., Recy, J., Taylor, F., Edwards, R.L. and Cabloch, G. 1997. Abrupt changes in early Holocene tropical sea surface temperature derived from coral records. *Nature,* 385: 705-7.

Broecker, W.S. 1986. Oxygen isotope constraints on surface ocean temperatures. *Quaternary Research,* 26: 121-34.

Broecker, W.S. and Denton, G.H. 1990. The role of ocean-atmosphere reorganizations in glacial cycles. *Quaternary Science Reviews,* 9: 305-41.

Clapperton, C.M., 1993, *Quaternary Geology and Geomorphology of South America.* Elsevier, Amsterdam: 796 pp.

CLIMAP Project Members. 1976. The surface of the Ice Age earth. *Science,* 191: 1131-44.

CLIMAP Project Members. 1981. *Seasonal Reconstructions of the Earth's Surface at the Last Glacial Maximum.* Geological Society of America Map and Chart Series No. MC-36. Geological Society of America, Boulder: 18 maps.

CLIMAP Project Members. 1984. The last interglacial ocean. *Quaternary Research,* 21: 123-234.

Colinvaux, P.A., de Oliveira, P.E. and Bush, M.B. 2000. Amazonian and Neotropical plant communities on glacial time-scales: the failure of the aridity and refuge hypothesis. *Quaternary Science Reviews,* 19: 141-69.

Dam, R.A.C., Fluin, J., Suparan, P. and van der Kaars, W.A. 2001. Palaeoenvironmental developments in the Lake Tondano area (North Sulawesi, Indonesia) since 33,000 BP. *Palaeogeography, Palaeoclimatology, Palaeoecology* 171: 147-183.

Farrera, I., Harrison, S.P., Prentice, I.C., Ramstein, G., Guiot, J. Bartlein, P.J., Bonnefille, R., Bush, M. Cramer, W. von Grafenstein, U., Holmgren, K., Hooghiemstra, H., Hope, G.S., Jolly, D., Lauritzen, S-E., Ono, Y., Pinot, S., Stute, M. and Yu, G. 1999. Tropical palaeoclimates at the Last Glacial Maximum: a new synthesis of terrestrial data. I. Vegetation, lake-levels and geochemistry. *Climate Dynamics,* 15(11): 823-56.

Flenley, J.R., 1993. Cloud forest, the Massenerhebung effect, and ultra-violet insolation. In: L.S. Hamilton, J.O. Juvik, and F.N. Scatena (Editors) *Tropical Montane Cloud Forests, Proceedings of an International Symposium.* East-West Center, Honolulu: 264pp.

Flenley, J.R. 1996. Problems of the Quaternary on mountains of the Sunda-Sahul region. *Quaternary Science Reviews,* 15: 549-55.

Flenley, J.R. 1997. The Quaternary in the tropics: an introduction. *Journal of Quaternary Science,* 12(5): 345-6.

Flenley, J.R. 1998. Tropical forests under climates of the last 30,000 years. *Climate Change,* 39: 177-97.

Gentilli, J. 1961. Quaternary Climates of the Australian Region. *Annals of the New York Academy of Sciences,* 95(1): 465-501.

Guilderson, T.P., Fairbanks, R.G. and Rubenstone, J.L. 1994. Tropical temperature variations since 20,000 years ago: modulating interhemispheric climate change. *Science,* 263: 663-5.

Harrison, S.P. and Dodson, J. 1993. Climates of Australia and New Guinea since 18,000 yrs BP. In: H.E. Wright Jr., J.E. Kutzbach, T. Webb III, W.F. Ruddiman, F.A. Street-

Perrott, and P.J. Bartlein (Editors) *Global Climates Since the Last Glacial Maximum.* University of Minnesota Press, Minneapolis: 265-93.

Hope, G.S. 1996. Quaternary change and historical biogeography of Pacific Islands. In: A. Keast and S.E. Miller (Editors) *The Origin and Evolution of Pacific Island Biotas, New Guinea to Eastern Polynesia: Patterns and Process.* SPB Publishing, Amsterdam: 165-90.

Hope, G.S. 2001. Environmental change in the Late Pleistocene and later Holocene at Wanda Site, Soroako, South Sulawesi, Indonesia. *Palaeogeography Palaeoclimatology, Palaeoecology* 171: 129-145.

Hope, G.S. and Peterson, J.A. 1975. Glaciation and vegetation in the high New Guinea mountains. *Royal Society of New Zealand Bulletin,* 13: 155-62.

Hope, G.S., Peterson, J.A., Allison, I., and Radok, U. 1976. *The Equatorial Glaciers of New Guinea.* Balkema, Rotterdam: 256 pp.

Hope, G.S. and Peterson, J.A. 1976. Palaeoenvironments. In: G.S. Hope, J.A. Peterson, I. Allison, and U. Radok.. *The Equatorial Glaciers of New Guinea.* Balkema, Rotterdam: 173-205.

Hope, G.S. and Tulip, J. 1994. A long vegetation history from lowland Irian Jaya, Indonesia. *Palaeogeography, Palaeoclimatology, Palaeoecology,* 109: 385-98.

Hostetler, S.W. and Mix, A.C. 1999. Reassessment of ice-age cooling of the tropical ocean and atmosphere. *Nature,* 399(6737): 673-6.

Klein, A. 1999. Modern and last local glacial maximum snowlines in the central Andes of Peru, Bolivia, and Northern Chile. *Quaternary Science Reviews,* 18: 63-84.

Kutzbach, J.E. and Ruddiman, W.F. 1993. Model description, external forcing and surface boundary conditions. In: H.E. Wright Jr., J.E. Kutzbach, T. Webb III, W.F. Ruddiman, F.A. Street-Perrott, and P.J. Bartlein (Editors) *Global Climates Since the Last Glacial Maximum.* University of Minnesota Press, Minneapolis: 12-23.

Kutzbach, J.E. and Guttner, P.J. 1986. The influence of changing orbital parameters and surface boundary conditions on climatic simulations for the past 18,000 years. *Journal of the Atmospheric Sciences,* 43(14): 1726-59.

Loeffler, E. 1972. Pleistocene glaciation in Papua and New Guinea. *Zeitschrift fur Geomorphologie,* 13: 32-58.

Martin, H. and Peterson, J.A. 1978. Eustatic sea level changes and environmental gradients. In: A.B. Pittock, L. Frakes, D. Jensen, J.A. Peterson and J. Zillman (Editors) *Climatic Change and Variability: A Southern Perspective.* Cambridge University Press, Cambridge : 108-24.

McCulloch, M.T., Gagan, M.K., Mortimer, G.E., Chivas, A.R., and Isdale, P. 1994. A high resolution Sr/Ca and δO^{18} coral record from the Great Barrier Reef, Australia, and the 1982-83 El Nino. *Geochimica Cosmochimica Acta,* 58: 2747-54.

McCulloch, M.T. Mortimer, G., Esat, T., Li, X., Pillans, B. and Chappell, J. 1996. High resolution windows into early Holocene climate: Sr/Ca coral records from the Huon Peninsula. *Earth and Planetary Science Letters,* 138(1-4): 169-78.

McCulloch, M.T., Tudhope, A.W., Esat, T.M, Mortimer, G.E., Chappel, J., Pillans, B., Chivas, A.R. and Omura, A. 1999 Coral record of equatorial sea-surface temperatures during the penultimate deglaciation at Huon Peninsula. *Science,* 283: 202-4.

Overpeck, J.T. Webb, T. III, and Prentice, I.C. 1985. Quantitative interpretation of fossil pollen spectra: Dissimilarity coefficients and the method of modern analogues for pollen data. *Quaternary Research,* 23: 87-108.

Read, J. and Hope, G.S. 1996. Ecology of - forests of New Guinea and New Caledonia. In T.T. Veblen, R.S. Hill and J. Read (Editors) *The Ecology and Biogeography of Nothofagus Forests.* Yale University Press, New Haven: 200-56.

Rhind, D. and Peteet, D. 1985. Terrestrial conditions at the last glacial maximum and CLIMAP sea surface temperature estimates: Are they consistent? *Quaternary Research,* 24: 1-22.

Schroeter, C. 1908. *Das Pflanzenleben der Alpen: Eine Schilderung der Hochgebirgs-flora.* Verlag von Albert Raustein, Zurich: 806pp.

Sonzogni, C., Bard, E. and Frauke, R. 1998. Tropical sea-surface temperatures during the last glacial period: a view based on alkenones in Indian Ocean sediments. *Quaternary Science Reviews,* 17: 1185-201.

Shackleton,N.J. 1967. Oxygen isotope analysis and Pleistocene temperatures reassessed. *Nature,* 215: 15-7.

Street-Perrott, F.A. 1994. Palaeo-Perspectives: changes in terrestrial ecosystems. *Ambio,* 23: 37-43.

Street-Perrott, F.A., Huang, H., Perrott, R.A., Eglinton, G., Barker, P., Ben Khelifer, L., Harkness, D.D., and Olago, D.O. 1997. Impact of lower atmospheric carbon dioxide on tropical mountain ecosystems. *Science,* 278: 1422-6.

Thompson, L.G., Mosley-Thompson, E., Davis, M.E., Lin, P-N., Henderson, K.A., Cole-Dai, J.F., Liu, K-B. 1995. Late Glacial Stage and Holocene tropical ice core records from Huascaran, Peru. *Science,* 269: 46-50.

Thompson, L.G. 1999. Ice core evidence for climate change in the tropics: implications for our future. *Quaternary Science Reviews,* 19:19-35.

Thunell, R., Anderson, D. Gellar, D. and Miao, Q. 1994. Sea surface temperature estimates for the tropical west Pacific during the last glaciation, and their implications for the West Pacific warm pool. *Quaternary Research,* 41: 255-61

van der Kaars, W.A., and Dam, M.A.C. 1995. A 135,000-year record of vegetational and climatic change from the Bandung area, West-Java, Indonesia. *Palaeogeography, Palaeoclimatology, Palaeoecology,* 117: 55-72.

van der Kaars, S. and Dam, R. 1997. Vegetation and climate change in West Java, Indonesia during the last 135,000 years. *Quaternary International,* 37: 67-71.

Walker, D. and Flenley, J. 1979. Late Quaternary vegetational history of the Enga Province of upland Papua New Guinea. *Philosophical Transactions Royal Society of London, B* 286: 265-344.

Webster, P.J. and Streten, N.A. 1978. Late Quaternary ice age climates of tropical Australasia: interpretations and reconstructions. *Quaternary Research,* 10: 279-309.

Late Quaternary Pollen and Associated Records from the Monsoonal Areas of Continental South and SE Asia

Andrew L. Maxwell and Kam-biu Liu

Introduction

Monsoonal circulation in southern and eastern Asia is one of the most important links in global-scale atmospheric circulation. Physically, the monsoon represents the largest south-north interhemispheric atmospheric exchange of energy. The monsoon also controls much of the natural resource base and agriculture for half of the population of the world. Despite its importance in global climate, past monsoon circulation is not understood well enough to permit accurate modeling of its role in global circulation. Although the Cooperative Holocene Mapping Project (COHMAP 1988) focuses on monsoon changes over the last 18,000 years, proxy data from Asia to confirm the model descriptions come only from a few sites in China and one site in India.

We review here terrestrial proxy records of late Pleistocene-Holocene climate change from South and SE Asia, including monsoonal China, where the Asian monsoon is the dominant control on precipitation regime (Figure 11.1; Table 11.1). Quality of the chronology is emphasized in discussing these studies, because the objective of this review is to clarify the pattern and timing of major changes in climate over the last 20,000 years, and to critically evaluate whether those changes confirm or contradict model predictions. We focus on four time periods which have been used as major markers in describing the sequence of changes in monsoon regime. These are: 1) the Late Glacial period, including, in a few cases, the Last Glacial Maximum (LGM); 2) the Pleistocene-Holocene climatic transition, usually a relatively abrupt change to more humid conditions that may signal a change to a stronger summer monsoon, *c.*10,000 years ago; 3) a Holocene precipitation maximum, here referred to as a 'monsoon maximum' because of its probable connection to a more intense or extensive summer monsoon; and 4) subsequent drier conditions, in some cases similar to present-day conditions, which may indicate a weakening of the monsoon. In this review, we are primarily discussing changes in the strength of the summer monsoon, not the winter monsoon, so the monsoon maximum is a period of maximum precipitation, not necessarily a period of maximum seasonality (contrast between summer and winter conditions).

ISBN 3-923381-47-6

Table 11.1. Sites discussed in the text and plotted in Figure 11.1, with references to articles discussing original proxy data and interpretations.

Site	Location	References
1	Bay of Bengal	Cullen 1981
2, 3	Eastern Arabian Sea (Malabar Coast)	van Campo 1986
4	Eastern Arabian Sea	Sarkar *et al.* 1990
5	Nilgiri Hills, southwestern India	Gupta 1971, 1990; Gupta and Prasad 1985; Sukumar *et al.* 1993
6	Son and Belan Valleys, central India	Williams and Royce 1982; Williams and Clarke 1984
7	Northern Madhya Pradesh, central India	Chauhan 1995
8	Southern Uttar Pradesh, central India	Gupta 1978;
9	Didwana Lake, Rajasthan, western India	Singh *et al.* 1990
10	Naukuchhiya Tal, northern India	Vishnu-Mittre *et al.* 1967; Rajagopalan *et al.* 1982
11	Butapathri bog, Kashmir	Dodia *et al.* 1984
12	Assam, northeastern India	Bhattacharya and Chanda 1992
13	Sumxi Co, Bangong Co, western Tibet	Gasse *et al.* 1991; van Campo and Gasse 1993; van Campo *et al.* 1996; Gasse *et al.* 1996
14	Dunde ice Cap, northern Qinghai, China	Thompson *et al.* 1989; Liu *et al.* 1998
15	Qinghai Lake, northern Qinghai, China	Lister *et al.* 1991; Kelts *et al.* 1989; Du *et al.* 1989
16	Baxie loess profile, Gansu, China	An *et al.* 1991
17	Roergai marsh, eastern Qinghai, China	Shen *et al.* 1996
18	Dahaizi Lake Lake Shayema, eastern Sichuan, China	Li and Liu 1988; Jarvis 1993
19	Xi Hu, northern Yunnan, China	Lin *et al.* 1986
20	Dianchi Lake, central Yunnan, China	Sun *et al.* 1986
21	Menghai, southern Yunnan, China	Liu *et al.* 1986
22a 22b	Ren Co, southern Tibet, and Hidden Lake, southern Tibet	Liu *et al.* 1996
23	Seling Co, southern central Tibet	Sun *et al.* 1993
24	Jianghan Plain, Hubei, China	Liu 1991
25	Nanping Bog, Lichuan, Hubei, China	Wang and Sun 1989; Gao 1988
26	Tianyang basin, Guangdong, South China	Lei and Zheng 1993
27a 27b	Core SO50-31KL, South China Sea, and Toushe Lake, central Taiwan,	Huang *et al* 1997
28	Yom Valley, central Thailand	Bishop and Godley 1994
29	Lake Kumphawapi, NE Thailand	Kealhofer and Penny 1997
30	Yeak Kara Lake, NE Cambodia	Maxwell 1996, 1999; Maxwell and Colm 1997

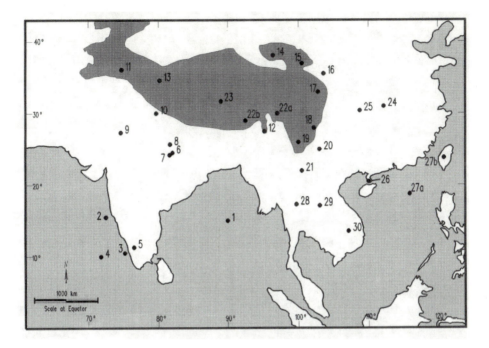

Fig. 11.1 Map of monsoon Asia showing the locations of sites discussed in the text, and cited in Table 11.1. The dark shaded area shows the approximate limits of the Tibet-Qinghai Plateau, roughly the 3000 m elevation isoline.

The Asian monsoon region covers part of Pakistan, almost all of India, Burma, Thailand, Laos, Cambodia, Vietnam, and much of central and southern China. We have plotted an inland limit for penetration of summer monsoon precipitation (Figure 11.2), based on available data for Tibet, Qinghai, and points across central China. Some of the sites discussed (e.g. Didwana, Dunde, Qinghai Lake) should be particularly sensitive to changes in monsoon strength, due to their positions near the present-day limit of monsoon influence. Implicit in this model is the assumption that a stronger summer monsoon has in the past not only brought more rain per year to sites within the present monsoon region but also covered a larger area, pushing farther inland over the Qinghai-Tibetan Plateau.

Discussions of the monsoon in China usually distinguish a 'SW' or Indian monsoon versus a 'SE' or East Asian monsoon. Some sources describe summer monsoon flow in SW China as coming from the Indian Ocean, and this generally west-southwesterly flow continues across China, south of 26-28° N latitude (Winkler and Wang 1993; Ramage 1971). Chinese literature frequently refers to a southeast monsoon prevailing across southern China, with flow originating in the western Pacific tropical easterlies. However, long term records of both marine and terrestrial July wind flow indicate that, although SW China is dominated by southwesterly Indian monsoon flow, South China is dominated by southerly flow from the South China Sea, as a continuation of cross-equatorial flow from the eastern Indian Ocean and the Sunda Shelf region (Figure 11.2; Hastenrath and Lamb 1979). Trewartha (1981: 211-3), citing Hsu (1958) says that precipitation in South China originates in the Indian Ocean during June and early July, and in the China Sea during late July and August, after the break-up of the Meiyu front. This bi-modal precipitation pattern in China should be considered when attempting to piece together past regional monsoon dynamics.

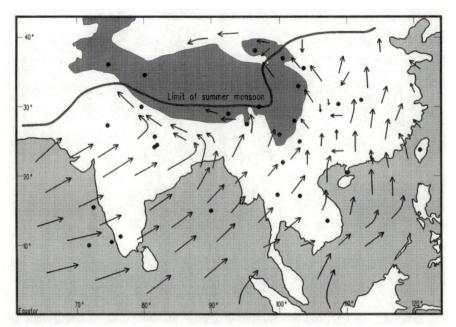

Fig. 11.2 July surface winds for monsoon Asia and adjacent marine areas, and generalized inland limit of summer monsoon wind penetration. Lengths of the arrows approximate directional frequency for July (after Ramage 1971; Hastenrath and Lamb 1979; Zhang and Lin 1985).

We begin this review by summarizing the results of pertinent modelling experiments, to establish a background on the global scale processes involved in monsoon circulation. Selected marine cores are then summarized to provide a late Quaternary chronology of changes in the upwind area for Asian summer monsoon winds. Reviews of terrestrial records then proceed geographically, starting in SW India, moving north to Tibet and Qinghai, through SW China, and, finally, to the few records available from Indochina and South China.

Ages referred to in discussions of the terrestrial sites are given in ^{14}C years before present (BP), or as thousands of ^{14}C years BP, unless otherwise noted. In the conclusion, conversion of the relevant dates to calendar years BP (cal years BP) is used in order to adapt the analysis to standard records used in model simulations and ocean oxygen-isotope chronologies. The distinction is significant, because ^{14}C dates are almost 2000 years younger than calendar ages at the Pleistocene-Holocene transition, and the difference is *c.*3000 years at the LGM.

Calendar years before present may be calibrated ^{14}C dates, (following Stuiver and Reimer 1993), or deduced from ice core laminations and/or $\delta^{18}O$ variations reflecting the SPECMAP curve (Imbrie *et al.* 1984). Table 11.2 includes a rough conversion (in 1000 year increments) from ^{14}C ages to calendar years.

Models and marine data

Atmospheric circulation models

Since the connection was described between changes in orbital parameters and the timing of glacial-interglacial periods (Hayes *et al.* 1976), various researchers have attempted to further clarify the role of solar radiation in long-term atmospheric circulation dynamics, and many efforts in that vein have concentrated on both regional and global factors affecting monsoon circulation (Clemens *et al.* 1991; Prell and Kutzbach 1987, 1992; de Menocal and Rind 1993). Kutzbach and Guetter (1986) present model runs later incorporated into the Cooperative Holocene Mapping Project (COHMAP 1988). The COHMAP models (1988) simulate temperature and precipitation change, every 3000 years between 18,000 years BP and the present. Tests of the sensitivity of South and East Asian regional atmospheric circulation to changes in radiation are based on the hypothesis that the LGM should have been a period of weak summer monsoons, because northern hemisphere summer solar radiation was significantly reduced relative to today's values. Conversely, 9000 cal years BP should have been a period of maximum monsoon intensity, because that is when the precessional cycle in the Earth's orbital variations brought maximum summer solar radiation and minimum winter solar radiation to the northern hemisphere. The radiation extremes exaggerate the land-sea pressure gradient compared to today's values, thus increasing monsoon intensity. The authors predict that the monsoon maximum, i.e. higher summer precipitation than present, prevailed in the eastern Eurasian sector over the period 12,000 to 6,000 years BP. This generalization has been borne out by proxy data including African lake levels (Street and Grove 1979), and pollen records, mostly from Australasia and a few from China.

Indian Ocean Cores

Bay of Bengal foraminifera assemblages (Cullen 1981; site 1 in Figure 11.1) indicate that salinity in the northern Bay was higher than present at the LGM, which, in turn, indicates reduced runoff and river discharge (Ganges and Brahmaputra) because of reduced precipitation. Glacial termination (*c.*12,500-10,500 years BP) brought a surge in low salinity waters to the northern parts of the Bay, increasing the equatorward salinity gradient to steeper-than-modern values. This change indicates markedly increased precipitation resulting from a strengthened southwest monsoon. Chronology in this study comes from a standard ocean oxygen isotope record (Broecker and van Donk 1970; Shackleton and Opdyke 1973), therefore the dates mentioned here are calendar years BP (or years ago), not ^{14}C years. Duplessy (1982) uses a similar method for cores throughout the northern Indian Ocean and Arabian Sea, and the record of monsoon strengthening at Glacial termination confirms Cullen's findings.

Peter Kershaw, Bruno David, Nigel Tapper, Dan Penny and Jonathan Brown (Eds): Bridging Wallace´s Line

Table 11.2. Chronology of inferred monsoonal precipitation changes for selected sites discussed in the text. Shaded areas represent periods of wetter-than-present climate, darker shading periods of maximum precipitation, relative to each site.

Site #		Oman Coast	(5) Nilgiris	(6) Son Valley	(9) Didwana
[14]C yrs.BP	Cal. yrs.BP				
1000	930				
2000	1940				dry
3000	3200			? ?	
4000	4400		driest		weaken
5000	5700	weaken			monsoon??
6000	6800				maximum
7000	7800		weaken		
8000	8800	monsoon			
9000	10,000	maximum		monsoon	
10,000	11,300		monsoon	maximum	
11,000	12,900				
12,000	14,000	strengthen	maximum		strengthen
13,000	15,400			dry	
		dry	strengthen		
14,000	16,800				dry
15,000	18,000		dry		

Van Campo (1986) presents two cores from the west coast of peninsular India (sites 2 and 3). Chronology again is determined by the oxygen-isotope changes, so the dates are in calendar years. Late Glacial monsoon strengthening came on gradually, beginning as early as 16,500 years ago at the southern site (site 3; Figure 11.3), and culminating at 11,000 years ago. In van Campo's isotope/pollen diagram levels of humid tropical pollen are relatively high through the late Pleistocene. In the southern core there is actually a relative maximum in humid tropical pollen just after 16,500 years ago. Even though this maximum may have been an artefact of synchronous low levels for Gramineae and mangrove taxa, still the data suggest that total influx of humid tropical taxa may have been fairly constant through the last 18,000 years, or at least not clearly reduced during the LGM.

Table 11.2 Continuation, right part

(13) Sumxi Co	(14) Dunde	(15) Qinghai L.	(16) Baxie Loess	(25) Hubei	(18) Sichuan
		weaken			
	driest				drier, more seasonal
Driest					
				monsoon	
Weaken		monsoon			monsoon
	monsoon				
Monsoon			monsoon	maximum	
Maximum		maximum			maximum
Reversal	maximum				
			maximum		
				strengthen	
Strengthen	dry		reversal		strengthen
		strengthen	strengthen		cool, wetter
				cool, wetter	dry
Dry		dry			
			dry	dry	

A core from the SE Arabian Sea (Sarkar *et al.* 1990; site 4), with ^{14}C chronology, indicates that transport of low salinity water (runoff from inland areas) from the Madras coast to the study site was enhanced during the LGM. This is taken to indicate a relatively strong winter monsoon, because the Madras coast receives the bulk of its annual precipitation from the northeast monsoon. Salinity levels show a distinct trough (negative δ^{18}O) at 18,500-18,000 years BP (21,000 cal years BP). However, maximum salinity values for the whole period (least input from Madras coast runoff) are found at 20,000 and 16,000 years BP (*c*.23,000 and 19,000 cal years BP, respectively), which implies a very tight temporal bracketing of the glacial maximum in that area.

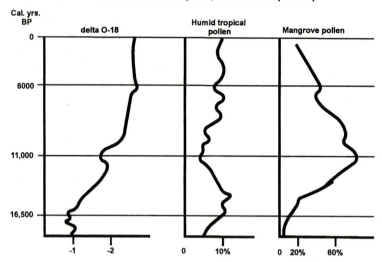

Malabar Coast core (site 3; from Van Campo 1986)

Fig. 11.3 Selected climate indicators from southwestern India: $\delta^{18}O$ PDB from tests of Globigerina ruber, composite humid tropical pollen types, and composite mangrove pollen types, from core MD77194 (site 3; from van Campo 1986).

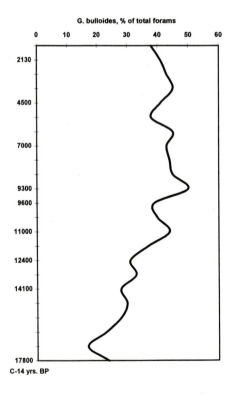

Fig. 11.4 Generalized percentage occurrence of the foraminifera Globigerina bulloides, an indicator of upwelling, in core RC27-23 from shelf off the coast of Oman (from Overpeck et al. 1996).

Higher Resolution Studies

Three recent analyses of proxy data, with higher temporal resolution, have added significantly to an understanding of the Pleistocene-Holocene monsoon transition. Sirocko *et al.* (1993), Overpeck *et al.* (1996) and Zonneveld *et al.* (1997) present well-dated cores from off the Arabian peninsula and the Somali coast (Figure 11.4; off the western edge of the map in Figure 11.1). Discreet steps in the transition from glacial to Holocene monsoon regime are identified and relatively precisely dated, using as proxies $\delta^{18}O$ values from *Globigerina bulloides* tests, dolomite content of dust, monsoon pollen indices, and relative abundances of upwelling-associated planktonic foraminifera and dinoflagellates. Although the findings are not as clear and precise as claimed in the articles, there is agreement on some major points (Overpeck *et al.* 1996; Figure 11.4). The monsoon strengthened in a series of discreet steps, not gradually as had previously been proposed. The first post-glacial surge in monsoon strength occurred *c.*13,100-12,500 years BP, the second surge at *ca.*10,000-9500 years BP[1], and Sirocko *et al.* (1993) find a third surge at 8800 years BP. These surges, particularly the latter two, were followed by a period of strong monsoons. The overall period of strengthened monsoon included some short-term oscillations, when relatively dry phases show up in all the records. One of these oscillations may coincide with the Younger Dryas event in the North Atlantic, i.e. *c.*11-10,000 years BP. The monsoon gradually weakened at around 5500 years BP, although Sirocko *et al.* (1993) find this weakening phase earlier, beginning *ca.*7300 years BP.

Overpeck *et al.* (1996) point out that solar radiation, putatively forcing the monsoon surge, did not simply peak at 9000 years BP. June radiation peaked at 11,300 cal BP (*ca.*10,000 years BP), and July radiation peaked at 9800 cal years BP (*c.*8800 years BP). Vegetation response lagged these radiation peaks, but, despite claims of a lag in monsoon response, the June peak at 11,300 cal years BP actually coincided well with the peak in monsoon-driven upwelling indicated in their cores. If August radiation peaked 1500 years later, at 8300 cal years BP, it would help in explaining the third surge mentioned by Sirocko *et al.* (1993), and also support the claim that the gradually shifting radiation peak led to the gradual waning of the monsoon, but, unfortunately, the data on August radiation is not presented. The initial surge in monsoon strength, coincident with peak radiation, still requires factors other than insolation to explain its abruptness, which the authors attribute to residual influence of Late Glacial Eurasian snow cover delaying the effect of gradually increasing insolation.

---

[1] Sirocko *et al.* (1993) find six stepped increases in monsoon strength, each lasting for less than 300 years. However, only three of these steps are really clear, at 13,100, 9900, and at 8800 years BP. The step at 14,300 years BP is not detected in $\delta^{18}O$, but only in dolomite content as a proxy for strength of northwesterlies; a step at 13,500 years BP is ambiguous, as the oxygen isotope value immediately rebounds to near its previous level before surging (negatively) again at 13,100 years BP; at 7300 years BP the change indicated by $\delta^{18}O$ is contradicted by the synchronous change shown in dolomite.

Peter Kershaw, Bruno David, Nigel Tapper, Dan Penny and Jonathan Brown (Eds): Bridging Wallace´s Line

Terrestrial data

South India

Gupta (1990) reviews studies from the Western Ghats (site 5) describing change in the shola, a type of tropical montane forest, through time. Of the pollen studies reviewed, the two sites at Colgrain and at Kakathope in the Nilgiri Hills have coarse chronologies which cover the last 20,000 years. The Kakathope cores (Gupta 1971) have only two dates, at 15,400 years BP and at 24,300 years BP, and these come from separate cores which the author feels are directly comparable. Despite the sparse dating, however, there is clearly only a short dry period associated with the LGM, which occurred *c.*16-15,000 years BP (19-18,000 cal years BP). Soon after that, an abrupt expansion in forest vegetation is indicated, which was the only period of significant forest cover for the sites. Forest extent at Kakathope decreased gradually after its late Pleistocene surge.

Gupta's review (1990) contains a version of the Colgrain pollen diagram which represents a liberal interpretation of the pollen data, not the data themselves, and the interpretation, included with the original article in Gupta and Prasad (1985), is not clearly explained. The review diagram shows maximum forest extent at 14,475 years BP, following a short non-arboreal phase around the LGM. The original data show very little change over the Pleistocene-Holocene transition other than a slight decrease in Gramineae and a re-establishment of *Ligustrum* after a late Pleistocene hiatus, which indicate change to a warmer moister climate. Tree taxa only show significant values after 7000 years BP. The extremely loose interpretation of the original data is puzzling, especially because no similar interpretation was applied to data from Kakathope. However, other data from the Nilgiris, plus van Campo's Malabar coast core (site 3; van Campo 1986), partially support the interpretation of an early surge in forest vegetation.

Sukumar *et al.* (1993) present variations in core sediment $\delta^{13}C$ for six sites in the Nilgiris (including Kakathope and Colgrain), reflecting past alternations between forest cover ($\delta^{13}C$ < -20 ‰) and tropical grass cover ($\delta^{13}C$ > -20 ‰) (Figure 11.5). A clearly arid period with grass dominating, occurred between 20,000 and 16,000 years BP, followed by an abrupt change to a wetter period by 14,000 years BP. Peak forest extent is inferred to have occurred at *c.*11,000 years BP. The lowest observed isotope value comes from Kakathope at 10,500 years BP, but the lowest value from Colgrain occurs at *c.*14,500 years BP. By just after 8000 years BP sites were becoming relatively arid, and peak aridity is inferred to have occurred between *c.*5000 and 3000 years BP. After this time the record is erratic but shows generally more humid conditions.

Data from the Nilgiris indicate two main points concerning late Pleistocene/Holocene climate: 1) the period of aridity associated with the LGM was of short duration, only about 2000-3000 years. However, the timing of the peak is not the same in all the records, varying between 20,000 years BP for the Nanjanad site (Sukumar *et al.* 1993) to 15,500 years BP at Kakathope; and 2) the period of maximum available moisture fell mostly before the Holocene, from 14,500 years BP at Colgrain to *c.*10,500 years BP for the Kakathope $\delta^{13}C$ signal, and Kakathope pollen indicates an even earlier surge in forest cover.

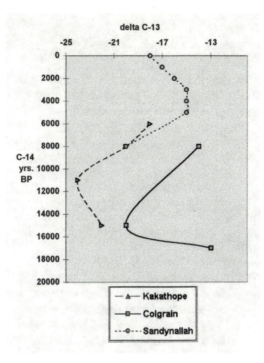

Fig. 11.5 $\delta^{13}C$ in peat profiles from the Nilgiri Hills, southern India (site 5; from Sukumar et al. 1993).

Finally, the pollen and $\delta^{13}C$ data indicating minimal forest cover during the Holocene may reflect human impacts, rather than climate change. The data from individual cultural indicators (large grass pollen, charcoal) are not conclusive, but the overall pattern of vegetation change may reflect long-term, persistent forest suppression. Gupta and Prasad (1985) mention archeological evidence which suggests that indigenous peoples have been active there throughout the Holocene.

Central India

Williams and Clarke (1984) and Williams and Royce (1982) present geomorphological and sedimentological data on the Son and Belan Rivers in Madhya Pradesh which indicate past changes in discharge and channel morphology (site 6). During the LGM, which in this study is taken to extend from *c.*25,000 to 17,000 years BP, the channel was braided, less sinuous and aggraded primarily by bedload deposition. This suggests dry conditions in the region, with precipitation more episodic than during a normal monsoon regime. There was also significant loess deposition in the Son and Belan Valleys during that period, implying a strong winter monsoon bringing dust from the plain of the Ganges

Peter Kershaw, Bruno David, Nigel Tapper, Dan Penny and Jonathan Brown (Eds): Bridging Wallace´s Line

and the foothills of the Himalayas. By 10,000 years BP the channel first began depositing finer-grained sediments, it then became more narrow and sinuous, incising into the late Pleistocene deposits. During that period the floodplains were subject to intermittent suspended load deposition. Regional conditions were more humid, evidenced by the reduced sediment load. The riverbed incision indicates relatively high discharge with low sediment load. The transition from bedload deposition to suspended load and channel incision dates to 13-10,000 years BP. A final period of aggradation of clays and loams, which may have signalled a precipitation maximum or more strongly seasonal climate, fell around 4500-3000 years BP in both valleys.

The Late Glacial sediments contain sporadic Upper Palaeolithic artefacts only at *ca.*18,000 years BP and at 26,000 years BP levels, and this is taken to indicate very sparse settlement along the river during the intervening period. Fossil animal bones are numerous, suggesting regional savanna woodland inhabited by deer, buffalo, elephant, and, at occasional deep pools along the river, hippo, crocodile and turtle. Epi-palaeolithic and mesolithic artefacts are numerous near the base of the Holocene fine sediments and at higher levels deposited during the period of river aggradation dated to 4500-3000 years BP. Neolithic materials from the latter period indicate farming at least 4000 years BP in these valleys (Williams and Clarke 1984).

Chauhan (1995) analyses Holocene soil profiles from three sites in NE Madhya Pradesh (site 7), to help elucidate the evolution of *sal* forests (dry deciduous forest dominated by *Shorea obtusa*) in that area. There are only two stratigraphically clustered dates per profile, and the pollen, lithology and setting suggests the possibility of depositional hiatuses. However, it is worth noting that three samples from one site, dated to *c.*8700 years BP, contain no tree pollen, being dominated by trilete pteridophyte spores. The top samples from this same site contain abundant pollen from the modern forest dominants (*Shorea, Terminalia, Anogeissus*) as do samples from another site dated to 1000 years BP. The third site shows sporadic tree pollen present at *c.*7000-5000 years BP. Although it is possible that these pollen records reflect very localized conditions (swamp succeeding to dry forest) there is also a suggestion that early Holocene climate could not support forests in this area. Gramineae pollen are present throughout the cores, which could indicate savanna, but Cyperaceae values (indicating wetlands?) roughly follow those of Gramineae. Sedge pollen achieves its peak value concurrent with the rise in tree pollen *c.*1000 years ago.

Gupta (1978) presents a pollen diagram from an ox-bow lake in the Ganga Valley, southeastern Uttar Pradesh (site 8). Again, interpretation is limited by inadequate chronology. One date of 4380 years BP in the lower half of the core gives no suggestion of the sedimentation rate. Chenopodiaceae are relatively abundant at levels between the dated sample and the base of the core, so the mid Holocene in this region may have been fairly dry, definitely drier than after 4000 years BP, when grasses, trees, shrubs and wetland pollen types dominate.

Attempting to pull together the records from Central India presents some difficulties, but the Son Valley record is used in Table 11.2 as some indication of change within the region, because its sequence is the longest and most clearly dated. Overbank flooding of fine-grain sediments in the immediate postglacial period, dated to *c.*10,000 years BP, indicates denser vegetation cover and a wetter climate than during the Glacial period. This floodplain accretion continued probably only for a thousand years, after which there was major downcutting through Pleistocene sediments by a high-energy, low sediment-yield flow. A major question arises as to whether the final stage of floodplain accretion by fine sediments, dated to 4500-3000 years BP, also represents increased regional precipitation as it does in the early Holocene. Arguments can be made for its representing either increased or decreased flow relative to the early Holocene downcutting stage. If the downcutting stage represents maximum Holocene flow, then the Son Valley record would fit into the monsoon regional pattern more neatly. However, other records from Central and Northern India, although neither complete nor dated precisely, all imply that the monsoon maximum occurred late in central India, roughly 6000-1000 years BP (Chauhan 1995; Gupta 1978; Vishnu-Mittre *et al.* 1967). If that is true, the late accretionary stage in the Son Valley could reflect the monsoon maximum, or a second stage in a double peak of monsoon strength.

Northwestern India

Singh *et al.* (1990) analyze a core from Lake Didwana in Rajasthan (site 9). The oldest date for the core is 12,800 years BP, so a time of peak LGM dryness cannot be identified. However, older sections of the core show arid steppe vegetation (*Aerva, Ephedra,* Gramineae, and, late in the period, *Artemisia*) dominant for the Late Glacial. Chenopodiaceae and *Calligonum* peak in the section dated 12,800-9400 years BP, while all other taxa are reduced. The authors interpret this as a change in lake conditions with higher precipitation, i.e. the rise in Chenopodiaceae reflects an increase in saline lake margin habitat as the lake initially expanded. The decrease in other taxa is taken as 'essentially statistical', so it would be useful here to see a concentration diagram to assess individual variations in Gramineae and *Artemisia*. After 9400 years BP there was a period of strongly fluctuating lake levels, followed by the establishment of savanna grassland in the basin, along with deep freshwater in the lake. The wettest period was *c.*7500-4000 years BP, after which the lake dried out and conditions did not support pollen preservation.

New data have recently become available from the same area. Enzel *et al.* (1999) discuss sedimentological changes in a core from Lake Lunkaransar, a site previously researched by Singh *et al.* (1972) and Bryson and Swain (1981). A higher-resolution chronology now puts the precipitation maximum between 6300 and 4800 years BP. Two points about the new chronology are important. First, they suggest that the period of maximum precipitation is not reflecting the period of strongest southwest monsoons. The authors attribute the Holocene precipitation maximum to a shift to predominant winter rains. One problem with this interpretation is that they seem to be using present-day monsoon precipitation (insufficient to support a permanent lake) as an analogue for previous periods. Second, the waning of precipitation, which this study finds occurring about 1000 years earlier than previous studies, occurred well before the rise of the Indus/Harappan civilization. This latter point is discussed below.

Peter Kershaw, Bruno David, Nigel Tapper, Dan Penny and Jonathan Brown (Eds): Bridging Wallace´s Line

Northern India

Naukuchhiya Tal in Kumaon Himalaya, is discussed by Vishnu-Mittre *et al.* (1967; site 10), although its chronology is published separately (Rajagopalan *et al.* 1982). At this site, the landscape was dominated by pine forest at 7570 years BP, but by 5800 years BP pine had given way to a more mixed oak forest assemblage. The oak forest remained at fairly constant levels up until the present day, with only a slight cooling evident at *c.*1000 years BP. This record implies that the early Holocene (pre 6000 years BP) was a time of neither maximum moisture availability nor maximum temperature.

Although at least 10 sites have been studied along the southern flank of the Himalayas, stretching from Ladakh in the northwest to Sikkim and Assam (Sharma 1992, 1993), either the length of the records or the quality of the chronologies is inadequate in all except the Naukuchhiya Tal study[2].

A baffling set of data is presented by Dodia *et al.* (1984), from two cores at Butapathri bog in Kashmir (site 11). One of the cores, with five radiocarbon dates, shows the cool, temperate conifers *Pinus, Abies,* and *Picea* increasing in importance both *c.*17,000 years BP, and more so *c.*9200 years BP, thus indicating at both times an increase in forest vegetation, presumably due to a change to warmer and/or moister climate. The only indication that the early Holocene may have been wetter than the Late Glacial is the loss of *Artemisia,* a shrub-steppe indicator, *c.*9000 years BP, although it is high at the LGM and after 2200 years BP. Whatever the interpretation given for individual taxa[3], core 1 reflectes the same trend (increasing conifer pollen) at both 17,000 years BP and at 9200 years BP, but this trend is not consistent with any other record from the monsoon region, unless the 9200 years BP section reflects a Younger Dryas-type reversal. The second core from Butapathri, taken from a site one km away from core 1, has only one date, near the base, of 10,950 years BP. If we interpolate ages going up-core, assuming present-day sediments at the top, this core shows the familiar pattern of a cold and dry Late Glacial changing at *c.*10,000 years BP

---

[2] A more recent study from Naukuchhiyatal (Kotlia *et al.* 1997) has problems with low pollen concentration and inadequate dating. A study from Ladakh (Bhattacharyya 1989) is hampered by a chronology probably subject to old carbon contamination and mixing of sediments. The area is tectonically active, and the core site is located on the edge of a lake, with limestone abundant in the catchment. Perusing just the dated sections still cannot clarify the picture, because too few levels are sampled for pollen, so comparison of time periods is difficult.
   Similar constraints limit the usefulness of a study from the Kathmandu Valley (Vishnu-Mittre and Sharma 1984). Five of the fourteen dates from four separate cores are inverted, and four are beyond ^{14}C range. The authors do not address the potential problems of old carbon contamination, sediment inversion resulting from tectonic movements, or sediment mixing at what seem to be essentially fluvial core sites.
[3] Early samples in the section dated to 17,000 years BP, with higher values for Gramineae and *Artemisia,* could reflect relatively cold and dry conditions, except that this same period also shows highest levels for *Carpinus* and *Corylus,* which reflect temperate broadleaf forest. If we take stronger representation of cool temperate conifers, at 3000 m elevation, as indicating relatively warm and/or wet conditions at the site, this would conflict with the interpretation of the second core.

to warmer and wetter conditions (no Younger Dryas-age reversal), which culminated *c.*4500 years BP in expansion of *Juglans*, Gramineae, and *Corylus*. After that, the inferred vegetation reflects cooler and drier conditions up to the present day. Although this interpretation is logical, the assumption of a constant sedimentation rate through the Holocene is risky, especially given the litho-logical changes in the core. Finding a clear climatic signal from Butapathri bog is confounded 1) by inconsistency between the two records for the Pleistocene-Holocene transition, and 2) by the fact that the record with five ^{14}C dates is still unclear chronologically because of depositional hiatuses and because the material used for dating was taken from relatively long sections (as much as 50 cm) of the length of the 280 cm core.

Finally, brief mention is made of a study from Assam (Bhattacharya and Chanda 1992; site 11). Pollen results from a 2.6 m section of peat and organic clay exposed in a quarry are presented in an illegible diagram with only one ^{14}C date. Because it was a floodplain deposit, extrapolation of age would be risky. However, the diagram does show that the date, 17,900 years BP, falls in the middle of a trend of rising arboreal pollen abundance. It is possible to interpret that trend as an indication that the LGM came early and had a muted signal. One metre up the core, arboreal pollen disappears abruptly. But there are several factors which may complicate an interpretation of just the arboreal pollen trend (e.g. strong dependence on an ambiguous curve for the shrub *Salix*), so it would be best to wait for more dates for this section, and for other material from Assam (Bhattacharya and Chanda 1987-88), before attempting a reconstruction of palaeoclimatic change in this subregion.

Western and Central Tibet

French-Chinese collaborative research has resulted in valuable records of climate change from Sumxi Co and Bangong Co in western Tibet (Gasse *et al.* 1991; van Campo and Gasse 1993; van Campo *et al.* 1996; Gasse *et al.* 1996) (site 13). The record at Sumxi Co covers the last 13,000 ^{14}C years, while that at Bangong Co, dated by various stratigraphical correlations with Sumxi Co, extends back to *c.*10,500 years BP. The records of the two lakes agree with each other with respect to climate changes. During the earliest period, before 10,000 years BP, the region was dry and the vegetation was dominated by the desertic elements Chenopodiaceae and *Ephedra*. Pulses of pedogenic calcite input during that period imply a general trend of increasing moisture until a short Younger Dryas-age reversion to colder drier conditions *c.*11,000-10,500 years BP. At 9900 years BP an abrupt rise in authigenic carbonate and *Artemisia* pollen and a decline in desert taxa signal a sudden change to a warmer, wetter climate, i.e., the strengthening of the monsoon. This is followed by a reversal at *c.*8000-7700 years BP when lake levels apparently fell. A second phase of warm humid climate and highest lake levels, indicating the strongest summer monsoon, occurred *c.*7500-6200 years BP. This period was followed by a dry period at *c.*6000 years BP, which may have been the end of summer monsoon influence in this area. There followed further gradual drying, culminating *ca* 3800 years BP., which seems to have been the driest time for the entire Holocene. After that, conditions fluctuated around a modern-day norm, except for a short arid pulse *c*.400 years BP.

In the centre of the Plateau, Seling Co (Lake) has furnished a 10,000 year record discussed in Sun *et al.* (1993) (site 23). The earliest period is marked by a pollen assemblage today found 1000 m higher, indicating that the Late Glacial was 4.5-5.5°C colder than today. Although there is a definite rise in pollen concentration at the onset of the Holocene, *c*.9600 years BP, the only indicator for a peak in available moisture at that time is dominance of pteridophyte spores. Sun *et al.* (1993) point out that, for this site, low total pollen concentration during the period *c*.8500-7500 years BP indicates wet conditions, as vegetation was slowly colonizing the bare terrain, and the core site, at mid-lake, was far away from pollen sources. This setting was followed by a gradual shift to denser vegetation by 6000-4000 years BP. Subsequent drier conditions are indicated by a rise in Cyperaceae as the lake dried up and lake edge habitat moved closer to the mid-lake core site. Pollen concentration rises very gradually, suggesting that much of the pollen in the core was wind-dispersed over intermediate to long distances. Tree pollen does not show up in abundance until relatively late, with *Abies* occurring with *Betula, Artemisia* and Compositae after 6000 years BP, *Picea* coming in at 4000 years BP, and *Pinus* surging with *Betula, Alnus,* and *Tsuga* between 2200 and 1000 years BP.

The most recent period at Seling Co, after *c*.1000 years BP, is difficult to interpret, with relatively high concentrations for *Tsuga, Betula* (cold temperate forest), Cyperaceae (alpine meadow or lake edge), and Chenopodiaceae (desert or saline lake edge).

An early Holocene monsoon surge could be interpreted by assuming that, although temperature and moisture may have been sufficient, colonization by trees anywhere near the site was postponed until the mid Holocene because of the difficulty of seed dispersal to, and establishment in, the centre of the Plateau. Alternatively, as mentioned above, the core site may have been in the middle of a very large lake, allowing only sparse pollen deposition at the site, even though vegetation around the plateau may have been relatively dense. After about 4000 years BP, the site became drier, indicated by the rise in Cyperaceae and the reduction in the *Artemisia*/Chenopodiaceae ratio. However, if we interpret the record more directly, i.e., taking low pollen concentration as an indicator of sparse regional vegetation, the site conditions could have been quite cold and dry before 9600 years BP, then cold with increased moisture from 9600 years BP until *c*.6000 years BP, maximum temperature and moisture between 6000 and 4000 years BP, followed by diminishing available moisture after *c*.4000 years BP. Although the Seling Co. record does not directly match other records from Western Tibet and Qinghai, the inferred late monsoon maximum, after 6000 years BP, echoes that found in the short record from Naukuchhiya Tal in northern India (site 10; Vishnu-Mittre *et al.* 1967; Rajagopalan *et al.* 1982), and perhaps also that of Lunkaransar (Enzel *et al.* 1999).

Northeastern Qinghai-Tibetan Plateau and Loess Plateau

Ice core data have recently become available from the Dunde Ice Cap in the northeastern Qinghai province (Thompson *et al.* 1989; Liu *et al.* 1998; site 14; Figure 11.6a). At 5300 m elevation, deposition of pollen on the ice cap has been sparse, but correlation of the recent ice core pollen record with local meteorological data shows that total pollen concentration serves as a useful proxy for regional moisture availability, as it affects vegetation density and productivity. A humid period, suggested by high pollen concentrations, between 10,000 and 4800 cal years BP implies the occurrence of an early Holocene summer monsoon maximum. A subsequent dry period, indicated by a surge in Chenopodiaceae and a drop in total concentration between *c.*4800 and 2700 cal years BP, appears to have been the driest for the Holocene. Several abrupt fluctuations followed this dry period, with relatively humid episodes occurring at 2700-2200, 1500-800, and 600-80 cal years BP (Liu *et al.* 1998).

Lake chemistry and palynological data from Qinghai Lake (site 15; Figure 11.6b) are presented in Du *et al.* (1989), Kong *et al.* (1990), Kelts *et al.* (1989) and Lister *et al.* (1991). The $\delta^{18}O$ and carbonate records begin at an extrapolated date of *c.*14,500 years BP, where the oxygen isotope signal indicates a general trend of increasing precipitation during the late Pleistocene, with two abrupt reversals, one of which may have been a Younger Dryas signal. The pollen record begins at *c.*11,000 years BP, with high values for the desert-steppe taxa *Artemisia*, Chenopodiaceae and *Ephedra*. A clear trend toward more negative $\delta^{18}O$ values, indicating higher precipitation, more run-off and lake-filling, takes about 700 years and culminates at 10,000 years BP. Strongest negative values for $\delta^{18}O$ correspond to distinct peaks in *Betula* pollen at 11,000 years BP and *c.*9600 years BP. A subsequent early Holocene trend toward less negative $\delta^{18}O$ values may indicate, as at Dunde (Liu *et al.* 1998), increasing monsoonal input as continental-type precipitation wanes. Although oxygen-isotope values are most negative by 10,000 years BP, the pollen shifts much more gradually as the herbaceous component gives way to trees, and *Betula* gradually gives way to *Pinus* and *Abies*. The early rise in *Betula* probably indicates initially increased humidity, but its partial replacement by *Pinus* signals gradual warming by 8000 years BP. The pollen record is somewhat erratic through the Holocene, but the average values indicate warmest conditions falling between 8000 and 3500 years BP. Lake level was highest at *c.*7000 years BP, and fell gradually about 10 m to its present level by *c.*3000 years BP. From 3000 to 1500 years BP arboreal pollen is still significant, but *Artemisia* recovers to pre-Holocene values. After 1500 years BP conifers almost completely disappear, *Betula* maintains a low but steady value, and Chenopodiaceae rises along with other herbs, indicating significantly drier conditions than those of the mid Holocene.

On the Loess Plateau, east of Qinghai Lake, one section from Baxie, in Gansu Province (site 16; Figure 11.6c) provides a well-dated record of the Pleistocene-Holocene transition (An *et al.* 1993a). Previous research at several sites in the Loess Plateau has established the methodology, using magnetic susceptibility of sediments as a proxy record of past changes in summer monsoon strength (An *et al.* 1991; Maher and Thompson 1992).

Peter Kershaw, Bruno David, Nigel Tapper, Dan Penny and Jonathan Brown (Eds): Bridging Wallace's Line

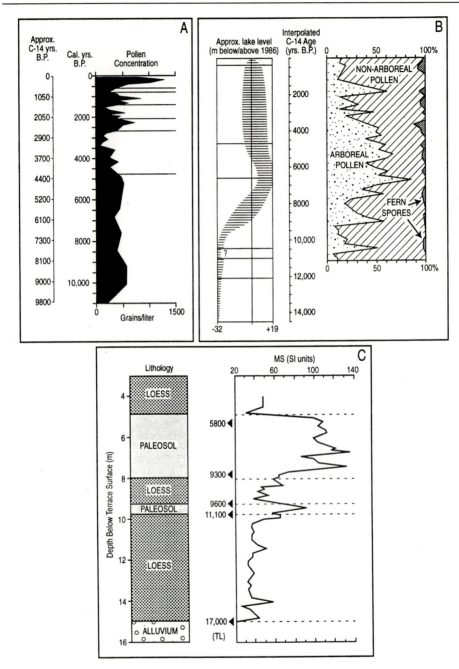

Fig. 11.6 Proxy data from sites in northeastern Qinghai and western Gansu Provinces, China: a) total pollen concentration in an ice core from Dunde Ice Cap (site 14; from Liu et al. 1998); b) inferred lake level and composite Holocene pollen percentages in core from Qinghai Lake (site 15; from Lister et al. 1991; Du et al. 1989); c) loess-paleosol sequence from Baxie, with magnetic susceptibility curve (site 16; from An et al. 1993a).

The Baxie profile shows a transition from dry Late Glacial loess deposition to humid-climate palaeosol development at *c*.11,000 years BP. In the case of the Baxie profile, the short Late Glacial period of palaeosol development seems to coincide with the timing for the Younger Dryas reversal in the North Atlantic region. In the North Atlantic this period was relatively dry and cold, but in central China the palaeosol development occurred under relatively humid conditions. This is clear evidence that events in the North Atlantic may have affected events in China, but not in a linear manner. There was a short reversal to drier conditions at *c*.9800-9300 years BP. This was followed by palaeosol development from 9300 years BP to 5500 years BP, which indicates a period of the strongest summer monsoons, with strong precipitation penetrating central China deeper than the present-day limit (this refers to precipitation, not necessarily to wind reversal, the limit of which is shown in Figure 11.2). After 5500 years BP (*c*.6300 cal years BP), loess deposition through the middle and late Holocene signals a return to relatively dry conditions.

These three records from Northern Qinghai have provided a fairly well resolved picture from a compact transect running across the margin of the Asian monsoon region. They show (Table 11.2) that the timing of major changes is close among the three sites, but not uniform. Concerning the chronology, the Dunde record is not finely resolved at the Pleistocene-Holocene transition, compaction of ice layers precluding a sampling resolution finer than 1000 years. At Qinghai Lake the Pleistocene-Holocene transition is documented and dated clearly, but the chronology more recent than 8400 years BP is questionable because its interpolation assumes no depositional hiatus, constant sedimentation rate, and present day deposition at the top. The Baxie record is clear concerning the early Holocene, but long periods of loess deposition reveal little about palaeoecological nuances in the late Holocene. However all three records clearly overlap in reflecting a monsoon maximum at least between *c*.8500 and 5500 years BP.

A long Pleistocene-Holocene record from Roergai (Zoige) Marsh, in adjacent northern Sichuan Province (site 17), has recently been published (Shen *et al.* 1996), including data from the LGM. The coldest period is found at 22,000-18,000 years BP, when the marsh, at *c*.3400 m above sea level, was covered with alpine meadow and desert-steppe vegetation. The late glacial brought gradually rising temperatures, with the introduction of subalpine shrubs. Alpine meadow disappeared completely by 11,000 years BP, and evergreen conifers came in. Further warming brought expansion of coniferous forest, and the monsoon seems to have strengthened significantly by 7000 years BP. After 3300 years BP Cyperaceae meadow persisted but coniferous forest faded out. A transfer function applied to the pollen data shows a bimodal temperature peak occurring at 6500 and 3500 years BP, and precipitation peaks at 8000-6000, 4500-3500, and the highest peak at 2000-1400 years BP. With the data taken together, the LGM was clearly very dry and cold, an early transition occurred at 12,500 years BP and the monsoon maximum fell between *c*.8000 and 3300 years BP, with a distinct dry period occurring for 1500 years in the middle of the monsoon maximum.

Records from the Qinghai-Tibetan Plateau are reviewed by Tang and Shen (1996). They find that the timing of major changes is roughly consistent across the Plateau. Distinct abrupt change to warmer wetter conditions occurred *c.*11,000-10,000 years BP. A monsoon maximum culminated at *c.*7000 years BP, after which time drying was gradual and progressive.

From southern Tibet two sites, Ren Co and Hidden Lake (sites 22a, 22b, respectively), presently under investigation, are anticipated to yield high-resolution records of Late Glacial-Holocene climate/vegetation changes (Liu *et al.* 1996). These sites are especially well situated to provide new data from the SE section of the Plateau, which will help in evaluating differences in long-term monsoon dynamics between the northwest of the Plateau and the southeast. This will improve our understanding of the importance of intra-plateau circulation and, more generally, of the overall role of terrain in Asian monsoon dynamics, a theme discussed by Tang (1979).

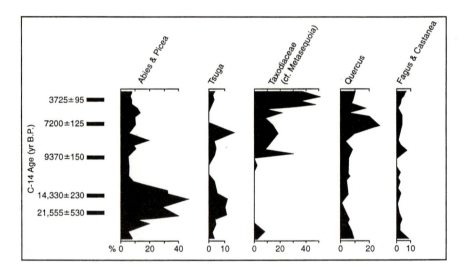

Fig. 11.7 Selected pollen percentage curves from Nanping Bog, Hubei Province, China (site 25; from Wang and Sun 1989; Gao 1988).

East-Central China

Two studies from Hubei province illustrate monsoon changes in the central portion of southern China. Wang and Sun (1989) and Gao (1988) discuss a four metre core from the Nanping peat bog (site 25; Figure 11.7). Before 22,000 years BP, a mixed broadleaf evergreen forest assemblage indicates conditions only slightly cooler than the present. LGM cooling set in by 22,000 years BP, with the forest dominated by *Abies, Picea, Pinus, Betula* and other temperate taxa. Around 12,000-13,000 years BP cool and increasingly humid conditions are indicated, when conifers began to diminish in importance as *Quercus* and *Salix* expanded along with Polypodiaceae. An abrupt change at 9300 years BP brought in *Metasequoia*, a warm temperate conifer, and more evergreen Fagaceae, signalling significant warming. Peak precipitation fell between 7200 and *c.*4000 years BP, indicated by increases in both forest and wetland taxa. After 4000 years BP, the site was slightly drier and cooler, dominated by *Metasequoia*. Liu (1991; site 24) finds a roughly similar pattern in the Jianghan Plain, although the chronology is not as clear as the one from Nanping bog. Late Pleistocene samples are dominated by the pollen of conifers, *Betula*, Rosaceae, Gramineae, and *Artemisia*, indicating clearly colder and drier conditions than present. All of these taxa diminish in importance by *c.*9300 years BP, although the timing of the change is not clear. Between 9000 years BP and 3500 years BP indicators of the forest composition are fairly steady, and indicate conditions warmer and more humid than present, which may be taken as the monsoon maximum. There are a few 'flickers' in the pollen signal, usually showing a reduction in some tree taxa, and a rise in grasses and ferns. There seems to be a slight cooling and drying after 3500 years BP, but the record is more erratic than before that date, accompanied by some sections with a marked dominance of ferns, which may indicate intensification of human influence.

The trends documented for Hubei are broadly similar to those established from the North China Plain and the lower Chang Jiang (Yangtze River) Valley. Pollen data from the latter regions also show maximum warmth and humidity during the middle Holocene, implying a summer monsoon maximum (Liu 1988; Liu *et al.* 1992).

Western Sichuan

Records from two cores on the southeastern edge of the plateau, in Sichuan Province (site 18), mirror the trends seen at Qinghai Lake. The record from Dahaizi Lake on Mt. Luoji (Li and Liu 1988) extends back to *c.*12,000 years, where that pollen spectrum is initially dominated by *Pinus* (only one sample), and *Artemisia* has a significant presence. In the subsequent pre-Holocene section of the core *Artemisia* drops out and *Pinus* decreases concurrent with increases in deciduous *Quercus*, *Betula*, and evergreen sclerophyllous Fagaceae (*Lithocarpus*, *Castanopsis* and *Cyclobalanopsis*). The record from nearby Lake Shayema (Jarvis 1993; Figure 11.8a) begins at 10,800 years BP and echoes the trends found at Lake Dahaizi. Moderately high levels for deciduous oaks before 10,000 years BP diminish by 9000 years BP, gradually at Shayema, abruptly at 10,000 years BP at Dahaizi. *Betula* was prevalent in the pre-Holocene and diminished rapidly to 9000 years BP, then more gradually until *c.*7000 years BP at Lake Shayema. *Betula* stays at roughly the same level until *c.*3800 years BP at Dahaizi. At Shayema there is a fairly abrupt rise in *Abies* and *Picea* pollen at 10,000 years BP, and a similarly abrupt decline at *c.*8500 years BP. Other than these few abrupt changes, both records are marked by gradual transitions in vegetation. Both cores show slight changes over the last 1000-1500 years, with sclerophyllous taxa being replaced by Gramineae, *Artemisia* and *Alnus* at Shayema, and by *Abies*, *Rhododendron* and *Alnus* at Dahaizi. The concentration diagram from Dahaizi suggests that this change, perhaps at both sites, results only from a loss of some evergreen oaks, and that the recovery of other taxa is

Peter Kershaw, Bruno David, Nigel Tapper, Dan Penny and Jonathan Brown (Eds): Bridging Wallace´s Line

an artefact of percentage calculations. The change is probably due to late clearance of forest for agriculture.

The overall picture for both sites shows late Pleistocene cool and humid conditions changing *c.*11,000 years BP to less cool and still humid conditions, followed by a change which was abrupt at Dahaizi and more gradual at Shayema to a warmer and more monsoonal climate beginning *c.*10,000 years BP, and then a gradual transition to established warm and seasonal climate by 7500 years BP. Shayema's record shows a further distinct shift to slightly warmer, drier conditions *c.*4000 years BP, similar to the modern climate.

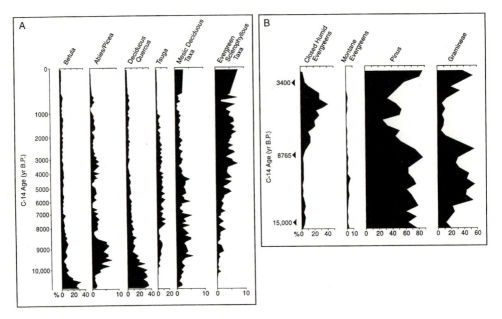

Fig. 11.8 Selected pollen percentage curves from a) Lake Shayema, Sichuan Province, China (site 18; from Jarvis 1993); b) Dianchi Lake, near Kunming, Yunnan Province, China (site 20; from Sun et al. 1986).

Yunnan

Three sites in Yunnan Province have provided the most complete records for SW China bordering on Indochina. Published together with a summary by Walker (1986), these studies fill in an important gap in the palaeomonsoon record.

The northernmost site at Xi Hu, near Er Yuan, yields a record beginning at 17,000 years BP, from a lake at 1980 m elevation (Lin *et al.* 1986; site 19). In the two cores used for pollen analysis, the top sediments date to 6340 and 4450 years BP. The oldest section of one core, 17,000-15,000 years BP, reflect conditions cooler and more humid than present, with a montane conifer assemblage occurring about 500-800 m lower than its current range. Mean annual temperature was depressed 2.5-4.0°C during that period. The section dated 14,000-

10,000 years BP saw very diverse assemblages fluctuating through the period, with montane conifers mixed with warm temperate deciduous elements (*Pterocarya, Juglans*, sclerophyllous *Quercus*), and with some Euphorbiaceae. That combination has no modern analogue, as it requires a downward altitudinal shift of the montane taxa and a simultaneous upward altitudinal shift of the subtropical/tropical taxa. The authors interpreted the Euphorbiaceae (genus *Euphorbia*) as indicating drier conditions, but also mention that the family types include pollen resembling *Baccaurea* and *Antidesma*, both of which genera are common in tropical semi-evergreen forests in Indochina today. Climate in this period probably had some periods of short warm summers and cold dry winters varying with periods which must have been relatively hot and dry, but the reconstruction is difficult because of the lack of an analogous modern pollen assemblage. By the end of that period, the Euphorbiaceae pollen are gone, along with all montane conifers other than *Tsuga*. The Xi Hu records after 10,500 years BP are fairly simple. *Pinus* dominates, particularly after displacing most evergreen Fagaceae by 7500 years BP. There are no major changes in the pollen assemblage above the 10,500 years BP level.

The central Yunnan site is located at Dianchi Lake near Kunming at an elevation of 1886 m (Sun *et al.* 1986; site 20). The record there, from two cores, begins at *c.*16,000 years BP. The oldest late Pleistocene sections indicate only slightly cooler (*c.*2°C) and wetter conditions than present. As at Xi Hu, the pollen assemblage contains elements which, at present, occur at both higher and lower elevations (e.g. simultaneous occurrence of montane *Tsuga, Betula*, and *Picea* with subtropical Fagaceae, Ulmaceae, *Carya, Juglans, Ilex, Pittosporum*, and Podocarpaceae) along with the more tropical Moraceae and *Mallotus*. This paradoxical assemblage is explained by the authors as reflecting a climate with cooler summers (allowing the montane taxa) and milder winters (allowing the subtropical taxa) than present, similar to the contemporary situation at Xi Hu. Wetter conditions are also reflected in the core lithology, as this is the only section with a relatively high inorganic (sand, silt, and clay) component to the sediments, indicating higher erosion rates. After about 9500 years BP the sediment deposition was mainly organic. The chronology is complicated by some reversals in one of the cores (DZ 18), but it appears as though there was a major change *c.*9000-9500 years BP from a cool, humid climate to a warmer one, still with significant year-round precipitation. Montane taxa disappeared along with the previously abundant deciduous oak, and *Pinus* became less abundant during that period. Seasonal difference in precipitation was least around 8000-7000 years BP. By 4000 years BP the climate became more clearly seasonal, approaching modern values, as *Pinus* recovered along with *Cyclobalanopsis* and sclerophyllous *Quercus*. Major changes late in the record, from *c.*1500 years BP, are tentatively attributed to increased agriculture and irrigation in the basin. However, the authors note that although the basin is known to have been settled as far back as 10,000 years BP, there is no evidence of elimination of forest and its replacement by scrub or herbs.

In the south, near the frontier with Laos and Burma, lies the southernmost Yunnan study site at Menghai (Liu *et al.* 1986; site 21). Chronology for these cores is a major problem. However, some observations proceed from the simple conclusion that the whole sediment profile dates to the period *c.*36,000-20,000 years BP. During that period there were several intervals when the local vegetation included Podocarpaceae, *Dacrycarpus* and *Dacrydium*, which are now completely absent near the site. By analogy with their present range which includes Hainan Island, it appears that the climate immediately preceding the LGM at this site was more humid than present and, if any cooler, only very slightly so. More specifically, there must have been more winter rain, and winter temperatures could have averaged as much as 5°C higher than today. Summer temperatures, on the other hand, could not have been higher than present, because a 5°C rise in summer temperature would have allowed the establishment of tropical evergreen rainforest at this site. Tropical rainforest pollen taxa other than Moraceae and Myrtaceae are largely absent from any level in the cores studied.

The Yunnan data, summarized in Walker (1986), show some surprising patterns. First, there is strong evidence that LGM climate was not drier than present, and there seems to have been a gradient of increasing mean annual precipitation to the south. This clearly suggests that a northeast winter monsoon was either not strongly felt in Yunnan, or that the duration of its maximum strength fell in the short period for which there is no record, *c.*20,000-17,000 years BP. Second, the Late Glacial climate supported, at Kunming and at Er Yuan, vegetation with no modern analogue, which was responding to less extreme temperatures both winter and summer, and more even precipitation through the year than is now the case. The evenness of precipitation continued into the Holocene, through a fairly clear increase in mean annual temperature. Seasonality, reflected in relative dominance of subtropical sclerophyllous Fagaceae and some *Pinus* spp., developed gradually, and reached its peak after 7000 years BP in the north, and by 4000 years BP at Kunming.

Before leaving these studies, it should be mentioned that some chronological problems remain unresolved, hampering interpretation of the palaeoclimate record. The chronology is recognized as being problematic at Menghai, but the other two studies do not address the potential problems of old carbonate contamination of dated materials, or of sediment mixing. Both Dianchi Lake and Xi Hu lie in geological settings dominated by limestone, and there is no reason to assume that ^{14}C ages from the cores would be unaffected by old carbon. The lakes also are set in one of the most tectonically active regions on Earth, so it is likely that lake sedimentation would be affected by slumps and earthflows. Either of these two conditions could account for the numerous age inversions in the main Dianchi core (DZ 18), but interpretation of the pollen record is confounded because we do not know if 1) the pollen deposition was continuous, but ages were inverted because of contamination, or 2) the ages are valid but the sediment column was subject to sporadic deposition of older material washed into the lakes from the hills. We cannot resolve these problems now. However, it is encouraging that certain cores from Xi Hu (No. 6) and from Dianchi (DZ 13; Figure 11.8b) do not show dating inversions and show less lithological evidence for sediment mixing, and it is these cores which support the discussion of inferred palaeoclimatic changes at those sites in Yunnan.

South China

From the Leizhou Peninsula near Hainan Island, Lei and Zheng (1993) describe a 230 m core, estimated to cover the last 350,000 years, from the Tianyang volcanic basin (site 26). An interval after 45,000 years BP sees a cool moist flora represented in the pollen record, until major change set in *c*.22,000 years BP. The section representing 22,000-12,000 years BP shows progressive drying, and this section actually shows the driest conditions for the entire 350,000 years represented in the core. The chronology in this section of core is problematic, as discussed below. There is first a fairly abrupt decline in the pollen of evergreen oaks, accompanied by an immediate surge in Gramineae and *Artemisia*. Both of these drop out by about the 20,000 years BP level, above which pteridophytes dominate the palynological spectrum. By 18,000-15,000 years BP tropical montane elements, particularly *Dacrydium*, had gone locally extinct. Some sub-tropical elements from the old assemblage persisted, representing a vegetation type which today occurs in areas with a mean annual temperature *c*.5°C cooler than now prevails at Tianyang. The progressive drying accelerated after 18,000-15,000 years BP until almost all moist forest elements were gone by the time of a depositional hiatus which is tentatively dated to the inception of the Holocene. By that time, vegetation was dominated by the ferns *Dicranopteris, Osmunda* and *Hicriopteris*. Above the depositional hiatus, a short, undated younger section, probably Holocene, indicates vegetation similar to the last described, but with higher values for Gramineae and *Artemisia*.

This record from Tianyang does not include charcoal counts, but it would be interesting to assess whether the latest Pleistocene changes could have been due to human modification. The progressive drying all the way through the Late Glacial is unexpected, especially the accelerated drying after 15,000 years BP, when other parts of the monsoon region were subject to increasingly humid conditions.

The Late Glacial section at Tianyang is dated by extrapolating from a 20,000 years BP ^{14}C date, using sedimentation rates from the part of the core representing subtropical-tropical vegetation. There is a possibility that the progressive drying described for the late Pleistocene may actually have occurred before and during the LGM. Progressively drier climate may have led to increased erosion and sedimentation rates, and, eventually, environmental changes which caused the depositional hiatus in the dry postglacial period. This interpretation allows the inferred vegetation changes to fit into the regional pattern more easily, but requires that some 10 m of sediment were deposited in the basin in the 4000 years between the last measured ^{14}C age and the LGM, which seems unlikely.

Two related records from this region, covering the last 25,000 years have recently become available in a study by Huang *et al.* (1997). A marine core (site 27a) from the South China Sea shows increased productivity in cooler (than present) waters during and after the LGM, until *c*.12,000 years BP. SST proxies show a decrease, relative to the present, of 4°C. The SST seems to reflect winter monsoon strength, which decreased first at 12,000 years BP and again at 6000 years BP. A complementary terrestrial record comes from highland Toushe Lake in central Taiwan, at 650 m elevation (site 27b). Here, the Late Glacial pollen assemblage is dominated first by *Alnus*, which succeeds to Gramineae,

Peter Kershaw, Bruno David, Nigel Tapper, Dan Penny and Jonathan Brown (Eds): Bridging Wallace´s Line

then *Artemisia* and the evergreen broad-leaved *Cyclobalanopsis*. The shift in vegetation compared to present indicates a 5°C lower temperature during the LGM. At 10,000 years BP there was clearly a transition to warmer and wetter conditions, signalled by a surge in pteridophyte spores and *Castanopsis* pollen. After that, *Castanopsis* decreases until *c.*8000 years BP, then maintains a fairly steady average value until it becomes negligible at the 3000 years BP level. Fern spores stay abundant in the early Holocene section, except for a major drop *c.*6000-7000 years BP, then decrease at 5500 years BP, and rebound again in the period 4000-2000 years BP. These data, combined with lithological changes in the sediments, indicate a weak summer monsoon until 12,000 years BP, a transitional phase until 10,000 years BP, and a strong monsoon through the Holocene. The signal for warm humid pollen types, however, is not steady through the Holocene, and it is difficult to demarcate a monsoon maximum.

Indochina Peninsula

Recently Kealhofer and Penny (1998) have produced the first adequately dated record from an upland site in mainland SE Asia (site 29). Cores from Lake Kumphawapi, in NE Thailand, cover the period from 12,200 years BP to *c.*5000-4000 years BP, and are subject to analysis of pollen, phytoliths and charcoal. Early in the record, at *c.*9000 years BP, a change from cool-dry conditions to warmer-wetter conditions is indicated by a sharp decline in *Celtis*, Compositae, and Amaranthaceae, concurrent with a rise in *Pinus*, Palmae, bambusoid phytoliths, Moraceae/Urticaceae and, slightly later, *Quercus*. This change dates to *c.*9000 years BP. Between the 9000 and 6500 years BP levels, several taxa indicate a floodplain forest vegetation, periodically inundated, which gradually changed to denser forest and, finally, to floating mat vegetation which now characterizes the site. Interpretation of climatic change is difficult for this section, as the arboreal element includes *Pinus* which, taken with the charcoal and pteridophyte curves, suggests either dry conditions or strong seasonality. Forest taxa begin to increase and diversify *c.*8500 years BP and reach peak diversity *c.*6500-5500 years BP, concurrent with a Cyperaceae and monolete fern spore peak representing the floating mat vegetation. At 5500 years BP there was an abrupt change to drier climate or more burning, signalled by a charcoal peak, an initial increase in Pinaceae, and a decrease in forest taxa diversity. This change could be due to increased human influence, although the phytolith record shows rice already occurring as early as 8000 years BP. Above the *c.*5000-4000 years BP level, the pollen and phytolith assemblages seem to indicate a reduction in the dominance of floating mat vegetation, less burning, a decrease in pine coverage, and a resurgence of forest cover. Data from this core, then, suggest a Pleistocene-Holocene transition occurring *c.*9000 years BP, with a gradual transition to maximum summer monsoon precipitation culminating at *c.*6500-5500 years BP. The late Holocene waning of monsoon strength is difficult to pin down because of the lack of dated material above the 5500 years BP level, and because anthropogenic burning may have disguised the climatic signal in the vegetation.

Bishop and Godley (1994; site 28) discuss channel morphological changes in the Yom river basin in central Thailand. Eight ^{14}C dates help with the chronology of channel changes, but the dates are clustered and young, only one older than 1700 years BP. The authors find indications that mid Holocene conditions were drier than those of either the early Holocene or the period after 1700 years BP. This observation conflicts with other evidence for a humid mid Holocene in Thailand (Kealhofer and Penny 1998). Lithological sections from NE Thailand summarized in Boonsener (1991), including a loess section (Udomchoke 1989), suggest very dry Holocene periods, with loess deposition dated to 8200 years BP and windblown sand dated to 3500 years BP. In those sections, however, there do occur intervening dated layers of alluvium and lake sediments, indicating wetter conditions. The suggestion of conditions dry enough to bring loess deposition to the Khorat Plateau through the mid Holocene does not fit well with the findings of Kealhofer and Penny (1998). A possible synthesis of the data, but requiring a liberal interpretation of the chronology, calls for a dry period as late as 8200 years BP, followed abruptly by a monsoon maximum falling between *c.*8200 and 4000 years BP, and then subsequent drying.

The remainder of late Quaternary palaeoclimate studies from Thailand focus on edaphically controlled vegetation changes associated with sea-level change (Pramojanee and Hastings 1983; Hastings and Liengsakul 1983; Somboon 1988), which indicate glacial-period regression beginning *c.*16,000 years BP and a Holocene sea-level high stand at *c.*4000 years BP. There is no clear signal for monsoon fluctuations in these records, even though upland taxa are represented in the mangrove section of the core discussed by Somboon (1988). Maloney (1992) provides a complete review of these works, including his own research at the coastal archaeological site, Khok Phanom Di. He finds that the few signals of vegetation change not attributable to sea-level fluctuations are reflecting human activity.

Sediments from volcanic lakes in NE Cambodia (site 30), with basal sediments dating to 9300 years BP, offer a continuous record of vegetation changes in the sub-region (Maxwell 1999, 2001). The remoteness of the study sites from major settlements suggests that Holocene vegetation changes have responded largely to climate. An abrupt change in sediment chemistry, pollen and charcoal abundance at 8500 years BP indicates the Pleistocene-Holocene transition from relatively cool, dry conditions to warmer, moister conditions. That transition is immediately followed by a 3000-year period of reduced fire activity and dense forest development that probably reflects the strengthened summer monsoon. After that, *c.*8500-5500 years BP, an increase in the abundance of disturbance taxa indicates response to increased fire activity. A final major change occurs at *c.*2500 years BP, after which dense forest recovers and fire activity is subdued. Although a strictly climatic interpretation of the record is possible, the sites are located in evergreen 'islands' embedded in dry deciduous forest, where anthropogenic burning is now an extensive and important control on vegetation structure and composition. It is unknown how long this control has been significant. Considering the scale and impact of anthropogenic burning on the vegetation in northern Australia (Kershaw 1986), it is important to develop a clearer picture of how human activities have shaped

Peter Kershaw, Bruno David, Nigel Tapper, Dan Penny and Jonathan Brown (Eds): Bridging Wallace's Line

forest evolution in continental SE Asia, the probable source area for major migrations to the islands of SE Asia and Australasia. Further research at the Cambodian sites should contribute to that picture.

Discussion

Late Glacial

Only sites in southern and central India, South China, and NE Qinghai-Tibetan Plateau have provided records datable to the LGM. In southern India, a short dry spell is indicated in the Nilgiris, and that condition is also suggested by an adjacent offshore core. In central India the regional signal indicates a long, dry LGM, roughly 24,000-17,000 years BP, with aridity continuing to *c.*11,000-10,000 years BP, but the timing of this latter period and subsequent changes is dimmed by the imprecision of the geomorphological signal. On the eastern margin of the Qinghai-Tibetan Plateau, at Roergai, conditions were coldest and driest at 22,000-18,000 years BP. In South China the Tianyang record shows climatic changes contemporaneous with the LGM beginning at 22,000 years BP, similarly to Nanping Bog in Hubei, but cooling and drying may have intensified gradually and taken 10,000 years to culminate in the local extinction of tropical taxa. This included a period after *c.*18,000-15,000 years BP when the climate continued to get drier, while most other sites in the region were beginning to experience increases in precipitation and warmer temperatures. The fact that the Tianyang record does not fit well with others in the region begs a confirmation of its Late Glacial chronology, as discussed above.

The majority of studies reviewed here have shown only a snapshot of Late Glacial (i.e. 15,000-10,000 years BP) conditions, usually only 1000-2000 years before the onset of a strong summer monsoon. The late Glacial was colder and drier than present at all sites except the Nilgiri Hills and Yunnan. The lack of data for the period before *c.*12,000 years BP implies that conditions at many sites were drier and/or colder to a degree that precluded lake deposition and/or pollen preservation. This is true of the sites on the Qinghai-Tibetan Plateau (too cold and dry), the Sichuan lakes and northern Yunnan, Didwana and, possibly, Kumphawapi (too dry). In these cases the suggestion of colder and drier conditions in the Late Glacial does not contradict the role of relatively low solar radiation at that time, but the lack of data obviously has told us little about the timing of changes which may or may not be responding to solar radiation.

Sites in central and southern Yunnan, although lacking a record of the LGM, were subject to precipitation equal to or greater than today's, with little temperature difference, in the periods immediately preceding and following the LGM (Walker 1986). These sites supported a vegetation with no modern analogue, an assemblage which implies markedly less seasonality, in both precipitation and temperature, than today. Walker (1986) suggests that the even precipitation at that time could have resulted from frontal disturbances, which may have occurred throughout year. Trewartha (1981) has described, in the modern circulation, the polar front in winter extending from Burma to Japan which accounts for significant frontal precipitation annually. With weakened monsoonal circulation during the Glacial period, this front may have been present throughout the year, or at least for a longer period before and after the northern winters.

Pleistocene-Holocene Transition

In attempting to clarify patterns of change in early Holocene monsoon strength, we should first point out that the records reviewed here do not support the idea that there was a uniform, synchronous response to increased summer insolation in the monsoon region. Almost all records have shown a precipitation maximum between 8000 and 6500 cal years BP, with notable exceptions, but the timing of onset, strengthening, and weakening vary significantly.

The timing of a major, relatively abrupt late Pleistocene/Holocene transition to wetter conditions falls at most sites between 11,000 and 9500 years BP (13,000-11,000 cal years BP) (Table 11.2). The exceptions to this pattern are found in records from southern India (Nilgiris, site 5) and western India (Didwana, site 9). Both the Malabar coast core described by van Campo (1986; site 3) and data from the Nilgiris (Sukumar *et al.* 1993; Gupta 1971) show an increase in monsoon strength in southern India beginning as early as 16,500 cal years BP, much earlier than any other sites reviewed here. The position of those sites at the southern edge of the monsoon region, on a peninsula, may have dampened the continentality which dominates sites farther north and farther inland. Perhaps the long term progression of monsoon change mimicked the annual 'march' of the monsoon across India, with monsoon onset beginning in the extreme southeast of the subcontinent and progressing toward the northwest (Ramage 1971). Another possible explanation involves the late Pleistocene winter monsoon. Sarkar *et al.* (1990) show that the strengthened northeast winter monsoon at the LGM caused enhanced precipitation on the southeast coast of India. With that monsoon stronger than present, it may have brought more rain during the LGM to the Nilgiris, which form the subcontinental divide about 300 km inland from the east coast.

The Didwana record fits well with the model of Overpeck *et al.* (1996), in terms of the timing of changes in the monsoon, but the connection with precipitation variations is not so clear. The Didwana site and the Thar Desert are marginal to monsoon circulation. Singh *et al.* (1990) point out that the site lies close to the modern zone of predominant winter rain. Furthermore, although present-day low level winds over the Thar Desert are monsoonal, precipitation during the southwest monsoon is not heavy. Ramage (1971) attributes this to subsidence dominating over monsoonal advection and precipitation in the area, a process which is not well understood. Bryson and Swain (1981) use a transfer function to show that at nearby Lunkaransar, precipitation in the mid Holocene was largely winter rain. Although their method produces a slightly different precipitation record at Didwana (there may be problems in correlating the two chronologies), it raises the possibility that precipitation and lake levels recorded in the later study of Didwana (Singh *et al.* 1990) may reflect non-monsoonal rain. Enzel *et al.* (1999) also suggest that the Holocene precipitation maximum (6300-4800 years BP) at Lunkaransar is due to an increase in winter precipitation, not to a Holocene monsoon maximum.

Peter Kershaw, Bruno David, Nigel Tapper, Dan Penny and Jonathan Brown (Eds): Bridging Wallace´s Line

Monsoon Maximum

At a temporal resolution of 3000-year increments, COHMAP (1988) describes stronger-than-present monsoons between 12,000 and 6000 years BP. Even at that coarse resolution terrestrial sites reveal two significant departures from the model descriptions. One is that the peak in monsoon strength occurred at most sites toward the end of this period, not at the beginning or middle as would be expected on the basis of orbitally induced insolation changes. This lag could be explained as a late response to radiation changes, delayed by residual Late Glacial boundary conditions (Eurasian snow cover and/or depressed sea-surface temperatures). The second exception is that several sites show maximum precipitation prevailing until as late as 5000-3500 years BP, 1000-2500 years later than model descriptions. This lag comes too late to invoke Late Glacial boundary conditions as an explanation.

At a finer temporal resolution, timing of changes outlined here from Asian terrestrial sites may be compared with high-resolution records from the Somali Coast and the Oman Coast (Zonneveld *et al.* 1997; Sirocko *et al.* 1993; Overpeck *et al.* 1996). The NE African records suggest a surge in monsoon strength as early as 15,000-14,500 cal years BP. The only Asian terrestrial site showing such an early change, other than the Nilgiris, is Lake Didwana, which is the closest site downwind of the Arabian Sea cores. All other terrestrial records show initial monsoon strengthening *c.*13,000-11,000 cal years BP (11,100-9800 ^{14}C years BP), followed by a further strengthening at 10,000-9000 cal years BP (9000-8100 ^{14}C years BP). The records not showing these two steps in monsoon increase, from the Son Valley and from Dunde, both have poor temporal resolution in the early Holocene. The Son Valley record, reflecting changes in a fairly large watershed, can only record major changes in river channel morphology and sedimentology occurring over several thousand years. At Dunde, compaction of the lower ice layers limits the ability to resolve early Holocene changes at less than 1000 year intervals, and the oldest sample dates only to 11,000 cal years BP.

Overall, then, most sites reviewed here reveal a two-stage increase in monsoon strength beginning 13,000-11,000 cal years BP, peaking after 10,000-9000 cal years BP, and finally weakening *c.*5000-3500 cal years BP (approximate corresponding ^{14}C age ranges are 11,100-9800 years BP, 9000-8100 years BP, and 4500-3300 years BP, respectively). This general pattern is also described in two reviews of Chinese data (Sun and Chen 1991; Fang 1991). It is plausible to assume that the 13,000-11,000 cal years BP climate change was due to the strengthening of the monsoon responding to orbital changes, as modelled by Kutzbach and Guetter (1986). The fact that there was a lag in the monsoon maximum confirms that other factors were also significant in forcing the monsoon.

The pattern found in terrestrial sites reviewed here is similar to, but lagging, that described by Overpeck *et al.* (1996). They suggest that the first surge in monsoon strength occurred at 15,500-14,500 cal years BP (13,100-12,500 ^{14}C years BP), followed by a second surge at *c.*11,300-10,600 cal years BP (10,000-9500 ^{14}C years BP). Comparison of their data with data presented in this review of terrestrial sites suggests that the influence of boundary conditions on monsoon dynamics may have delayed changes in terrestrial expression of the Asian monsoon relative to the NE African monsoon.

Model experiments by de Menocal and Rind (1993) also indicate that Asian and African monsoons have responded differently to forcing factors, but in their study northeastern Africa is considered as part of the Asian subregion, because of the dominant influence of the Indian Ocean. The NW African monsoon is more strongly influenced by North Atlantic sea-surface temperature (SST) than is the Asian monsoon; and the Asian monsoon is influenced more by terrain, particularly the Qinghai-Tibetan Plateau (de Menocal and Rind 1993). Overpeck *et al.* (1996) outline correlations between the timing of certain events in the North Atlantic and changes in the Indian monsoon, and argue that Late Glacial-early Holocene changes in the North Atlantic affected snow cover and albedo in Central Asia, causing the gradually diminishing dominance of glacial boundary conditions and the increasing dominance of solar radiation changes as forcing factors. However, the generalized timing of the sequence of changes they describe for the NE Africa/Arabian Sea region do not clearly follow the specific site data presented.

Recent research has drawn on high-resolution data to draw a clear picture of the timing of both major and minor changes in regional monsoon circulation (Gasse and van Campo 1994; Overpeck *et al.* 1996). These syntheses have depended heavily on two sites in Asia, Rajasthan (Didwana and Lunkaransar) and Western Tibet (Sumxi Co, Banggong Co), because these studies are detailed multi-proxy records with good chronological control. It is still not clear, however, how representative these two areas are of monsoonal precipitation. Problems with direct interpretation of the Didwana record have been discussed above. Sumxi Co, presently outside the range of monsoonal precipitation, has yielded a record which does fit fairly well with other records from farther east on the plateau (Liu *et al.* 1998; Du *et al.* 1989), but what is missing now is an explanation of the mechanism by which Sumxi Co falls under the influence of the Indian Ocean monsoon. Attributing change at that site simply to a strengthening of the monsoon skirts the problem of how a steeper sea-land pressure gradient, caused by more intense insolation, forced the Tibetan surface low either to shift north, west, or to shrink in size to the point where western Tibet received precipitation from the Indian Ocean (or from the China Sea?). That Sumxi Co may have received the bulk of precipitation in the past from another, non-monsoonal, source is at least suggested by the fact that sites much farther north, off the plateau, showed a similar late Pleistocene-Holocene precipitation record (Rhodes *et al.* 1996).

The idea that monsoon changes, especially shifts in the monsoon front, could have been time-transgressive through the region has been broached by An *et al.* (1993b, 2000). Their review of Chinese sites reveals that the monsoon reached maximum strength at different times in different zones within China. Of particular interest here is that SW China, including Yunnan and western Sichuan, experienced a monsoon peak earliest, at *c.*12,000 years BP. The peak in East Central China (e.g. around Hubei) came in at 6000 years BP, and along the South China coast as late as 3000 years BP. Although our review reveals different timing for the monsoon maximum in SW China, and data from South

Peter Kershaw, Bruno David, Nigel Tapper, Dan Penny and Jonathan Brown (Eds): Bridging Wallace´s Line

China are as yet inadequate to pin down a maximum, still the idea of a time-transgressive monsoon maximum is helpful for illustrating how the Indian Ocean monsoon is not necessarily one system, especially not as it affects land areas. Tang (1979) discusses the idea of a plateau monsoon, as distinct from the Indian (SW) monsoon and the East Asian (SE) monsoon. The plateau monsoon may behave differently from monsoons driven by land-sea pressure gradients, because plateau monsoon flow responds to subregional lowland-highland pressure gradients. Much of the moisture from this monsoon is recycled from eastern China's humid lowlands. This subregional circulation, mentioned in Trewartha (1981: 184-185), results in maximum upward heat flow in the southeastern part of the plateau and subsidence in the northwest. Intra-plateau patterns of change in the monsoon regime are discussed by Shen and Tang, based on their review of Plateau sites (Tang and Shen 1996). Details of this circulation are beyond the scope of this review, but the idea that past monsoon changes may have been time-transgressive, as this review has indicated, may be more easily explained by considering variations in terrestrial subregional circulation.

Monsoon Weakening

The timing of monsoon weakening in the records reviewed here varies with about the same chronological range as monsoon onset, but the pattern of change does not reveal any clear geographical trend (see Table 11.2). Weakening began earliest (other than in the Nilgiris) in the northwest at Sumxi Co/Bangong Co at *c.*7000 cal years BP. The record from Qinghai Lake ostensibly reveals a late transition to drier conditions (*c.*1500 years BP) but, as mentioned above, the chronology for that time period at Qinghai Lake is not confirmed by any absolute dating. Almost all of the sites reviewed here experienced a relatively abrupt change to drier conditions *c.*4000-3500 years BP (4400-3800 cal years BP). At Dunde and Didwana the ensuing 1000 year period was actually drier than present, and driest for the entire Holocene. A similar period of maximum aridity in western Tibet (Sumxi Co, Bangong Co) culminated *c.*4300 cal years BP. At Roergai a major dry period fell in the middle of the monsoon maximum, for about 1500 years, centred on *c.*5000 years BP (5700 cal years BP). This period of mid-late Holocene aridity was one of the widespread changes identified and discussed by Gasse and van Campo (1994). Monsoon weakening could be viewed as a gradual process if one looks at the region as a whole and finds the variance in timing for that event among all the sites (Overpeck *et al.* 1996). However, taking each site record individually, the inferred change from wetter-than-present conditions to drier-than-present conditions is often distinct and occurs over only 500-800 years. That type of change is especially clear at sites on the margins of present-day monsoon flow, although highland sites in the heart of monsoon flow (Sichuan, Yunnan) have changed more gradually. It is difficult to reconcile a change of that character with the dominance of gradually waning summer insolation as a forcing factor, because for any one of the sites on the margin, the change was not gradual.

Human Environments

The emphasis of this review has been on inferred climate changes, so we cannot deal in detail with archaeological evidence for demographic change or settlement density in the region. However we mention a few observations from the articles pertaining to the interplay between climate dynamics and the environmental conditions which may have encouraged or discouraged human settlement in this region during the late Quaterny. Bryson and Swain (1981), in discussing conditions in Rajasthan, around Lake Didwana, note that the last period of desiccation there (*c*.3500 years BP) coincided with the decline and eventual disappearance of the Indus Culture. Whether that collapse was a result of climate change or whether the peak of the civilization brought on local vegetation changes which may have exacerbated the climate-forced drying has been debated for years, especially in light of roughly similar environmental changes in the Mayan area. However, the new data from Enzel *et al.* (1999) clearly indicate that the Indus/Harappan civilization arose *after* the major decrease in regional precipitation. The timing of that major dry period coincided with the general pattern described by Gasse and van Campo (1994), which suggests the dominance of climatic influence.

During the late Pleistocene Central India was apparently too dry for riparian settlements in the Son and Belan Valleys, but sporadic finds of artefacts in more recent strata suggest not only settlement there by 10,000 years BP but, by 4000 years BP, incipient agriculture (Williams and Royce 1982; Williams and Clarke 1984). There, again, the late wet period ended by 3000 years BP. In southern India, unambiguous cultural indicators are not found in the pollen signal, but researchers suspect that swidden cultivation may have been an important influence on the highland vegetation for as much as 10,000 years (Gupta and Prasad 1985).

In Yunnan Province, China, Sun *et al.* (1986) remark that although archae-ological research has shown settlement in the Lake Dianchi basin for 10,000 years, the pollen record shows very little evidence of major disruption of the vegetation until *c*.1500 years BP. Interpreting records like this may continue to be a problem as new data come in from heavily forested sites in the region, because swidden cultivation, a prevailing mode of subsistence, may be practi-cally undetectable in the pollen record. A similar situation seems likely in north-ern Thailand, where Spirit Cave deposits show early domestication of plants, in the absence of any evidence for widespread landscape change (Gorman 1970; Solheim 1970). Data from NE Thailand show a charcoal peak associated with a surge in *Pinus* and monolete fern spores *c*.5500 years BP, but that peak was not synchronous with a rise in any unambiguous indicators of agriculture. It could be reflecting change to a drier, more fire-prone climate. The Tianyang core in Guangdong Province shows a major surge in pteridophyte spores occurring in the Late Glacial, a change which, elsewhere in South China, is taken to indicate human disturbance (Liu and Qiu 1994). But the chronology at Tianyang leaves open the possibility that such a change could be attributed simply to cold, dry LGM conditions, not to early (Late Glacial) human modification of the landscape.

Peter Kershaw, Bruno David, Nigel Tapper, Dan Penny and Jonathan Brown (Eds): Bridging Wallace´s Line

It would be useful to determine a more nearly standard interpretation of distinct surges in pteridophyte spores in the SE Asian records, or to determine whether more taxonomic detail is needed in interpreting spore records. Van der Kaars (1991) has already suggested that they indicate relatively humid conditions in the Sahul Shelf region, but in the Asian monsoon region it may not be appropriate to take pteridophyte spores simply as indicators of moist conditions. Palaeovegetation studies from South China have found a close association between periods of fern dominance (mostly *Dicranopteris* spp.) and historically confirmed human disturbance (Liu and Qiu 1994). In this review, interpretation of the pteridophyte surge is critical in a few studies (Sun *et al.* 1993; Lei and Zheng 1993; Huang *et al.* 1997; Kealhofer and Penny 1998), and as mentioned above for Tianyang, pteridophytes may indicate cold and/or dry conditions, or response to fire.

Neither the geographic representation of these pieces of evidence nor the difficulty in interpreting subtle changes in the pollen and charcoal records allow sweeping conclusions about the environmental conditions for human settlement and migration. There are no solid data yet from the part of the monsoon region which should have served as the source area for out-migration to Australasia. And there are no clear indications of early (e.g. mid Holocene) human-induced modification of the forests of mainland SE Asia. Eventually, when the monsoon and archaeological records become clearer, it would be useful to incorporate studies of monsoon change into a discussion of environments affecting settlement and migration in the region. Throughout the region, sea level was relatively stable by 4000 years BP, and in most sites covered here, climate at that time was receding from an early to mid Holocene monsoon maximum. How did the waning monsoon and drier climate affect populations at the brink of agriculture?

Conclusion

We have presented a critical review of palaeoecological studies addressing late Quaternary monsoon dynamics in South and SE Asia, including monsoonal China. Because of the complexity of global and regional, atmospheric, marine and terrestrial factors forcing long term changes in the monsoon, it is extremely difficult to present a clear explanation of past changes. In addition to this theoretical complexity is the more mundane problem that available data suffer either from imprecise chronologies or poor geographical representation. Generally, it is possible to discern a two-step increase in monsoon strength, at 13,000-11,000 cal years BP and 10,000-9000 cal years BP, followed by a period of maximum monsoon precipitation lasting until *c.*5000-3500 cal years BP. This regime, best reflected in records from the eastern margins of the Qinghai-Tibetan Plateau, lagged by about 1500 years the pattern described for NE Africa and the Arabian Sea summarized by Overpeck *et al.* (1996). Whether this time-transgressive pattern may be following a west-east gradient, a coastal-inland gradient, or a lowland-highland gradient is not yet clear from available data.

Timing of changes does not appear to reflect linear response to global-scale or extraterrestrial forcing factors. Many changes were abrupt, and show surprising variations within the few subregions which have well-dated records. Based on available data, it is difficult to identify more than just a few gradual changes in monsoon circulation, but the sites in the middle of the region reviewed here seem to show more gradual Holocene changes, i.e. at Roergai, Sichuan, and Yunnan. Although regional patterns can be induced by averaging data from numerous sites, local differences in the timing of changes are significant, probably owing to the predominant influence of the complex Asian terrain on regional atmospheric circulation.

The major persisting problem is that we do not have long, clear, well-dated records from land sites well within the Asian monsoon region. Specifically, the areas poorly represented by good data are central and NE India, southern Tibet, the Indochina peninsula, and South China. Some studies described in this review, and a few not mentioned, could be useful with improved chronologies; i.e. more radiocarbon dates from cores, especially from India, which have already been analysed for pollen and lithology. Although records from sites at the margins of monsoon influence are valuable for their sensitivity to changes in monsoon extent, we cannot really be sure what regional changes those sites reflect until we know what is happening in the core area of the monsoon.

Acknowledgements

Authors' research on sites in Cambodia and Tibet was funded by the National Science Foundation (for Tibet, grant nos. SBR-9410411 and ATM-9410491; for Cambodia, grant no. SBR-9506344), and further support for the Cambodian research came from the Geological Society of America, Association of American Geographers, the LSU Department of Geography and Anthropology, and Sigma Xi. The LSU Graduate School and the Department of Geography and Anthropology also provided partial funding for travel to the Monash conference. We are very grateful for beneficial discussions of Qinghai-Tibetan data with Tang Lingyu, Shen Caiming, and Yao Zuju.

References

An, Z., Kukla, G.J., Porter, S.C. and Xiao, J. 1991. Magnetic susceptibility evidence of monsoon variation on the loess plateau of central China during the last 130,000 years. *Quaternary Research,* 36: 29-36.

An, Z., Porter, S.C., Zhou, W., Lu, Y., Donahue, D.J., Head, M.J., Wu, X., Ren, J. and Zheng, H. 1993(a). Episode of strengthened monsoon climate of Younger Dryas Age on the Loess Plateau of Central China. *Quaternary Research,* 39: 45-54.

An, Z., Porter, S.C., Wu, X., Kutzbach, J., Wang, S., Liu, X., Li, X., Wang, J., Zhou, W., Xiao, J., Liu, J. and Lu, J. 1993(b). The Holocene Climatic Optimum in central and eastern China and variations in the East Asian summer monsoon. *Chinese Science Bulletin,* 14: 1302-6. (in Chinese).

An, Z., Porter, S.C., Kutzbach, J.E., Wu, X.H., Wang, S.M., Liu, X.D., Li, X.Q. and Zhou, W.J. 2000. Asynchronous Holocene optimum of the East Asian monsoon. *Quaternary Science Reviews,* 19: 743-762.

Bhattacharya, K. and Chanda, S. 1987-1988. Late Quaternary vegetational history, palaeoecology and biostratigraphy of some deposits of Brahmaputra Basin, Upper Assam, India. *Journal of Palynology,* 23-24: 225-37.

Bhattacharya, K. and Chanda, S. 1992. Late Quaternary vegetational history of Upper Assam, India. *Review of Palaeobotany and Palynology,* 72: 325-33.

Bhattacharyya, A. 1989. Vegetation and climate during the last 30,000 years in Ladakh. *Palaeogeography, Palaeoclimatology, Palaeoecology,* 73: 25-38.

Bishop, P. and Godley, D. 1994. Holocene palaeochannels at SiSatchanalai, north-central Thailand: ages, significance and palaeoenvironmental indications. *Holocene,* 4: 32-41.

Boonsener, M. 1991. The Quaternary stratigraphy of northeast Thailand. *Journal of Thai Geosciences,* 1: 23-32.

Broecker, W.S. and van Donk, J. 1970. Insolation changes, ice volumes, and the ^{18}O record in deep sea cores. *Journal of Geophysical Research,* 8: 169-96.

Bryson, R.A. and Swain, A.M. 1981. Holocene variations of monsoon rainfall in Rajasthan. *Quaternary Research,* 16: 135-45

Chauhan, M.S. 1995. Origin and history of tropical deciduous Sal (*Shorea robusta* Gaertn.) forests in Madhya Pradesh, India. *Palaeobotanist,* 43(1): 89-101.

Clemens, S., Prell, W., Murray, D., Shimmield, G. and Weedon, G. 1991. Forcing mechanisms of the Indian Ocean monsoon. *Nature,* 353: 720-5.

COHMAP 1988. Climate changes of the last 18,000 years: Observations and model simulations. *Science,* 241: 1043-52.

Cullen, J.L. 1981. Microfossil evidence for changing salinity patterns in the Bay of Bengal over the last 20,000 years. *Palaeogeography, Palaeoclimatology, Palaeo-ecology,* 35: 315-56.

de Menocal, P.B. and Rind, D. 1993. Sensitivity of Asian and African climate to variations in seasonal insolation, glacial ice cover, sea surface temperature, and Asian orography. *Journal of Geophysical Research,* 98(D4): 7265-87.

Dodia, R., Agrawal, D.P. and Vora, A.B. 1984. New Pollen Data from Kashmir Bogs. In: R.O. Whyte (Editor) *The Evolution of the East Asian Environment.* Center of Asian Studies, University of Hong Kong, Hong Kong: 569-77.

Du, N., Kong, Z. and Shan, F. 1989. A preliminary investigation on the vegetational and climatic changes since 11,000 years in Qinghai Lake - an analysis based on palynology in core QH85-14C. *Acta Botanica Sinica,* 31: 803-14. (in Chinese).

Duplessy, J.C. 1982. Glacial to interglacial contrasts in the northern Indian Ocean. *Nature,* 295: 494-8.

Enzel, Y., Ely, L.L., Mishra, S., Ramesh, R., Amit, R., Lazar, B., Rajaguru, S.N, Baker, V.R. and Sandler, A. 1999. High-resolution Holocene environmental changes in the Thar Desert, Northwestern India. *Science,* 284: 125-8.

Fang, J-Q. 1991. Lake evolution druing the past 30,000 years in China, and its implications for environmental change. *Quaternary Research,* 36: 37-60.

Gao, F. 1988. Analyses of the formation environment of peat and rotted wood in Nanping of Lichuan. *Geographical Research,* 7: 59-66. (in Chinese, with English abstract)

Gasse, F. and van Campo, E. 1994. Abrupt post-glacial climate events in West Asia and North Africa monsoon domains. *Earth and Planetary Science Letters,* 126: 435-56.

Gasse, F., Fontes, J.C., van Campo, E and Wei, K. 1996. Holocene environmental changes in Bangong Co basin (Western Tibet): Part 4: Discussion and conclusions. *Palaeogeography, Palaeoclimatology, Palaeoecology,* 120: 79-92.

Gasse, F., Arnold, M., Fontes, J.C., Fort, M., Gilbert, E., Huc, A., Li, B., Li, Y., Liu, Q., Mélières, F., van Campo, E., Wang, F. and Zhang, Q. 1991. A 13,000-year climate record from western Tibet. *Nature,* 353: 742-5.

Gorman, C.F. 1970. Excavations at Spirit Cave, North Thailand, 1966: some interim interpretations. *Asian Perspectives,* 13: 79-107.

Gupta, H.P. 1971. Quaternary vegetational history of Ootacamund, Nilgiris, south India 1: Kakathope and Rees-corner. *Palaeobotanist,* 20(1): 74-90.

Gupta, H.P. 1978. Holocene palynology from Meander Lake in the Ganga Valley, district Pratapgarh, U.P. *Palaeobotanist,* 25: 109-119.

Gupta, H.P. 1990. Sholas in south Indian montane: Past, present, and future. *Palaeobotanist,* 38: 394-403.

Gupta, H.P. and Prasad, K. 1985. The vegetational development during 30,000 years BP at Colgrain, Ootacamund, Nilgiris. *Journal of Palynology,* 21: 174-87.

Hastenrath, S. and Lamb, P.J. 1979. *Climatic Atlas of the Indian Ocean. Part 1: Surface Climate and Atmospheric Circulation.* University of Wisconsin Press, Madison: 97 charts.

Hastings, P. and Liengsakul, M. 1983. Chronology of Late Quaternary Climatic Changes in Thailand. In: N. Thiramongkol and V. Pisutha-Arnold (Editors). *Proceedings of the First Symposium on Geomorphology and Quaternary Geology of Thailand.* Department of Geology, Chulalongkorn University, Bangkok: 24-34.

Hayes, J.D., Imbrie, J. and Shackleton, N.J. 1976. Variations in the Earth's orbit: Pacemaker of the ice ages. *Science,* 194: 1121-32.

Huang, C-Y., Liew, P.M., Zhao, M., Chang, T-C., Kuo, C-M., Chen, M-T., Wang, C-H. and Zheng, L-F. 1997. Deep sea and lake records of the Southeast Asian palaeo-monsoons for the last 25 thousand years. *Earth and Planetary Science Letters,* 146: 59-72.

Imbrie, J., Hayes, J.D., Martinson, D.G., McIntyre, A., Mix, A.C., Morley, J.J, Pisias, N.G., Prell, W.L., and Shackleton, N.J. 1984. The Orbital Theory of Pleistocene Climate: Support for a Revised Chronology of the Marine $\delta^{18}O$ Record. In: A. Berger, J. Imbrie, J. Hays, G. Kukla and B. Saltzman (Editors), *Milankovitch and Climate.* Reidel, Dordrecht, Holland: 269-305.

Jarvis, D.I. 1993. Pollen evidence of changing Holocene monsoon climate in Sichuan Province, China. *Quaternary Research,* 39(3): 325-37.

Kealhofer, L.K. and Penny, D. 1998. A combined pollen and phytolith record for fourteen thousand years of vegetation change in northeastern Thailand. *Review of Palaeobotany and Palynology,* 103: 83-93.

Kelts, K., Chen K.Z., Lister, G.S, Yu, J.Q., Gao, Z., Niessen, F. and Bonani, G. 1989. Geological fingerprints of climate history: a cooperative study of Qinghai Lake, China. *Eclogae geologicae Helvetiae,* 82(1): 167-82.

Kershaw, A.P. 1986. Climatic change and Aboriginal burning in north-east Australia during the last two glacial/interglacial cycles. *Nature,* 322: 47-9.

Kong, Z., Du, N., Shan, F., Tong, G., Luo, S. and Fan, S. 1990. Vegetational and climatic changes in the last 11,000 years in Qinghai Lake - Numerical analysis based on palynology in core QH85-14C. *Marine Geology and Quaternary Geology,* 10(3): 79-90. (in Chinese, with English abstract)

Kotlia, B.S., Bhalla, M.S., Sharma, C., Rajagopalan, G., Ramesh, R., Chauhan, M.S., Mathur, P.D, Bhandari, S. and Chacko, S.T. 1997. Palaeo climatic conditions in the upper Pleistocene and Holocene Bhimtal-Naukuchiatal lake basin in south-central Kumaun, North India. *Palaeogeography, Palaeoclimatology, Palaeoecology,* 130: 307-22.

Kutzbach, J.E. and Guetter, P.J. 1986. The influence of changing orbital parameters and surface boundary conditions on climate simulations for the past 18 000 years. *Journal of Atmospheric Science,* 43: 1726-59.

Lei, Z-Q. and Zheng, Z. 1993. Quaternary sporo-pollen flora and palaeoclimate of the Tianyang volcanic lake basin, Leizhou Peninsula. *Acta Botanica Sinica,* 35: 128-38. (in Chinese, with English abstract)

Li, X. and Liu, J. 1988. Holocene vegetational and environmental changes at Mt. Luoji, Sichuan. *Acta Geographica Sinica,* 43(1): 44-51. (in Chinese, with English abstract)

Lin, S., Qiao, Y. and Walker, D. 1986. Late Pleistocene and Holocene vegetation history at Xi Hu, Er Yuan, Yunnan Province, southwest China. *Journal of Biogeography,* 13: 419-40.

Lister, G.S, Kelts, K., Chen, K.Z., Yu, J.Q. and Niessen, F. 1991. Lake Qinghai, China: closed-basin lake levels and the oxygen isotope record for ostracoda since the latest Pleistocene. *Palaeogeography, Palaeoclimatology, Palaeoecology,* 84: 141-62.

Liu, G-X. 1991. Late-glacial and postglacial vegetation and associated environment in Jianghan Plain. *Acta Botanica Sinica,* 33(8): 581-8. (in Chinese, with English abstract)

Liu, J., Tang, L., Qiao, Y., Head, M.J. and Walker, D. 1986. Late Quaternary vegetation history at Menghai, Yunnan Province, southwest China. *Journal of Biogeography,* 13: 399-418.

Liu, K-B. 1988. Quaternary history of the temperate forests of China. *Quaternary Science Review,* 7: 1-20.

Liu, K-B., Overpeck, J.T., Tang, L., Xu, X., Yang, S., Cole, J.E., Trumbore, S.E. and Shen, C. 1996. Monsoon climate changes and vegetation dynamics in eastern Tibet since the Last Glacial Maximum. *Association of American Geographers 92nd Annual Meeting – Abstracts.* Charlotte, North Carolina.

Liu, K-B. and Qiu, H-L. 1994. Late Holocene pollen records of vegetational changes in China: Climate or human disturbance? *Terrestrial, Atmospheric and Oceanic Sciences,* 5(3): 393-410.

Liu, K-B, Sun, S. and Jiang, X. 1992. Environmental change in the Yangtze River delta since 12,000 years B.P. *Quaternary Research,* 38: 32-45.

Liu, K-B., Yao, Z. and Thompson, L. 1998. A pollen record of Holocene climatic changes from the Dunde Ice Cap, Qinghai-Tibetan Plateau. *Geology,* 26(2): 135-8.

Maher, B.A. and Thompson, R. 1992. Plaaeoclimatic significance of the mineral magnetic record of the Chinese loess and palaeosols. *Quaternary Research,* 37: 155-70.

Maloney, B.K. 1992. Late Holocene climatic change in Southeast Asia: the palynological evidence and its implications for archaeology. *World Archaeology,* 24(1): 25-34.

Maxwell, A.L. 1996. Palynological analysis of Holocene lake sediments from northeastern Cambodia. *Association of American Geographers 92nd Annual Meeting – Abstracts.* Charlotte, North Carolina.

Maxwell, A.L. 1999. *Holocene Environmental Change in Mainland Southeast Asia: Pollen and Charcoal Records from Cambodia.* Unpublished Ph.D. Thesis. Louisiana State University, Baton Rouge.

Maxwell, A.L. 2001. Holocene monsoon changes inferred from lake sediment pollen and carbonate records. *Quaternary Research,* 56: 390-400.

Maxwell, A.L. and Colm, S.E. 1997. Lake sediment charcoal records of land-use change in northeastern Cambodia. *Association of American Geographers 93rd Annual Meeting – Abstracts.* Ft. Worth, Texas.

Overpeck, J., Anderson, D. Trumbore, S. and Prell, W. 1996. The southwest Indian Monsoon over the last 18 000 years. *Climate Dynamics,* 12: 213-25.

Pramojanee, P. and Hastings, P.J. 1983. Geomorphological and palynological investigation of sea level changes in Chanthaburi, S.E. Thailand. In: N. Thiramongkol and V. Pisutha-Arnond (Editors) *Proceedings of the 1st Symposium on the Geomorphology and Quaternary Geology of Thailand.* Department of Geology, Chulalongkorn University, Bangkok: 35-51.

Prell, W.L. and Kutzbach, J.E. 1987. Monsoon variability over the past 150,000 years. *Journal of Geophysical Research* 92(D7): 8411-25.

Prell, W.L. and Kutzbach, J.E. 1992. Sensitivity of the Indian monsoon to forcing parameters and implications for its evolution. *Nature,* 360: 647-52.

Rajagopalan, G., Vishnu-Mittre, B. Sekar, and T.K. Mandal. 1982. Birbal Sahni Institute radiocarbon measurements III. *Radiocarbon,* 24: 45-53.

Ramage, C.S. 1971. *Monsoon Meteorology.* Academic Press, New York: 253 pp.

Rhodes, T.E., Gasse, F., Lin., R., Fontes, J-C., Wei, K., Bertrand, P., Gibert, E., Melieres, F., Tucholka, P., Wang, Z., and Cheng Z-Y. 1996. A late Pleistocene-Holocene lacustrine record from Lake Manas, Zunggar (northern Xinjiang, western China). *Palaeogeography, Palaeoclimatology, Palaeoecology,* 120: 105-21.

Sarkar, A., Ramesh, R., Bhattacharyya, S.K. and Rajagopalan, G. 1990. Oxygen isotope evidence for a stronger winter monsoon current during the last glaciation. *Nature,* 343: 549-51.

Shackleton, N.J. and Opdyke, N.D. 1973. Oxygen isotope and palaeomagnetic stratigraphy of equatorial Pacific core V28-238: oxygen isotope temperatures and ice volumes on a 10^5 and 10^6 year scale. *Quaternary Research,* 3: 39-55.

Sharma, C. 1992. Palaeoclimatic oscillations since the last deglaciation in western Himalaya: a palynological assay. *Palaeobotanist,* 40: 374-82.

Sharma, C. 1993. Palynostratigraphy of lake deposits of Himalaya and palaeoclimate in Quaternary period. *Current Science,* 64(11-12): 930-2.

Shen, C-M., Tang L-Y. and Wang, S-M. 1996. Vegetation and climate during the last 22,000 years in Zoige Region. *Acta Micropalaeontologica Sinica* 13(4): 401-6. (in Chinese, with English abstract)

Shen, C-M. and Tang L-Y. 1996. Pollen evidence of changing Holocene monsoon on Qinghai-Xizang Plateau. *Acta Micropalaeontologica Sinica* 13(4): 433-6. (in Chinese, with English abstract)

Singh, G., Joshi, R.D. and Singh, A.B. 1972. Stratigraphic and radiocarbon evidence for the age and development of three salt lake deposits in Rajasthan, India. *Quaternary Research,* 2: 496-505.

Singh, G., Wasson, R.J. and Agrawal, D.P. 1990. Vegetational and seasonal climate changes since the last full glacial in the Thar Desert, northwestern India. *Review of Palaeobotany and Palynology,* 64: 351-8

Sirocko, F., Sarnthein, M., Erlenkeuser, H., Lange, H., Arnold, M. and Duplessy, J.C. 1993. Century-scale events in monsoonal climate over the past 24,000 years. *Nature,* 364: 322-4.

Solheim, W.G.II. 1970. Northern Thailand, Southeast Asia, and World Prehistory. *Asian Perspectives,* 13: 145-62.

Somboon, J.R.P. 1988. Palaeontological study of the recent marine sediments in the lower central plain, Thailand. *Journal of Southeast Asian Earth Sciences,* 2: 201-10.

Street, F.A. and Grove, A.T. 1979. Global maps of lake-level fluctuations since 30,000 yr. B.P. *Quaternary Research* 12: 83-118.

Stuiver, M. and Reimer, P.J. 1993. Extended 14C data base and revised Calib 3.0 14C age calibration program. *Radiocarbon,* 35: 215-30

Sukumar, R., Ramesh, R., Pant, R.K. and Rajagopalan, G. 1993. A $\delta^{13}C$ record of late Quaternary climate change from tropical peats in southern India. *Nature,* 364: 703-6.

Sun X. and Chen Y. 1991. Palynological records of the last 11,000 years in China. *Quaternary Science Review,* 10: 537-44.

Sun X., Du, N., Chen, Y., Gu, Z., Liu, J. and Yuan, B. 1993. Holocene palynological records in Lake Selingcuo, Northern Xizang. *Acta Botanica Sinica,* 35(12): 943-50. (in Chinese, with English abstract)

Sun X., Wu, Y., Qiao, Y. and Walker, D. 1986. Late Pleistocene and Holocene vegetation history at Kunming, Yunnan Province, southwest China. *Journal of Biogeography,* 13: 441-76.

Tang, L-Y. and Shen, C-M. 1996. Holocene pollen records of the Qinghai-Xizang Plateau. *Acta Palaeontologica Sinica,* 13(4): 407-22. (in Chinese, with English abstract)

Tang M. 1979. The average characteristics of the Plateau monsoon climate. *Acta Geographica* 34. (in Chinese)

Thompson, L.G., Mosley-Thompson, E., Davis, M., Bolzan, J., Dai, J., Yao, T., Gundestrap, N., Wu, X., Klein, L. and Xie, Z. 1989. Holocene-Late Pleistocene climate ice records from Qinghai-Tibetan Plateau. *Science* 246: 474-7.

Trewartha, G. 1981. *The Earth's Problem Climates.* University of Wisconsin, Madison: 371 pp.

Udomchoke, V. 1989. Quaternary stratigraphy of the Khorat Plateau area, northeastern Thailand. In: N. Thiramongkol (Editor) *Proceedings of the Workshop on Correlation of Quaternary Successions in South, East, and Southeast Asia.* Department of Geology, Chulalongkorn University, Bangkok: 69-94.

van Campo, E. 1986 Monsoon fluctuations in two 20,000-yr. oxygen-isotope/pollen records off southwest India. *Quaternary Research,* 26: 376-88.

van Campo, E. and Gasse, F. 1993. Pollen and diatom-inferred climatic and hydrological changes in Sumxi Co basin (western Tibet) since 13,000 yr. B.P. *Quaternary Research,* 39(3): 300-13.

van Campo, E., Cour, P. and Hang, S. 1996. Holocene environmental changes in Bangong Co basin (Western Tibet): Part 2: The pollen record. *Palaeogeography, Palaeoclimatology, Palaeoecology,* 120: 49-63.

van der Kaars, W.A. Palynology of eastern Indonesian marine piston-cores: A Late Quaternary vegetational and climatic record for Australasia. *Palaeogeography, Palaeoclimatology, Palaeoecology,* 85: 239-302.

Vishnu-Mittre, Gupta, H.P. and Robert, R. 1967. Studies of the Late Quaternary vegetational history of Kumaon Himalaya. *Current Science,* 36(20): 539-40.

Vishnu-Mittre and Sharma, C. 1984. Vegetation and climate during the last glaciation in the Kathmandu Valley, Nepal. *Pollen et Spores,* 26: 69-94.

Walker, D. 1986. Late Pleistocene - early Holocene vegetational and climatic changes in Yunnan Province, southwest China. *Journal of Biogeography,* 13: 477-86.

Wang, K. and Sun, L. 1989. Changes of palaeovegetation and palaeoclimate in Lichuan, Hubei Province, since 20,000 B.P. *Geographical Research,* 8(3): 61-5. (in Chinese, with English abstract)

Williams, M.A.J. and Royce, K. 1982. Quaternary geology of the middle Son Valley, north central India: Implications for prehistoric archaeology. *Palaeogeography, Palaeoclimatology, Palaeoecology,* 38: 139-62.

Williams, M.A.J. and Clarke, M.F. 1984. Late Quaternary environments in north-central India. *Nature,* 308: 633-5.

Winkler, M.J. and Wang, P.K. 1993. The Late-Quaternary vegetation and climate of China. In: H.E. Wright, J.E. Kutzbach, T. Webb III, W.F. Ruddiman, F.A. Street-Perrott and P.J. Bartlein (Editors) *Global Climates since the Last Glacial Maximum.* University of Minnesota Press, Minneapolis: 221-64.

Zonneveld, K.A.F., Ganssen, G., Troelstra, S., Versteegh, G.J.M. and Visscher, H. 1997. Mechanisms forcing abrupt fluctuations of the Indian Ocean summer monsoon during the last deglaciation. *Quaternary Science Review,* 16: 187-201.

The Reconstruction of Flood Regimes in SE Asia from El Niño-Southern Oscillation (ENSO) Related Records

David Godley

Introduction

It has been estimated that the Asia-Pacific region experiences around 60 percent of all major natural disasters, resulting in economic losses of between US $5 and $10 billion per year (Inter Press Service 1998). The largest proportion of this damage is flood and drought related. The scope of the damage and the socio-economic impact on a nation and its people can be seen in a few recent figures from Thailand. During the 1996 wet season, flooding caused an estimated $88 million damage, affected 5 million people and submerged 480,000 acres of arable land (Xinhua News Agency 1996; Lloyds List 1996). This contrasts with the situation in the 1987 'wet season' when only 9.1 percent of the nation's paddy fields were planted on time due to poor or delayed rains (Xinhua News Agency 1987). It is clear that this cycle of extreme flood and drought events has had a significant impact on the Thai way of life and it seems reasonable to assume that this has long been the case. Khanittana (1989) suggests that water played a critical role in the location of early Thai settlements whilst Jumsai (1989a, 1989b) also theorised that cultural adaptation to regular flooding has remained an integral part of the modern Thai mentality. If such assumptions are correct, the record of monsoonal flooding is absolutely critical to any understanding of the cultural as well as economic history of the region.

This chapter explores the relationships that exist between the modern flow regimes of northern Thai rivers and ENSO related phenomena from around the Pacific Rim as a basis for interpretation of archaeological and historical features of the region. The extensive exploration of ENSO and its relationship to rainfall in the region (Walker 1910; Tanaka 1978; Yamamoto 1978; Parthasarathy *et al.* 1990; Khandekar 1991; Nicholls 1993; Chang and King 1994; Vijayakumar and Kulkarni 1995 amongst numerous others) indirectly suggest that the ENSO record may contain an indication of any changes in the character of SE Asian flood regimes. Despite the perceived importance of cycles of flood and drought in SE Asia, very little is known on historical, *let alo*ne archaeological, timescales about the interaction between the rivers of SE Asia and the communities which remain dependent on them. Extant records, which are at best piece-meal, provide evidence of individual flood and drought events in select locations but do little to help us understand the broader flood regimes that have so influenced

ISBN 3-923381-47-6
© 2002 by CATENA VERLAG, 35447 Reiskirchen

the context of the region's cultural development. I attempt to remedy this by constructing a proxy flood-drought chronology using ENSO data from around the Pacific rim, supplemented by other long-term records, especially those from China.

Flooding and ENSO

The hypothesis that El Niño events are closely related to a decrease in the annual number of tropical cyclones formed in the Western North Pacific has gained wide acceptance (Li 1988; Chan 1985; Wu and Lau 1992; Bao and Xiang 1993; Chan 1995) although there are those who would argue against it (e.g. Ramage and Hori 1981; Lander 1994). Statistically-significant connections between El Niño and the location of cyclone genesis in the Western Pacific (Aiko, 1985; Lander, 1994) and the South China Sea (Chan 1995; McGregor 1995) have been found. During El Niño years there appears to be an extension to the East of the axis of the monsoon trough. Because the majority of cyclones form in this trough, the mean location of cyclone genesis also shifts to the East. What has yet to be established is the impact on annual floods in the interior of mainland SE Asia. However a few broad generalizations seem in order.

The peak of the annual flood in inland Thailand usually occurs at the end of their wet season, beginning in the North in August and gradually moving to the South over the following months. The timing, severity and rate of southerly progression of this flooding is related to the withdrawal of the summer monsoon and the southerly migration of the Inter-Tropical Convergence Zone (ITCZ) (Kong Phumi'akat and Kong Utuniyomwitthaya Uthok 1989). Associated with the withdrawing ITCZ is the passage of typhoons, cyclones, depressions and storms. These are among the prime causes of heavy rainfall late in the wet season (Phiansap *et al.* 1990; Wongwitat and Uprasitwong 1991). It has been suggested that both the frequency of occurrence and the speed with which these systems pass play an important role in controlling the severity of the annual flooding (Phiansap *et al.* 1990; Wiraphan 1994; Watchari Wiraphan pers. comm.). In short, the greater the number of tropical systems and the slower their movement, the greater the potential rainfall in any given area. With heavy rainfall comes an increased flood risk implying a relational connection between tropical cyclone activity and flooding. If this is indeed the case, any factor controlling cyclone formation can, in turn, be used as a proxy flood record.

The synoptic conditions associated with El Niño years also cause anomalous Westerlies, which result in a greater number of tropical cyclones recurving East of the Philippines archipelago. Consequently the number penetrating into SE Asia is reduced (Aiko 1985; McGregor 1995). Therefore it appears that El Niño causes a reduction in the total number of cyclones formed, an Eastward shift in the location of genesis and a greater degree of recurve. An outcome of these phenomena is a decrease in the number of cyclones and typhoons landing in SE Asia. This is likely to lead to a reduction in the severity of the flood peak for that year. Therefore it can be argued that a decrease in the number of cyclones making landfall results in a drier than normal wet season, and that these drier years tend to correspond with El Niño years or the presence of El Niño-like conditions. If this is the case, the ENSO cycle should be in phase with the flood-drought cycle and this pattern can be exploited by using ENSO proxies as indicators of flood patterns.

Constructing an ENSO-based Flood-Drought Proxy

With the exception of the Javanese tree-ring chronology (Murphy and Whetton 1989), there are no long ENSO records from SE Asia that can be used as a flood proxy. In order to create such a record, a composite ENSO index can be constructed from the various existing Pacific Rim chronologies of, and proxy records for, ENSO. Eight 500 year long chronologies of ENSO, including one from India, outside of the Pacific rim, have been selected.

Northern Chinese Rainfall

Fluctuations in rainfall in northern China have been shown to be associated with variations in ENSO (Huang and Wang 1985; Zhang and Chen 1987; Yang and Xu 1994; Xu 1995). It has been argued that up to 82% of the variance in drought occurrence can be explained by the Southern Oscillation, with 80% of drought years occurring within 2 years of an El Niño event (Tang and Guo 1993). 'Wetness-Dryness' data from 35 stations in northern China were used by Whetton *et al.* (1990) to construct a proxy chronology of El Niño between 1470 and 1979 from several thousands of historical sources (Chung yang ch'i hsiang ch'u ch'i hsiang k'o hsüeh yen chiu yüan 1981).

Javanese Tree-Ring Chronology

As with the Chinese data, the relationship between El Niño and rainfall in the Indonesian archipelago is well established (Berlage 1957; Nicholls 1981; Hastenrath 1987; Cole *et al.* 1992). Rainfall on Java is highly seasonal and this affects the growth cycle of Teak (*Tectona grandis*) (De Boer 1951). Tree ring chronologies show a significant correlation with the ENSO cycle (Murphy and Whetton 1989) and can therefore be used as a proxy signal for El Niño.

Indian Drought Record

The impact of broad climatological and circulatory phenomena on Indian rainfall has been the subject of study for most of this century, and the connection between Indian rainfall, the monsoon and ENSO has been well documented (Parthasarathy *et al.* 1990; Khandekar 1991; Kiladis and Sinha 1991; Vijaya-kumar and Kulkarni 1995). The occurrence of below-average rainfall across much of India is associated with El Niño. Whetton and Rutherfurd (1994) have compiled records of widespread drought and famine from across India and argue that given the correspondence between poor rainfall and El Niño this record can be used as a proxy for El Niño. Because Kripalani *et al.* (1995) have tentatively identified a correlation between the occurrence of drought in India and in Thailand, the Indian data are likely to reflect conditions in SE Asia despite their location beyond the Pacific rim.

Quelccaya Ice Cap Ice Accumulation and $\delta^{18}O$ Records (Two Data Series)

Approximately 80% of the annual snowfall on the Peruvian Quelccaya ice cap occurs during the November to April wet season (Thompson 1992). During a strong El Niño year there is a significant decline in rainfall, resulting in a substantial decrease in ice accumulation on the ice cap (Thompson *et al.* 1984a; 1984b). This reduction in rainfall in the mountains of southern Peru constrasts with the high high rainfall received in the usually arid, adjacent coastal regions within El Niño years. The changes in SSTs which are characteristic of an El Niño event cause a fractionation of oxygen isotope abundances in sea water. Because the snow has an oceanic origin, this fractionation affects the isotopic abundances in the snow that forms the ice caps. Rates of ice accumulation and δO^{18} values are therefore both indicators of El Niño.

Typhoon Landings

The link between typhoon formation, typhoon landfall and El Niño has already been discussed. In short, the average number of typhoons landing in SE Asia decreases during El Niño years (Aiko 1985; McGregor 1995). Wang (1990) has used the nadir in the number of typhoons landing in southern China to compile a chronology of likely El Niño years.

South American Historical Records

There are several chronologies of El Niño events based on South American historical records. Those of Quinn and Neal (1992) and Hamilton and Garcia (1986) have been selected. These chronologies are based on historical accounts of coastal flooding, heavy rain in Peru and the productivity of coastal fisheries. The Quinn (1992) chronology was specifically excluded as it makes use of ENSO-related climatic anomalies from outside of the Pacific basin (the Nile River). The only extra-Pacific data source used here is the Indian drought chronology.

Discussion

The above data fall into two categories: those records which consist of an annual series of anomalies ($\delta^{18}O$, ice accumulation, Javanese tree rings and the record of northern Chinese rainfall) and those records of individual El Niño events (typhoon landings and the two South American El Niño event sequences). While the annual data series have values for every year, the El Niño event chronologies consist solely of the years that the various authors identified as probable El Niño years. Accordingly, the two groups were processed slightly differently and then combined to form two sequences that reflect aspects of the likelihood of flood and drought.

To remove long period variations in each of the annual series, the values were smoothed using a 10-year Gaussian filter. Each yearly value was then expressed as an anomaly from the smoothed curve[1]. The northern Chinese flood-drought data were inverted so that in each case a positive value in each of the data sets is associated with El Niño or El Niño-like conditions. El Niño years in the event chronologies were given a positive value to follow the same pattern. The years that were not identified as El Niño years in those chronologies were assumed to have no information.

The values of the anomalies were then reduced to an indication of its sign only (positive, negative or zero). The sign of each anomaly was then added to form a single value with a possible range from +8 to –4 (Figure 12.1). This process was carried out for the entire 1480-1980 sequence. The purpose behind the summing of the data is two-fold. Firstly, it reduces the possible impact of a change in the El Niño-proxy relationship in a single data series. Whetton and Rutherfurd (1994) and Whetton *et al.* (1996) noted that over short periods a contrary relationship sometimes appeared between the Chinese and Javanese proxy series and the Quinn (1992) El Niño chronology. While it is likely that the contrary periods were caused by the influence of the Nile data on Quinn's (1992) chronology (Peter Whetton, pers. comm.), by combining the eight data sets described above, a change to a contrary relationship in any one series should be compensated for by the remaining stable ones. Secondly, the composite El Niño index gives an indication of the uniformity of the anomalies across the Pacific Rim. It seems reasonable to assume that, under the influence of strong SST or SOI anomalies, there would be a greater chance that all of the Pacific rim proxy locations will be influenced by these conditions, and therefore their records would be in unison. The uniformity of the sign of the proxy anomalies across the eight data series (and therefore the absolute value of the composite index derived here) can be viewed as an indirect measure of the intensity of the SST and pressure anomalies that they are recording.

As a test of this El Niño intensity-Proxy Unison model, values from 1900 to 1980 have been compared against the record of major floods and droughts in Thailand (Figure 12.2). In this test all but three of the known severe drought years have positive values of the ENSO composite index and, although weaker, all but two of the severe flood years have values less than one. When compared against known El Niño and La Niña events the pattern also holds, although it appears that the measure works slightly better for El Niño than it does for La Niña (Figure 12.2). This is to be expected as La Nina events are characterized by much more subtle changes than are El Niño events (Trenberth 1997).

---

[1] This data-set was kindly provided by Ian Rutherfurd, Centre for Catchment Hydrology, Monash University.

Peter Kershaw, Bruno David, Nigel Tapper, Dan Penny and Jonathan Brown (Eds): Bridging Wallace´s Line

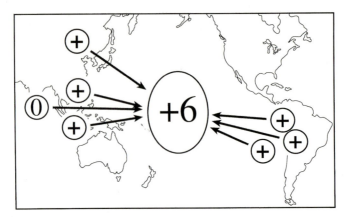

Fig. 12.1 The ENSO proxy pattern for a hypothetical year. In this year, all of the proxy records used indicate positive anomalies except the Indian record. Because the Indian proxy only records positive anomalies it is given a value of zero. The sum of the various proxy values is therefore +6. With such unison between the proxies it would be highly likely that these records document a year in which there were very strongly anomalous conditions characteristic of an El Niño year.

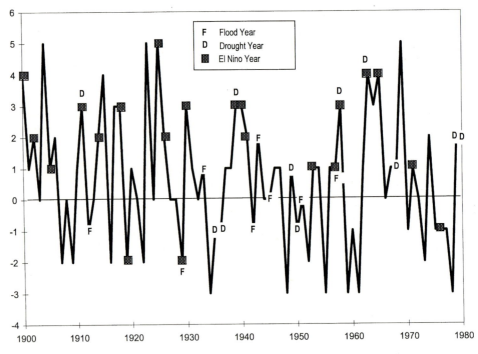

Fig. 12.2 The intensity of the ENSO composite index for the period 1900-1980. Flood, Drought and El Niño years (after Quinn et al. 1978) are marked. The record of flood and drought is based on the gauging records from seven stations in northern and central Thailand (after Godley 1997)

It has been noted (Godley 1997) that the impact of SST and SOI anomalies on Thai peak discharges appears to be strongest when these anomalies have persisted for several years. Given that this pattern appears on the inland rivers of Thailand, it seems reasonable to conclude that the same will hold on the other rivers of the region. Consequently, when SST and SOI anomalies have persisted for an extended period, there is a greater chance that the flood hydrology across inland SE Asia will be accurately represented by the composite ENSO record. It is therefore informative to examine conditions over extended periods rather than individual years. A five-year period was chosen and, with this in mind, the composite of the eight El Niño chronologies has been used in two ways.

Firstly, a five-year running 'Drought Index' (I) has been constructed using the sign of the composite data, whereby:

$$I = \frac{(D-F)/(F+D)}{N_0+1}$$

I   =   Drought index
D   =   Number of positive values of the composite index over five years.
F   =   Number of negative values of the composite index over five years.
N_0 =   Number of zero anomaly years in the composite index over five years.

The use of proportional indices is quite common in the analysis of climatic data, both in its own right and as a means to compensate for changes in the total number of records. Chu (1926), for example, used the index $I = D/F$. Chu's index is, however, poorly defined when F is zero or when there is a large difference between the values of D and F. To compensate for this, Brooks (1949) and Gong and Hameed (1991) have used $I = F/(F+D)$ multiplied by a constant. While formulae of this type work for binary functions, they are not suited to the tertiary function (positive, negative and zero) being used here.

The index, I, is a reflection of the continuity of anomalies over successive, running five year periods. If five successive years have positive or negative values, the index value is 1 or -1 respectively. Under such conditions it is highly likely that the period in question represents a spell of more flood- or drought-prone conditions. A modern analogy to this is the prolonged El Niño event of the early 1990s. Between 1990 and 1995 SSTs in the central Pacific remained high and this led to a persistence of the Southern Oscillation (Trenberth and Hoar 1996). Between 1990 and 1993 gauging data show that the peak discharges of all of the major Thai rivers were among the lowest this century (Godley unpublished data). Smaller values of I, on the other hand, reflect the presence of both signs in the five-year period. A spell of abnormally dry or wet weather under such conditions seems less likely. As a result, the drought index I can be viewed as an indication of the probability that a spell of abnormal hydrological conditions will occur, with the sign indicating flooding (-) or drought (+). The absolute value of I does not, however, give an indication of the severity or intensity of the flood- or drought-prone spells thus identified, only indicating that such a spell is likely.

In order to extract the intensity of these abnormal spells, it is necessary to use the original eight data series in a slightly different way. In years when there is

Peter Kershaw, Bruno David, Nigel Tapper, Dan Penny and Jonathan Brown (Eds): Bridging Wallace´s Line

agreement across the eight data series, the intensity of the SOI and SST anomalies should be greater than is the case when there is disagreement (i.e. some series having positive anomalies while others have negative anomalies). If this is the case, and it seems reasonable to assume that it is, then the degree to which the sign of the anomalies are uniform will be an indicator of the strength or intensity of the ENSO event. The sum of the eight proxy anomaly signs for a given year can therefore be used as an indicator of the intensity of SOI and SST anomalies. The same five-year running mean has been applied to the 'intensity' data as was applied to the drought index.

Because a five-year running mean has been applied to the data and it is difficult to identify the timing and statistical significance of trends in the data thus modified, the ENSO composite is expressed here as a cumulative deviation from the mean value. This is a well accepted statistical tool used to counter the difficulties associated with running means. It is important to point out that the actual values of this deviation are meaningless. It is the direction, slope and points of inflection of the cumulative deviation curve that are important to the following discussion.

A 500 Year Proxy Flood/Drought Chronology

The cumulative deviation from the mean annual sign shows quite clearly that the composite 500 year ENSO series can be divided into three distinct phases (Figure 12.3) suggesting three dominant flood-drought regimes during the past 500 years. The transition from one phase to another is marked by an inflection of the slope on the cumulative deviation curve. The first phase, from the early 16th to the mid 17th centuries, corresponds to a period of positive slope of the cumulative deviation from the mean value of I. The second, from the mid 17th to the late 18th centuries, corresponds to a period of sustained negative values of I. In the final phase, from the early 19th century to the end of the series, there is a return to a positive slope in the cumulative deviation from the mean value of I.

Phase 1 (Early 16th to Mid 17th Centuries)

During the first phase, corresponding to a positive slope in the curve of the cumulative deviation from the long term mean value of I, the rate of increase in the cumulative deviation gradually increases because it contains more positively anomalous (dry) years than the 500 year average. For the slope to increase as it does, the relative proportion of positive to negative years must also be increasing. This trend is clearly visible in the drought index (I) values through this period (Figure 12.3). There are four points when the value of I reaches 1.00. Each of these points corresponds with peaks in the intensity (Figure 12.4) of the El Niño signal (unison between all of the proxies) suggesting four very dry spells. While the sign of the drought Index (I) for the most part remains positive, the intensity of the El Niño signal throughout this phase tends to fluctuate between highly positive and highly negative (Figure 12.4) suggesting that short-lived wet spells lasting a few years occurred during this period dominated by low flood conditions.

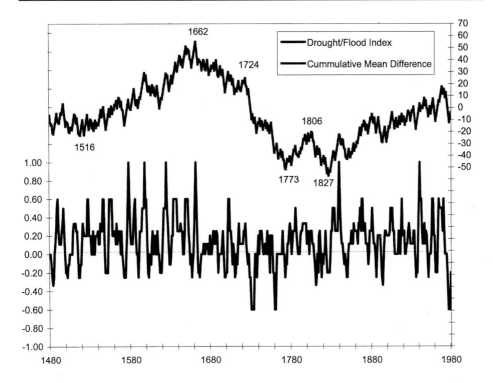

Fig. 12.3 The Drought Index 'I' for the period 1480 to 1980 showing the cumulative deviation from the mean value of I (top) and values of I based on a running five year period (bottom). Dates mark significant points of inflection in the cumulative deviation.

This initial phase is characterised by the dominance of years with positive anomalies across most of the El Niño proxies suggesting generally dry summer monsoon seasons and consequently drought-prone conditions throughout SE Asia. There were probably regular periods during which these positive anomalies persisted for several years. As these periods of persistent positive anomalies coincide with the intensification of the El Niño signal it is likely that during this first phase there were regular failures of the monsoon rains which would have resulted in a significant decrease in the peak stages of the rivers across the region. This overall drought-prone regime was interspersed with relatively short-lived wet and more flood-prone spells lasting probably no more than a few years at a time.

This first phase ends abruptly in the mid 17th century with a point of inflection in the cumulative deviation from the mean (Figure 12.3). The sharp change from positive to negative slopes marks a rapid, rather than gradual, shift in the sign of the annual anomalies. T-tests have been applied to the intensity data to confirm the significance of this change following the approaches of Yan *et al.* (1992 and 1993) and Shi and Zhu (1996) (Figure 12.4). The test has been applied as two 50-year sliding windows. The student's t-test applied in this fashion is an expression of the probability that two consecutive 50-year periods come from the same population. The point at which the test value first drops below the 5% level marks the boundary between two statistically distinguishable populations. The year 1652 marks one such statistically significant change (Figure 12.4) and this date falls between the two shoulders (1650 and 1662) which mark the point of inflection in the cumulative deviation from the annual mean El Niño intensity (Figure 12.3). The close correspondence between the point of inflection and the t-test results strongly suggests that the mid 17th century marks a major change in ENSO regimes and, by inference, hydrological regimes.

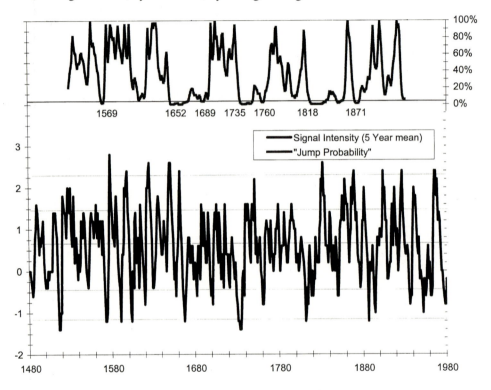

Fig. 12.4 The intensity of the constructed El Niño signal 1480 to 1980 showing the five year running mean intensity of the combined proxy signal (bottom). The 500-year mean and 1σ and 2σ values based on the actual (non-normal) probability distributions are marked. The results of sliding t-test values on successive 50-year periods are used to test the statistical likelihood that the two periods come from a population with the same mean (top). When the t-value initially falls below 5% (p>0.05), it marks the boundary between two statistically different populations.

Phase 2 (Mid 17th to Late 18th Centuries)

The data imply that a sharp and rapid change to flood-prone conditions began in the mid 17th century and lasted until the late 18th century. This change is expressed by a general decline in both the values of I and the intensity of positive anomalies. There are two possible explanations for this pattern. First, it is possible that it represents a period of rapid fluctuation between strongly positive and strongly negative anomalies. If the period of fluctuation is less than five years, the averaging effect of the five-year means would result in weakly positive values because the range and distribution of possible values of I is skewed slightly to the positive. Second, it is possible that the distributions represent a genuine absence of strong positive anomalies and an increase in the number of negative anomaly years. If this second explanation is true, the mid 17th to late 18th centuries should coincide with a reduction in the number and relative strength of El Niño events. Quinn *et al.*'s (1978) ranked chronology (which was not used in the construction of the chronology series) shows an overall reduction in both the number and intensity of El Niño events between 1650 and 1800 AD (Figure 12.5). Thus, it supports the supposition that this phase was characterized by a general reduction in the frequency and intensity of El Niño events. While it is difficult to conclude from the combined chronology presented here that conditions were particularly flood-prone, the reduction in the number of El Niño events taken in conjunction with the lack of persistent positive anomalies suggests that conditions are unlikely to have been particularly dry. There is, however, some circumstantial evidence which implies that this second phase, particularly the mid 18th century, may well have been more flood-prone than the preceding period.

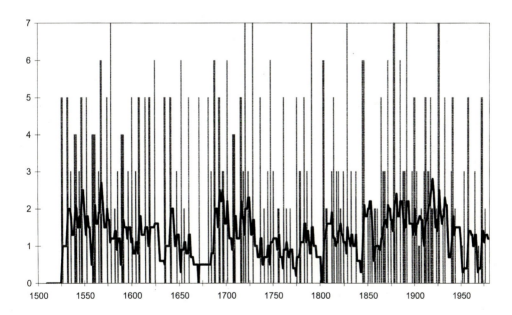

Fig. 12.5 Intensities of El Niño events between 1500 AD to 1978 AD (0 = no event to 7 = very strong event) with the five year mean intensity (black line). El Niño event intensities are those of Quinn et al. (1978). A decrease in the number and intensity of El Niño events is apparent during the mid 17th and mid 18th centuries.

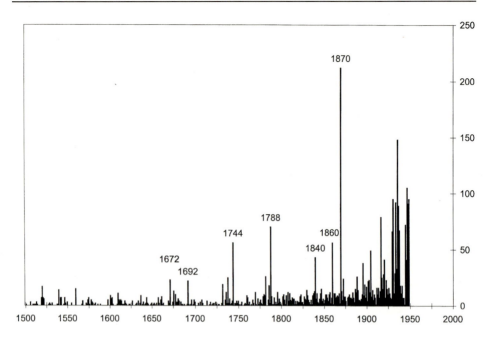

Fig. 12.6 The number of records of flood 1500-1949 AD from the Chiangjiang, Yalong, Minjian, Tuojiang, Jialing and Wujiang Rivers of Sichuan Province (southern China) shows that between the mid 17ᵗʰ and early 19ᵗʰ centuries there was an increase in very well recorded (and presumably very large) floods in the Province. The increase in the number of records of flood from the late 19ᵗʰ century onwards makes the later part of the data difficult to interpret (collated from descriptions of flood events in Shui li pu Ch'ang-chiang shui li wei yüan hui et al. 1993).

Between 1724 and 1773 AD there is a sharp decrease in the cumulative deviation from the mean, reflecting a period of strongly negative I values (Figure 12.3). Approximately one third of all negative values of I for the whole series occur in the hundred years between 1680 and 1780 AD. The intensity signal shows a similar over-representation of negative values (Figure 12.4). While the low I and intensity values taken by themselves are not conclusive, when combined with the increased occurrence of very well recorded floods in Sichuan Province (which suggests either extreme or widespread flood events) (Figure 12.6), peaks in Guangdong Province of the number typhoon landings and bumper harvests[2] and a decline in famine (Figure 12.7), there appears to be

---

[2] The correlation between the numbers of both typhoon landings and bumper harvests is significant to the 5% level (Zheng 1988). While typhoons can be associated with massive damage and loss of life, they are also associated with abundant rainfall. In years when typhoon landings are high, the destruction of crops in the direct path of the typhoon is countered by prolific rainfall over a much greater area. The increased rainfalls mean that poor harvests are less likely, resulting in decreased occurrence of famine, and good harvests over a much greater area.

at least *prima facie* evidence that this middle phase was considerably wetter than the preceding one. Further supportive evidence comes via dendro-climatological studies of the Hengduan and Tianshan Mountains of SW China which again suggests that the mid 17th to mid 18th centuries was a major wet period (Wu *et al.* 1988; Wu and Zhan 1991). In short, the bulk of the evidence clearly implies that this second phase was wetter than the previous one. If the wet seasons of this phase were considerably moister, as these records suggest, it is likely that flooding was also more frequent.

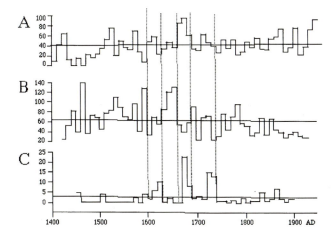

Fig. 12.7 Decadal numbers of
a) Typhoon landings,
b) Famine records and
c) Bumper harvests from Guangdong Province (coastal southern China) between 1400 and 1900 (after Zheng 1988).

During the overall depression of the values of I which is characteristic of this phase, there are two short periods during which low values of I correspond with an intensely negative signal. The concurrence of both implies a strong and persistent La Niña-like signal during the 1730s and 1830s (Figures 12.3 and 12.4). Under such conditions severe flooding would have been both more likely and more severe. While appropriate historical documents from Thailand during this period are scarce, there are several well-dated descriptions of flood events in this phase. The Nan chronicle records that the city of Nan was subjected to severe flooding in August 1817 (Wyatt 1994) and early Chakri Dynasty (1781 to present) records include accounts of devastating floods during 1785 (van Beek 1995) and 1831 AD (Hallet 1890; United States Department of State 1969). Descriptions of the 1785 flood suggest stage heights that have never been surpassed (Sapphaisan and Siwattana 1986).

Phase 3 (Early 19th Century to Present)

The start of the third phase marks the transition to the present day flood regimes. Two inflection points in the cumulative deviation from the annual mean in 1773 and 1827 AD (Figure 12.3) form shoulders around a significant change in the running t-test which occurs in 1818 (Figure 12.4). This range of dates combined with the fact that the change in slope of the cumulative deviation is more gradual, suggest that the end of the very wet second phase was a gradual one which probably started some time in the late 18th century and was complete by the early 19th century. Like the first phase, the final phase exhibits a predominance of positive values of I, indicating more positively anomalous (dry) years. Unlike the first phase however, there was less persistence of the dry spells suggesting that this final phase is probably wetter than the first one.

The ENSO-based reconstruction suggests that during the past 500 years the hydrological regimes of SE Asia can be divided into three broad phases during which the wet season rains were either more likely to have been particularly strong or, alternatively more likely to have failed. These changes in the long term character of the wet season rains would have led to corresponding changes in the frequency of flood events. The combined Pacific rim proxy created here suggests that prior to the mid 17th century, El Niño events dominated. This would have led to a reduction in wet season rainfall and following from this, a general decline in the occurrence of large magnitude floods. The overall predominance of El Niño-like conditions ended rapidly in the mid 17th century when a change to fewer El Niño and a greater number of La Niña events occurred. It seems highly likely that this change resulted in regular and probably severe flooding of the inland rivers. On time-scales longer than 500 years, data with both the resolution of the Pacific rim proxies and a strong connection to ENSO are not readily available. In order to reconstruct flood and drought regimes over longer time scales it is necessary to turn to other sources. The most obvious source applicable to SE Asia is the corpus of Chinese historical records.

The Longer Record of Flooding from Chinese Historical Sources

The Chinese recorded weather patterns cosmologically, as well as economically-important events and therefore it is possible to examine long chronologies of both climatological and hydrological phenomena. Like ENSO records, these can also be used to suggest a chronology of SE Asian hydrological regimes. In keeping with the ENSO theme, the focus remains on those phenomena and regions that have been shown to exhibit a teleconnection to ENSO, thereby exploiting the ENSO-Indochina flood hydrology link which has already been established. Although the distribution of flood and drought across China is not uniform, there are three patterns that dominate. These are a) wet (dry) across all of China, b) wet (dry) in the north and dry (wet) in the south, and c) dry (wet) in the north and south and wet (dry) in the middle and deltaic reaches of the Chiangjiang and Huanghe River basins. All three of these patterns appear to be linked to hemispherical circulation The strongest links have been found between pattern B and the strength of the monsoon (Shi and Zhu 1996) and between pattern C and the El Niño cycle (Zhang *et al.* 1989; Huang 1992; Tang and Guo 1993; Shi and Zhu 1996), although both patterns B and C can be indirectly associated with aspects of the ENSO cycle. In short, during El Niño years there is a greater likelihood of extreme flooding in the Chiangjiang and Huanghe basins and extreme drought in the north and south of the country. During the opposite ENSO phase the reverse is true. This suggests that flood-drought extremes in Indochina will tend to occur in phase with those of northern and southern China and out-of-phase with those of the Chiangjiang and Huanghe basins. Clearly, this is not likely to be accurate on an annual level as there are many other variables involved, but the 500 year record created above suggests a long period oscillation in the frequency of the two ENSO phases, and it is on this time-scale that these Chinese records may reflect flood and drought regimes in inland SE Asia.

Utilizing over 50,000 reports of flood and drought spanning the past 2000 years, Gong and Hameed (1991) examined the relative proportions of flood and drought records from three discrete climatic sub-regions of eastern China. While the semi-arid north-western region used by Gong and Hameed (1991) falls outside of the area of northern China which exhibits the strongest out-of-phase correlation with ENSO (Whetton *et al.* 1990), it provides the best data currently available over the last millennium and the area does at least partially overlap with the area of strong correlation with ENSO. The wet and semi-wet sub-regions, on the other-hand, include those reaches of the Huanghe and Chiangjiang basins within which the occurrence of severe rainfall exhibits an in-phase relationship with ENSO (Zhang *et al.* 1989; Huang 1992; Tang and Guo 1993; Shi and Zhu 1996).

An indication of flooding for Southern China comes from the accounts of flooding in Sichuan Province from the Chiangjiang, Yalong, Minjian, Tuojiang, Jialing and Wujiang Rivers (Shui li pu Ch'ang-chiang shui li wei yüan hui *et al.* 1993). Extracted from this collection is the number of records describing flooding during each year and the number of rivers recorded as being in flood. The flood record from this part of China is likely to be a good parallel for Indochina for two reasons. Firstly, it is the closest of the Chinese Provinces that also has an extensively preserved literature. Secondly, the basins are located well inland and at the edge of the reach of tropical cyclones (Meteorological Department 1983; McGregor 1995) so that changes in cyclone frequency and intensity are likely to influence flood levels in the same manner as has been proposed for inland SE Asia.

There is one key problem with the use of historical accounts of flood and drought such as those described above. Exaggeration of the extent or severity of disasters in official reports could be used as a means of gaining tax relief from the central government (Edkins 1903; Wang 1973; Zelin 1984). Frequently such requests were used to cover-up corruption and self-interest among the same local administrators who were reporting the 'disaster' and therefore the reports do not always provide an accurate reflection of the true damage or extent of a disaster (Zelin 1984; Smith 1991). It is expected that the long term averaging of the Gong and Hameed (1991) data will minimize the affects of any exaggeration, but a useful test of the reliability of the flood/drought data is to use other meteorological phenomena which are not typically subject to 'institutionalized' exaggeration. The records of two phenomena that can be linked indirectly to variability in winter pressure systems are therefore also used. These are dust-fall and winter thunderstorms.

In extracting over 1000 accounts of dust-fall since 300 AD, Zhang (1984) showed that there is a highly significant negative correlation between the frequency of dust-fall and Chu's (1973) reconstruction of temperatures. Unusually cold weather, associated with increases in dust-fall, can be caused by a strengthening and southward migration of the Siberian high pressure system (Bao 1987; Domrös and Peng Gongbing 1988). This migration creates steeper pressure gradients and increased wind strengths, which are more favourable for the transfer of dust. It is possible, at several levels, to suggest a link between the position and strength of the pressure systems over China and the ENSO cycle. First, inter-annual changes in the intensity of Chinese winters suggest a hemispherical oscillation of pressure systems. Second, the dust-falls tend to be restricted to the Chiangjiang and Huanghe basins (Zhang 1984). This implies that the area is dry because the dust would be washed out of the air if this was not the case. Third, Tang *et al.* (1989) have found that, in the summer following one of these unusually cold winters, there is an increased likelihood of severe floods occurring in the north of China. An oscillatory pattern which produces unusually wet conditions in the north of China and unusually dry conditions in middle and deltaic reaches of the Chiangjiang-Huanghe rivers is highly suggestive of ENSO (Shi and Zhu 1996). In short, this suggests that dust-fall occurrence will be in phase with wetter conditions in Indochina because the dust-fall is controlled by variability in the regional pressure systems which are, in turn, related to ENSO.

Winter thunderstorms, on the other hand, are relatively rare events and were seen by Chinese astrologers as abnormal or evil (Wang 1980). Because winter thunder is both a rare event and was seen as a portent of things to come, all occurrences of winter thunder that were heard are believed to have been recorded (Wang 1980). As for dust-fall, there is an apparent connection between temperature and the occurrence of winter thunder. High frequencies of winter thunder tend to be associated with colder weather (Wang 1980). Strong, quickly moving cold fronts cause instability in the air mass, leading to an increased chance of thunderstorm formation. The colder and more defined the body of air is, the greater the likelihood of thunderstorms. Wang suggests (1980) that the position and strength of the Siberian high, through its impact on temperatures, is a key factor in the occurrence of winter thunder.

Most of records of winter thunder were written in the lower Huanghe and Chiangjiang basins (Wang 1980) and this is simply a factor of the greater population in this area. Indirectly at least, records of winter thunder from the Huanghe and Chiangjiang basins imply that the strong and quickly moving cold fronts which cause the thunder have moved south into areas with higher population densities. If this was not the case, the thunder would not be heard and therefore would not be recorded. In short, records of winter thunder appear to be an indirect measure of the strength and position of the pressure systems that dominate winter weather in China. Tentatively at least, the occurrence of both winter thunder and dust-fall can be viewed as a reflection of the strength and position of the winter pressure systems over China. It seems reasonable to assume, albeit tentatively, that long-term variation in the numbers of recorded dust-falls and winter thunder through time should reflect variation in these pressure systems and hence flood regimes in SE Asia.

Inferred SE Asian Flood/Drought Regimes between 500 and 1500 AD

It is clearly possible to identify several long-term trends across all of the long Chinese records discussed above (Figure 12.8). While it would be easy to dismiss a single record, the general concordance across all of these records supports the inference that they are all being controlled by broad and regionally pervasive phenomena. Three multi-century phase changes are apparent across the data between 500 AD and 1500 AD. These phases are discussed below along with tentative implications for the wet season hydrology in SE Asia.

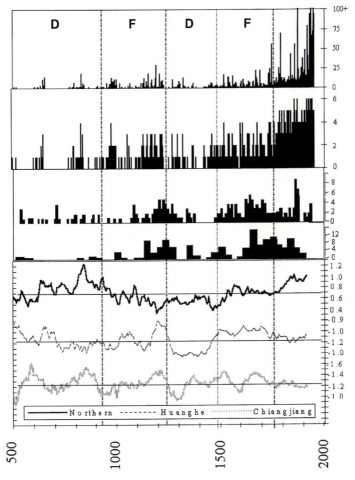

Fig. 12.8 The correspondence between the Chinese flood/drought indicators for the past 2000 years showing century scale periods of higher likelihood of flood (F) and drought (D) in Southeast Asia. From top to bottom: yearly number of flood records from Sichuan, yearly number of Sichuan rivers recorded as being in flood, decadal dust-fall occurrence (after Wang 1980), 30-yearly winter thunder occurrence (after Zhang 1984) and the seventy-five year average moisture indices for eastern China (after Gong and Hameed 1991).

6th to 10th Centuries

From 500 to 1000 AD there are comparatively few records available and so it is difficult to separate the effects of poor document preservation from true changes in the flood/drought trend. Even with limited data, a few trends emerge. First, and most obvious, is the general paucity of records that would suggest heightened flooding in Indochina. While it is quite possible that this is due solely to the poor preservation of records, Gong and Hameed's data from the semi-wet Huanghe reaches (which has the strongest known ENSO component of the various series used) show that there are proportionally more records of flood than there are of drought (Figure 12.8). If El Niño is the phenomena driving this flooding it implies that conditions in Indochina would have been dry.

Within this tentatively identified dry phase, a number of relatively short peaks, lasting between 50 and 100 years, represent periods of increased flooding. If it can be assumed that the Sichuan record bears the strongest relationship with the true conditions in Indochina given its proximity to the region, then the data suggest that the 7th and 9th centuries could have been wetter than the average for the 500 years of this phase as a whole. This is at least partially supported by Gong and Hameed's (1991) series (Figure 12.8). Similarly fluvial gravels at the base of a six meter alluvial sequence in Central Thailand (Bishop *et al.* 1994, 1996) have been dated to the 8th century and, given the resolution and the limited nature of the available flood data, these two records appear to correspond suggesting an increase in flooding between the 7th and 9th centuries AD.

10th to Mid 13th Centuries

With the exception of the semi-arid series (Figure 12.8), all the records begin to show an increase in the likelihood of increased flooding from the mid to late 10th century. The exact timing of this increase varies between the series, but is generally bound by the 10th and 14th centuries. If the political situation in China is taken into account, the peaks in dust-fall, winter thunder and Sichuan records of flood may, in fact, under-represent the true conditions, as this period corresponds with the Mongol invasion of northern China. Consequently, record keeping may have been poorer in the effected areas, resulting in fewer documents being written and a greater proportion being destroyed. If northern records have been destroyed then the original number of records could have been considerably greater than is implied by the number of occurrences of winter thunder and dust-fall that are preserved. The implication is that this period could have been one of regular and extreme flood events.

Mid 13th to 16th Centuries

Although the timing of an onset of drier conditions varies slightly across the series, by the start of the 14th century all of the Chinese data series are indicative of dry conditions in SE Asia. This change is marked by a rapid decline between 1250 and 1300 AD in the number of records of dust-fall and winter thunder. At the same time there is a corresponding increase in the proportion of records of flood in the Huanghe and Chiangjiang basins. This change is most pronounced in the semi-wet Huanghe delta sub-region (Figure 12.8) which has the strongest in-phase relationship with ENSO. While there is an apparent increase in the number of flood reports in Sichuan Province during the early 14th century, only 32 years between 1250 and 1450 AD have records of flood, and in all but seven of those years the records come from a single river (Figure 12.8). The preservation of multiple records of a single river in flood may suggest good preservation of documents rather than widespread flooding.

At the very least, these Chinese data suggest that a major and striking change to the climatological regimes of the region occurred at the start of this period. The general paucity of flood records from Sichuan and semi-arid northeastern China in combination with the high proportion of floods in the Huanghe delta strongly suggests that this change was ENSO related and therefore should correspond to a general desiccation of Indochina. As the records of dust-fall and winter thunder also show this pattern, it seems unlikely that the changes observed in the flood and drought based records are artifacts of either preservation or 'institutionalized' exaggeration. This dry phase appears to end during the 16th century. Given the coarser resolution of this series of records, a 16th century end to this period of increased desiccation is not inconsistent with the patterns identified in the 500 year ENSO records presented above.

Discussion

The reconstruction of regional hydrological regimes presented here suggests several long term cycles of flood-dominated and drought-dominated regimes in the SE Asia region since 500 AD. The data are inferred from a wide variety of records that all indirectly record the character of ENSO. The impact of these regime changes is likely to have produced drier-than-normal wet seasons during the 6th to 10th centuries and wetter-than-normal wet seasons between the 10th and 13th centuries and corresponding changes to flood frequencies. From the 13th to 15th centuries there is strong evidence for a significant and long-term desiccation of the wet season in Indochina which ended in the 16th century. Such an interpretation has a number of implications for our understanding of SE Asian history, a few of which are now outlined.

The distribution of Mon-Khmer and Thai sites, particularly in Thailand and Cambodia, has been used to suggest that these two ethnic groups preferred different physical environments (Vella 1986; van Liere 1989). The reconstruction of flood regimes presented here may help explain this pattern of settlement location. The general consensus has been that the Mon and Khmer preferred higher ground away from flood-prone areas. Indeed this has been explicitly used to explain why the Mon and Khmer apparently avoided the marshy plains of the lower Chao Phraya River (Vella 1986) and why Sukhothai is placed so far away from a major river (van Liere 1989). The most extreme form of this position is held by van Liere (1989) who contends that 'the banks of major rivers ... were not occupied at all by the Mon or Khmer peoples' and that 'this is not so remarkable, since monsoon rivers are unpredictable and their banks unfit for early settlement'. While this statement seems over-dramatic, during a period of

Peter Kershaw, Bruno David, Nigel Tapper, Dan Penny and Jonathan Brown (Eds): Bridging Wallace´s Line

regular extreme flooding many riverbank sites might have posed an unacceptable risk and might therefore have been avoided. The reconstruction of broad
flood frequencies carries with it an implicit increase in rainfall during those
periods of increased flood risk. As a result upland Khmer settlements like
Sukhothai, in what is currently one of the driest Provinces in Thailand, may well
have been far wetter than is implied by the present meteorological regime.

This image of Khmer-era flood regimes is starkly contrasted with the reconstructed hydrological record for the 13th to 15th centuries. During this phase,
monsoon rains were likely to have been regularly poor resulting in conditions
which were predominantly dry and only infrequently subjected to severe flooding. Such a reconstruction paints a considerably different picture of the hydrological environment than does Ramkhamhaeng's late 13th century statement
about his Kingdom of Sukhothai on the central Thai plain. While the line 'There
is rice in the fields and fish in the water' (Chamberlain 1991) is often quoted,
even under present hydrological conditions the nearby Yom River regularly runs
dry during part of the year (David Godley, unpublished data). While Sukhothai
is not actually located on the Yom River, the city is only fed by two small
streams both of which have only limited catchment areas and, depending on the
availability of groundwater, the city may have faced regular shortages of water.
It has been suggested that Sukhothai had to import rice (Kasetsiri 1973), and this
may be a reflection of crop failure caused by poor or unreliable rainfall. The
lack of a reliable water supply is implied and under these conditions it may have
been difficult to keep rice in the fields and fish in the water every year. The
lines from Ramkhamhaeng's inscription have often been read as a statement of
the prosperity and bounty of his Kingdom at Sukhothai. Alternatively of course,
the contradiction between hydrological reconstruction and inscription could be
used to argue against the authenticity of the inscription[3].

A general 13th to 15th-16th century desiccation of the SE Asian landscape
may, in part, help account for the demise of the Sukhothai Kingdom. Unlike
most other major cities that are all located on the banks of large rivers,
Sukhothai is quite isolated from a permanent and reliable year-round source of
water. Although their ultimate purpose is not clear, there are a number of major
engineering works in the Sukhothai hinterland that would have funneled water
towards the city (Godley *et al.* 1993) and therefore could have functioned to
supplement the water supply of the city. Under the later weak Kings, the source
areas for what amounts to a network of canals had fallen to 'foreign' powers,
leaving water supplies at Sukhothai at their most vulnerable. While it is always

---

[3] There has been quite a lively debate on the matter of the authenticity of the
Ramkhamhaeng Inscription (cf. Samakhom 1990; Chamberlain 1991). While the
veracity of the arguments for and against are far beyond the scope of this work, of
particular interest is the opinion that the inscription is an imitation that was composed
during the early 19th century (Krairiksh 1991a, 1991b). If it is indeed an early 19th
century composition, the author of the inscription would have had no means of
knowing that the late 13th century was, in all probability, extremely dry. An early 19th
century visitor to the region would have been there at the very end of a La Niña
dominated wet phase. At this time, the conditions around Sukhothai may well have
included abundant rice in the fields and fish in the water.

precarious to invoke simplistic environmental determinism, the lack of a reliable and locally controlled source of water during this particularly dry phase may have at least contributed to the decline of Sukhothai. Similar might be said of the late 13th century decline of Pagan in Upper Burma. It has long been argued that Pagan's prosperity depended on a climate wetter than the modern one (Mackenzie 1913) although it must be added that there have been those who have argued against it (Luce 1940; Cooler 1997). Again, like at Sukhothai, an extensive system of weirs, dams and diversionary barricades (Stargardt 1990) built and designed during the earlier wetter phase may have been insufficient after an onset of desiccation in the 13th century.

Stott (1992) recently concluded that 'South East Asia [is a place] where human-water, not human-land, relationships define geography'. The correspondence between broad shifts in flood regimes with the rise and fall of Kingdoms suggests that like geography, the history of the region has been influenced by those very same human-water relationships. Expanding on this idea Rigg points out that 'Water has long been recognized as an important linking theme in the study of South East Asia. The history of the region has been moulded by the accessibility afforded by the sea and waterways. ... Water is also the pivotal element in the cultivation of ... rice. The rituals and songs which dwell upon the vagaries of the monsoon ... all indicate the central role that the gift of water plays in the life of farming families in the region' (Rigg 1992: 1). Given this pattern it should not be a surprise that the cycle of flooding could become an integral part of the Thai psyche as Jumsai would have it for of the features of the Thai environment, it is water that closely resembles their spiritual ideals: "Of all the natural elements, water manifests the essence of change and unpredictability. For the Thai it is an unpredictability that lies within the limits of reasonable expectation ... While the natural environment would seem to be permanent, the Buddhist concepts of impermanence and the universal cycle of change are ever present in the depths of Thai consciousness." (Rutnin 1989: 243)

Acknowledgements

The research reported here is based on work supported by the Australian Research Council, Monash University, the Hydrology Division of the Royal Irrigation Department (Thailand) and the National Research Council of Thailand. Ian Rutherfurd and Sultan Hameed both kindly shared data with me. Paul Bishop suggested numerous improvements to the text.

References

Aiko, T. 1985 A climatological study of typhoon formation and typhoon visits to Japan. *Meteorology and Geophysics,* 36: 61-118.

Bao, C. 1987. *Synoptic meteorology in China*. China Ocean Press, Beijing: 269 pp.

Bao, C. and Xiang, Y. 1993. Relationship between El Niño event and atmospheric circulation, typhoon activity and flooding. In: W. J. Kyle and C. P. Chang (Editors) *Proceedings of the 2nd International Conference on East Asia and Western Pacific Meteorology and Climate*. World Scientific, Singapore: 239-49.

Berlage, H.P. 1957 Fluctuations of the general atmospheric circulation of more than one year, their nature and prognostic value. *Koninklijk Nederlands Meteorologisch Instituut Mededelingen en verhandelingen,* 69: 152.

Bishop, P., Hein, D. and Sutthinet, T. 1994. Twelve centuries of occupation of a river-bank setting: old Sisatchanalai, northern Thailand. *Antiquity,* 68: 745-57.

Bishop, P., Hein, D. and Godley, D. 1996 Was medieval Sawankhalok like modern Bangkok: Flooded every few years but an economic powerhouse nonetheless? *Asian Perspectives,* 352: 119-54.

Brooks, C.E.P. 1949. *Climate through the ages.* Benn, London: 395 pp.

Chamberlain, J.R. (Editor) 1991. *The Ram Khamhaeng Controversy: Collected Papers.* The Siam Society, Bangkok: 575 pp.

Chan, J.C.L. 1985. Tropical cyclone activity in the northwestern Pacific in relation to the El Niño/Southern Oscillation phenomenon. *Monthly Weather Review,* 113: 599-606.

Chan, J.C.L. 1995. Prediction of annual tropical cyclone activity over the western north Pacific and the south China Sea. *International Journal of Climate,* 15: 1011-9.

Chang, W.Y.B. and King, G. 1994. Centennial climatic changes and their global associations in the Yangtze River Chang Jiang Delta, China and subtropical Asia. *Climatic Research,* 4(2): 95-103.

Chu, C-C. 1926. Climatic pulsations during historical times in China. *Geographical Review,* 162: 274-83.

Chu K-C. 1973. A preliminary study on the climatic fluctuations during the last 5,000 years in China. *Scientia Sinica B,* 16: 226-56.

Chung yang ch'i hsiang ch'u ch'i hsiang k'o hsüeh yen chiu yüan 1981. *Chung-kuo chin wu pai nien han lao fen pu t'u chi,* Pei-ching: Ti t'u ch'u pan she (Yearly charts of wetness and dryness for the last 500 year period) (In Chinese).

Cole, J.E., Shen, G.T., Fairbanks, R.G. and Moore, M. 1992. Coral monitors of El Niño/Southern Oscillation dynamics across the equatorial Pacific, in H. F. Diaz and V. Markgraf (Editors) *El Niño, Historical and Paleoclimatic Aspects of the Southern Oscillation.* Cambridge University Press, Cambridge: 349-375.

Cooler, R. 1997. Sacred Buildings for an Arid Climate: Architectural evidence for low rainfall in ancient Pagan, *The Journal of Burmese Studies,* 1: 19-44.

De Boer, H.J. 1951. Tree ring measurements and weather fluctuations in Java from A.D. 1514, *Proceedings of the Koninklijke Nederlandse Akademie van Wetenschappen,* B54: 194-209.

Domrös, M. and Peng Gongbing. 1988. *The Climate of China.* Springer-Verlag, Berlin: 361 pp.

Edkins, J. 1903 (Reprinted 1980). *The Revenue and Taxation of the Chinese Empire.* Garland Publishing, New York: 249 pp.

Godley, D. 1997. *Flood Regimes in Northern Thailand: An Inter-disciplinary Approach* Unpublished M.Sc. Thesis, School of Geography and Environmental Science, Monash University.

Godley, D., Bishop, P. and Supajanya, T. 1993. Recent data on Thanon Phra Ruang between Sukhothai and Sisatchanalai: road or canal? *Journal of the Siam Society,* 81(2): 98-108.

Gong, G. and Hameed, S. 1991. The variation of moisture conditions in China during the last 2000 years. *International Journal of Climatology,* 11: 271-83.

Hallett, H.S. 1890. *Thousand Miles on an Elephant in the Shan States.* William Black-wood and Sons, Edinburgh: 484 pp.

Hamilton, K. and Garcia, R.R. 1986. El Niño/Southern Oscillation events and their associated midlatitude teleconnections. *Bulletin of the American Meteorological Society,* 67: 1354-61.

Hastenrath, S. 1987 Predictability of Java monsoon rainfall anomalies: a case study. *Journal of Climate and Applied Meteorology,* 26: 133-41.

Huang, J-Y. and Wang, S-W. 1985. Investigations on variations of the subtropical high in the western Pacific during historical times. *Climatic Change,* 7: 427-40.

Huang, R. 1992. The East Asia/Pacific pattern teleconnection of summer circulation and climate anomaly in East Asia. *Acta Meteorologica Sinica,* 61: 25-37.

Inter Press Service. 1998. *Nature: Thai Drought Prediction Shows Need for Disaster Planning.* (By-line: Teena Gill, February 4, 1998)

Jumsai, S. 1989(a). *Naga: Cultural Origins in Siam and the West Pacific.* Oxford University Press, Singapore: 183 pp.

Jumsai, S. 1989(b) Oceanic origins of Thai culture. In: *Culture and Environment in Thailand: A Symposium of the Siam Society.* The Siam Society, Bangkok: 7-21.

Kasetsiri, C. 1973. *The Rise of Ayudhya: A History of Siam in the fourteenth and fifteenth Centuries.* Ph.D. thesis, Cornell University. Oxford University Press, Kuala Lumpur: 254 pp.

Khandekar M.L. 1991. Eurasian snow cover, Indian monsoon and El Niño/Southern Oscillation – a synthesis. *Atmosphere-Ocean,* 294: 636-47.

Khanittana, W. 1989. The order of the natural world as recorded in Tai languages, In: *Culture and Environment in Thailand: a Symposium of the Siam Society.* The Siam Society, Bangkok: 233-42.

Krairiksh, P. 1991(a) Towards a revised history of Sukhothai Art: A reassessment of the Inscription of King Ram Khamhaeng. In: J.R. Chamberlain (Editor) *The Ram Khamhaeng Controversy – Collected Papers.* The Siam Society, Bangkok: 53-159.

Krairiksh, P 1991(b) An epilogue to the Ram Khamhaeng inscription. In: J.R Chamberlain (Editor) *The Ram Khamhaeng Controversy – Collected Papers.* The Siam Society, Bangkok: 553-65.

Kiladis, G. and Sinha, S.K. 1991. ENSO, monsoon and drought in India. In: M.H. Glantz, W. Katz and N. Nicholls (Editors) *Teleconnections Linking Worldwide Climate Anomalies.* Cambridge University Press, Cambridge: 431-58.

Kong Phumi´akat and Kong Utuniyomwitthaya Uthok 1989. Ekkasan Wichikan Kann-suksa pariman nam fon thi mi phon to kankoet uthakphai nai Prathet Thai (Study of the effect of rainfall amount on flood occurrence in Thailand). Krom Utuniyom-witthaya, Bangkok (in Thai).

Kripalani, R.H., Singh, S.V., Panchawagh, N. and Brikshavana, M. 1995. Variability of the summer monsoon rainfall over Thailand: comparison with features over India. *International Journal of Climatology,* 15: 657-72.

Lander, M.A. 1994. An exploratory analysis of the relationship between tropical storm formation in the Western North Pacific and ENSO. *Monthly Weather Review,* 122: 636-51.

Li, C. 1988. Actions of typhoons over the western Pacific Including the South China Sea and El Niño. *Advances in Atmospheric Sciences,* 5: 107-15.

Lloyds List. 1996. *Frequency of Flooding Likely to Increase in the Future; Extensive Floods are Projected to Continue in the Future.* Lloyd's Information Casualty Report. October 18, 1996.

Luce, G.H. 1940. Economic Life of the Early Burman. *Journal of the Burma Research Society,* 30: 283-335.

Mackenzie, J.C. 1913. Climate in Burmese History. *Journal of the Burmese Research Society,* 3: 4-46.

McGregor, G.R. 1995 The tropical cyclone hazard over the South China Sea 1970-1989. *Applied Geography,* 151: 35-52.

Meteorological Department. 1983. *Thangdoen khong phayu mun khet ron nai Prathet Thai lae boriwen klaikhiang nai khap 30 pi Pho. So. 2494-2523* (Tracks of tropical cyclones over Thailand and its neighboring area during the 30 year period 1951-1980). Ministry of Communications, Indonesia. (In Thai)

Murphy, J.O. and Whetton, P. 1989. A reanalysis of a tree ring chronology from Java. *Proceedings of the Koninklijke Nederlandse Akade,* B92: 241-57.

Nicholls, N. 1981. Air-Sea interaction and the possibility of long range weather prediction in the Indonesian Archipelago. *Monthly Weather Review,* 109: 2435-43.

Nicholls, N. 1993. ENSO, drought and flooding rain in South-East Asia. In: H. Brookfield and Y. Byron (Editors) *South-East Asia's Environmental Future: The Search for Sustainability.* Oxford University Press, Kuala Lumpur: 154-75.

Parthasarathy, B., Sontakke, N.A., Munot, A.A. and Kothawale, D.R. 1990. Vagaries of Indian Monsoon rainfall and its relationships with regional/global circulations. *Mausam,* 41: 301-8.

Phiansap, P., Thosangkhahathisalun, S., Phengyai, T. and Insawang, S. 1990. *Utthokphai an nuang ma chak Phayu Son Ron Ira rawang wan thi 3-5 Tulakhom 2533,* Krom Utuniyomwitthaya (Floods from Tropical Storm Ira between 3-5 October 1990). (In Thai)

Quinn, W.H. 1992. A study of Southern Oscillation-related climatic activity for A.D. 622-1900 incorporating Nile River flood data. In: H.F. Diaz and V. Markgraf (Editors) *El Niño: Historical and Paleoclimatic Aspects of the Southern Oscillation.* Cambridge University Press, Cambridge: 119-49.

Quinn, W.H. and Neal, V.T. 1992. The historical record of El Niño events. In: R.S. Bradley and P.D. Jones (Editors) *Climate since AD 1500.* Routledge, London: 623-48.

Quinn, W.H., Zopf, D.O., Short, K.S. and Yang, K. 1978. Historical trends and statistics of the Southern Oscillation, El Niño and Indonesian droughts. *Fisheries Bulletin,* 76: 663-78.

Ramage, C.S. and Hori, A.M. 1981. Meteorological aspects of El Niño. *Monthly Weather Review,* 109: 1827-35.

Rigg, J. 1992. The gift of water. In: J. Rigg (Editor) *The Gift of Water: Water Management, Cosmology and the State in South East Asia.* School of Oriental and African Studies, London: 1-6.

Sapphaisan, C. and Siwattana, T. 1986. *Kanpongkun namthuam lae kanrabai nam khong Mahanakhon.* Mahawitthayalai Kasetsat (Flood prevention and water retention in Bangkok). (In Thai)

Samakhom, S. 1990. *Khamaphiprai ruang silacharuk Sukhothai lak thi 1* (Discussions on Sukhothai inscription number 1). The Siam Society, Bangkok. (In Thai)

Shi, N. and Zhu, Q. 1996. An abrupt change in the intensity of the east Asian summer monsoon index and its relationship with temperature and precipitation over east China. *International Journal of Climatology,* 16: 757-64.

Shui li pu Ch'ang-chiang shui li wei yüan hui, Ch'ung-ch'ing shih wen hua chü and Ch'ung-ch'ing shih po wu kuan pien. 1993. *Ssu-ch'uan lian ch'ien nien hung tsai shih liao hui pien,* Pei-ching: Wen wu ch'u pan she (Historical materials on floods in Sichuan for 2000 years). (In Chinese)

Smith, P.J. 1991. *Taxing Heaven's Storehouse: Horses, Bureaucrats, and the Destruction of the Sichuan Tea Industry, 1074-1224.* Council on East Asian Studies, Harvard University, Cambridge: 489 pp.

Stargardt, J. 1990. *The Ancient Pyu of Burma (Volume 1).* PACSEA (Cambridge) in association with the Institute of SE Asian Studies (Singapore), Cambridge: 462 pp.

Stott, P. 1992. Angkor: Shifting the hydraulic paradigm. In: J. Rigg (Editor) *The Gift of Water: Water Management, Cosmology and the State in South East Asia.* School of Oriental and African Studies, London: 47-58.

Tanaka, M. 1978. Synoptic study on recent climatic change in monsoon Asia and its influence on agricultural production. In: K. Takahashi and M. Yoshino (Editors) *Climate Change and Food Production.* University of Tokyo Press, Tokyo: 81-100.

Tang, M, Tianshi, L., Jian, Z. and Cunqiang, L. 1989. The operational forecasting of total precipitation in flood seasons April to September of 5 years 1983-1987. *Advances in Atmospheric Sciences,* 63: 289-300.

Tang, Y. and Guo, L. 1993. Research on drought/flood influence factors in China. *Chinese Geographical Science,* 31: 34-43.

Thompson, L.G. 1992, Ice core evidence from Peru and China. In: R.S. Bradley and P.D. Jones (Editors) *Climate since A.D. 1500.* Routledge, London: 517-48.

Thompson, L.G., Mosley-Thompson, E., Grootes, P.M., Pourchet, M. and Hastenrath, S. 1984(a). Tropical glaciers: potential for ice core paleoclimate reconstructions. *Journal of Geophysical Research,* 893: 4638-46.

Thompson, L.G., Mosley-Thompson, E. and Morales Arnao, B. 1984(b). Major El Niño/Southern Oscillation events recorded in stratigraphy of the tropical Quelccaya ice cap. *Science,* 226(4670): 50-2.

Trenberth, K.E. 1997. The definition of El Niño. *Bulletin of the American Meteorological Society,* 78: 2771-7.

Trenberth, K.E. and Hoar, T.J. 1996. The 1990-1995 El Niño-Southern Oscillation Event: longest on record. *Geophysical Research Letters,* 231: 57-60.

United States Department of State. 1969. *Records of the Department of State Relating to Internal Affairs of Siam 1910-1929.* National Archives and Records Service, Washington D.C.

van Beek, S. 1995. *The Chao Phya: River in Transition.* Oxford University Press, Kuala Lumpur: 210 pp.

van Liere, W. J. 1989. Mon-Khmer approaches to the environment. In: *Culture and Environment in Thailand: A Symposium of the Siam Society.* The Siam Society, Bangkok: 143-60.

Vella, W.F. 1986. The cultural boundaries of water usage: reservoirs versus canals in the Chao Phraya Plain. In: R.D. Renard (Editor) *Anuson Walter Vella.* University of Hawaii, Honolulu: 92-7.

Vijayakumar, R. and Kulkarni, J.R. 1995. The variability of inter-annual oscillations of the Indian summer monsoon. *Advances in Atmospheric Sciences,* 12: 95-102.

Walker, G.T. 1910. On the meteorological evidence for supposed changes of climate in India. *Memoirs of the Indian Meteorological Department,* 21: 1-21.

Wang, P-K. 1980. On the relationship between winter thunder and the climatic change in China in the past 2,200 years. *Climatic Change,* 31: 37-46.

Wang, S. 1990. El Niño events from 1470 to 1988. *Meteorological Monthly,* 154: 15-20. (In Chinese)

Wang, Y-C. 1974. *Land taxation in Imperial China, 1750-1911.* Harvard University Press, Cambridge: 192 pp.

Whetton P. and Rutherfurd, I. 1994. ENSO teleconnections in the Eastern hemisphere over the last 500 years. *Climatic Change,* 28: 221-53.

Whetton P., Allan, R. and Rutherfurd, I. 1996. Historical ENSO teleconnections in the eastern hemisphere: Comparisons with latest El Nino series of Quinn. *Climatic Change,* 321: 103-9.

Whetton, P., Adamson, D. and Williams, M. 1990. Rainfall and river flow variability in Africa, Australian and East Asia linked to El Niño-Southern Oscillation events. In: P. Bishop (Editor) *Lessons for Human Survival: Nature's Records from the Quaternary.* Geological Society of Australia Symposium Proceedings No. 1: 71-82.

Wiraphan, W. 1994. *Kanpramoen phon kanphayakon nam pi 2535,* Krom Utuniyom-witthaya (Flood forecasting evaluation in 1992), Bangkok. (In Thai)

Wongwitwat, P. and Nongnat, U. 1991. *Fonlaeng nai Prathet Thai*. Krom Utthunniyom Witthaya (Drought in Thailand), Bangkok. (In Thai)

Wu, G. and Lau, N.C. 1992. A GCM simulation of the relationship between tropical storm formation and ENSO. *Monthly Weather Review*, 120: 958-77.

Wu, X. and Zhan, X. 1991. Tree ring width and climatic change in China. *Quaternary Science Reviews*, 10: 545-9.

Wu, X., Lin, Z and Sun, L. 1988. A preliminary study on the climatic change of the Hengduan mountain area since 1600 AD. *Advances in Atmospheric Sciences*, 54: 437-3.

Wyat,, D.K. 1994. *The Nan Chronicle*. Cornell University Press, Ithaca.

Xinhua News Agency. 1987. *Thailand Hit by Worst Drought in Decade*. July 24, 1987.

Xinhua News Agency. 1996. *Floods Death Toll Rises to 88 in Thailand*. October 22, 1996.

Xu, Q. 1995. Analysis of causes and seasonal prediction of severe floods in Yangtze/Huaihe Basins during summer 1991. *Advances in Atmospheric Sciences*, 12: 215-24.

Yamamoto, T. 1978. Some remarks on the relationship between the climate of Japan and that of tropical Asia, Australia and East Africa. In: K. Takahashi and M. Yoshino (Editors) *Climate Change and Food Production*. University of Tokyo Press, Tokyo: 47-62.

Yan, Z., Li, Z.Y. and Wang, X.C. 1993. An analysis of decade- and century- scales climatic jumps in history. *Scientia Atmospherica Sinica*, 174: 359-67.

Yan, Z., Ye, D. and Wang, C. 1992. Climatic jumps in flood/drought historical chronology of Central China. *Climate Dynamics*, 6: 153-60.

Yang, S. and Xu, L. 1994. Linkage between Eurasian winter snow cover and regional Chinese Summer Rainfall, *International Journal of Climatology*, 14: 739-50.

Zelin, M. 1984. *The Magistrate's Tael*. University of California Press, Berkeley: 385 pp.

Zhang, D. 1984. Synoptic-climatic studies of dust fall in China since historic times. *Scientia Sinica B*, 278: 825-36.

Zhang, J. and Chen, Z. 1987. The operational seasonal forecasting of the summer rainfall in China. *Advances in Atmospheric Sciences*, 4: 349-62.

Zhang, X., Song, J. and Zhao, Z. 1989. The Southern Oscillation reconstruction and Drought/Flood in China. *Acta Meteorologica Sinica*, 33: 290-301.

Zheng, S. 1988. Climatic change and its effect on food production during the period 1400-1949 in Guangdong Province In: Z. Jiacheng (Editor) *The Reconstruction of Climate in China for Historical Times*. Science Press, Beijing: 138-144.

PART III

THE PEOPLING OF SUNDA AND SAHUL

Early Hominids in SE Asia:
Older, Younger, Smarter and More

Ian Walters

Introduction

Just over a hundred years ago, Eugene Dubois, a medical anatomist graduate from the Netherlands, set out to find the fossil 'missing link' between modern humans and what had come to be thought of, in 19th century intellectual circles, as 'our animal past'. Charles Darwin was of course the giant influence, but Dubois had been as much influenced by his mentor and teacher, Ernst Haeckel. Unlike Darwin, Haeckel had speculated that any missing link in the evolution to modern humans would be found in the South East or South Asian region. From comparative anatomy, and using what was known, archaeologically and palaeo-environmentally, as a tree of life, Haeckel placed human ancestry in a category of, as translated in the English language edition of his work, Ape-man. To this classification he gave the Linnaean nomenclature: ape-human without speech, *Pithecanthropus alalus*. Dubois aimed to find this missing link and sought funds to do so. Without luck in that regard, he joined the Netherlands East Indies army as a medical officer and was duly sent to Sumatra. Immediately he began fossil hunting. Dubois then applied for, and was granted a transfer to Java. It was there, between 1889 and 1891, that he found the fossils which were to make him famous.

Dubois' original discoveries at Trinil were followed by finds on the Solo River at Sangiran, Sambungmachan and other sites in central and east Java. These finds were described by a group of scholars: Weidenreich, von Koenigs-wald, Marks, and the Indonesian students Jacob and Sartono. The last two were to become the doyens of indigenous Indonesian scholarship in this field, training more recent analysts who now carry, with international collaborators, the burden of work on these sites and collections.

ISBN 3-923381-47-6
© 2002 by CATENA VERLAG, 35447 Reiskirchen

Morphology of early hominoids

Mainstream descriptive analyses of hominoid anatomy are made from a variable, but generally similar, set of fossils. These fossils are seen as diverse, yet possessing an identifiable set of attributes. Rightmire (1990) identifies traits which are clearly derived from earlier *Homo* and Australopithecines. These attributes include 'a heavy brow, midline keeling and parietal tori, strong flexion of the occiput and development of a prominent transverse torus, features of the cranial base and expansion of cranial capacity' (Rightmire 1990: 201). Specimens are seen as having a cranial capacity a little over 1000 ml, and have a longish, thick boned vault with marked post-orbital constriction, its greatest breadth lying at the base (Howells 1980). Skulls possess a low frontal with a pronounced and continuous supraorbital torus. Sagittal keeling is sometimes present. The occipital is sharply angled, with the upper plate relatively small. The face is short but heavily constructed, and relatively wide (Howells 1980). They are upright walkers with relatively robust post-cranial skeletons. There are a variety of features such as a P_4 diastema and presence or absence of sagittal keeling, which distinguish some specimens. However, much of the diversity of form is related to size, a situation which causes single lineage advocates to reject implications in these differences for specific or generic status other than a generalised *Homo erectus*. Nevertheless, as Rightmire (1990: 178) says, 'there is a lot of variation'.

Taxonomy

Rightmire (1990) presents a case for *H. erectus* being a real taxon. He and other colleagues, adhering to the mainstream interpretation of pre-modern hominids in SE Asia, believe occupation to have involved a single species. It was Dubois who established the idea of such a species, believing he had discovered Haeckel's missing link. Weidenreich and many of those who followed him were of a view that the assemblage of skeletal remains they found and worked on were from one taxon. Weidenreich (e.g. 1945) however, complicated things by conflating his single species with several Linnaean names such as *Homo*, *Pithecanthropus* and *Meganthropus*, which would normally be ascribed to genera and the species they contained. Oddly, he seems to have intended these as mere descriptors, rather than any implication of a diverse suite of taxa. Despite his intentions, this seems to have left something of a legacy for his colleagues and students, Jacob, Sartono and others, to continue this idea of ascribing multiple Latin names to the array of individuals and specimens that constitute the Javanese fossil assemblage (e.g. Marks 1953; Sartono 1961; von Koenigswald 1973). For some, these variable descriptions did imply separate genera or 'subgeneric status' (Tobias and von Koenigswald 1964: 517). Jacob and Sartono, and after them their own students, have shifted ground many times over the years, coming up with several different interpretations of hominid taxonomy in the region, often with an implied evolutionary sequence in mind (e.g. Jacob 1973).

Such interpretations have fitted nicely with the single species ideas of hominid taxonomists elsewhere. Prominent figures in hominid evolutionary theory and taxonomy, such as Thorne, Wolpoff, Frayer and Lovejoy, have all used the SE Asian material to support their single species interpretation of pre-modern hominid taxonomy and evolution (e.g. Lovejoy 1970; Thorne and Wolpoff 1981; Pope and Cronin 1984; Wolpoff *et al.* 1984).

There have been other views proposed. For example, Franzen, Tyler, this author, as well as the Indonesian scholars, have all put forward arguments for the possible presence of multiple taxa in the SE Asian record prior to the presence of modern hominids in the region (Franzen 1985; Tyler 1991, 1996; Walters 1996). Tattersall (e.g. 1986; see also 1993) has been a strong proponent of the important and interesting conclusion that we have to pay more attention to the possibility of multiple hominid species being present until quite late in the Pleistocene. Tattersall and his co-workers have shown, in one key example of careful descriptions of material from the SE Asian region, that this is the case for hominoids (Schwartz *et al.* 1994, 1995). My own quantitative work has recently supported this conclusion (Walters *et al.* 1998) and raised the inference that it could well be the case for hominids in the region as well (Walters 1998). Led by scholars such as Stringer and Tattersall, such views of a hominid taxonomy containing far greater diversity than previously thought now seem to be gaining more widespread acceptance (e.g. see Gibbons 1996, 1997 for a popular documentation of this).

Taxonomic classification of early hominoids comes down to the old arguments between splitters and lumpers. For the lumpers, there is but one world-wide species of pre-modern hominid. These scholars have to see Neanderthals as part of modern *H. sapiens'* variety, ignoring certain key archaeological indicators to the contrary. They also sometimes see sub-species, a level of classification abhorred by others as biologically meaningless. Some well-known splitters, such as Groves, are content with one *Homo* species prior to moderns, but with the caveat that it contains a number of sub-species (Groves 1989). At the extreme of the lumpers are scholars like Thorne who advocate the sinking altogether of *H. erectus* as a taxon, collapsing the entire schema into an evolving *H. sapiens,* which itself moved once out of Africa maybe as much as two million years ago (Thorne and Wolpoff 1981).

Martin and others (e.g. Martin 1991; Cope 1993; Martin and Andrews 1993; Tattersall 1993) have expended a deal of analytical and creative energy trying to sort the splitting lumping problems, suggesting control of some measure of baseline diversity is the way to proceed. This very interesting methodological approach involves the use of reference maxima, utilising collections of known and agreed upon taxonomy, and therefore measurable diversity. Tests are then run on quantitative measures from the fossil samples and tested against these reference sets. This allows for testing of the hypotheses that more than a single species exists, or does not exist, in the fossil samples. I have applied these methods to hominoids and pre-modern hominids from SE Asia and found the single species hypothesis wanting in each case. I remain convinced that the Martin methods lead to support of the ideas of scholars like Tattersall that more than one pre-modern hominid species existed, perhaps until quite late in the SE

Asian record. In contrast, leading scholars like Thorne remain as convinced as ever of the correctness of their position. To contradict Martin and his colleagues, other scholars have suggested that the reference maxima are exceeded because there was greater morphological diversity in the past due to factors such as more pronounced sexual dimorphism, as well as methodological inadequacies in the statistical tests of the single species hypothesis (Kelley and Plavcan 1998).

   This brings us to evolution, and the way taxonomy links with evolutionary schemes to characterise early hominids in the SE Asian region.

Chronology and evolution

Essentially, the evolutionary debate is a global one, juxtaposing the 'Out of Africa' theory with 'Multi-regional Evolution'. This point, in turn, must lead to a discussion of dating and chronology. The earliest attempts at constructing chronologies depended on comparative anatomy and correlation of strata. Following chronometric advances after World War II, the number of ascribed dates began to gradually increase. The latest scene is one of a vast array of dates from vastly different sources using a litany of methods. In a 1980s review of these, Pope (1983) treated older dates with suspicion, scrutinising them with a hygienic rigor not always attributed to younger ones. He concluded that no hominids in the region dated to older than about a million years before present and that the great bulk of them are considerably younger. Some of these old dates, however, have been recently re-assessed by one of the original workers who had produced the oldest date some decades ago. This new laboratory analysis by Swisher *et al.* (1994) indicates that the reality of a more ancient heritage of hominids may have to be acknowledged. Not only do a number of dates extend beyond the cluster at just less than a million years, but some appear to be twice as old as that. The establishment of dates at nearly 2 million (1.9 Ma and 1.8 Ma: Swisher *et al.* 1994) years before present is intriguing to say the least.

   Dates for the most recent pre-modern hominids in the region have been difficult to determine. There has been no shortage of conjecture about continuity of these pre-modern hominids with anatomically modern humans or of replacement by them, yet no clear dates for these events have emerged. Some relative dating has been inferred by use of comparative anatomy or morphological similarities. However, Swisher and his colleagues have again troubled us with a set of startling dates for the uppermost pre-modern specimens from Java (Swisher *et al.* 1996). These results place the latest pre-modern hominids in Java somewhere between 53,000 and 27,000 years BP, making them contemporaneous with *H. sapiens* in the region. This has very important implications for theories of replacement by moderns, as well as for gradual evolution of pre-moderns into moderns. It also makes the relationship of these most recent SE Asian early hominids to Pleistocene Australians a more complex one than may have previously been thought, especially since it seems Australia's oldest inhabitants may pre-date this considerably (Thorne *et al.* 1999).

Culture and material culture

There have been difficulties linking stone artefact use and early fossil hominids in SE Asia. Pope sought to overcome this by means of a hypothesis suggesting *H. erectus* in SE Asia was a bamboo tool user, not requiring stone because of the utility of the plant, and its distribution in the karst landscapes of the region where *H. erectus* had been found. There was a widespread belief, retained by many, that a lack of watercraft or ability to move beyond the shores of Sunda-land implied a lack of other human attributes in these early creatures. Lack of water crossing ability also had implications for the colonisation of Australia, suggesting the accepted view that only *H. sapiens* had the ability to do so, and crossed into greater Australia during the late Pleistocene.

There is, however, Morwood's and colleagues' new evidence from Flores (Morwood *et al.* 1998). This must have implications for what we know and previously believed about early SE Asian hominids. On Flores, strata, dated to some 800,000 years BP, contain what appear to be *in situ* stone artefacts. That the site's occupants appear to have been using stone artefacts on an island offshore of the Sundaland continent this early may well cause earlier notions that these early hominids lacked equivalent modern human capacities in language and material culture to be rejected. For them to have been there in that context, it is argued, implies the undertaking of significant water crossings from nearby Sundaland. It must therefore also have implications for conceptualisation and organisation at a level not generally attributed to hominids of that form. It has long been argued that water crossings of this magnitude were an attribute of modern *H. sapiens*, who colonised Australia and moved to other such places by virtue of their superior linguistic and intellectual abilities. Morwood's sites are sedimentary deposits near the shores of ancient lakes, so there may be arguments from taphonomists about movement of relatively recent stone artefacts downwards into older silts not yet petrified at the time, according to standard taphonomic processes. But this seems unlikely, and after full scrutiny it may well be sustained that Morwood and his colleagues have come across some quite exciting new material which forces significant re-thinking of the abilities and evolutionary place of these hominids.

Conclusion

The fossil material, including its provenance, biological and cultural status, is fraught with all sorts of uncertainties. Dating of the region's early hominids has not been sorted out to everyone's satisfaction. For early dates the issue is dependent on whether the new results of Swisher and his colleagues are accepted or not. But perhaps pre-modern hominids in SE Asia are older than we thought. The same may be said of the new terminal dates. The hominids may have survived until quite recently. Evidence now strongly suggests that they entered the region early, some two million years ago. It also seems at least some of their number retained a more robust ancestral morphology until more recently than 50,000 years BP, within the time frame of anatomically modern occupation. This in turn seems to imply cohabitation of at least two different species, one

being *H. sapiens*. The early dates, and maybe even the late dates, give prota-
gonists every reason to continue their arguments. However, while a two million
year time scale gives more time for evolution in the region, it also allows the
numerical evidence to indicate not simply a transition to gracility within a single
lineage, but rather a burst of speciation on the fertile shores of Sundaland.

There is still no agreement on the taxonomy of SE Asia's pre-modern
hominids. But then, there probably never will be. The single lineage adherents
remain convinced one species moved out of Africa and evolved in form through
to the present. Those who reject this view see at least a replacement of the
earlier species by anatomically modern humans, and at an even greater extreme,
the possible co-existence in the region of multiple species of hominid prior to the
eventual situation of *H. sapiens* prevailing. Models of evolution have to address
the complexity of the SE Asian material. But to date, none of them seem to have
given any significant ground, preferring instead to interpret the data in ways
which remain consistent with their personal evolutionary schemes. If we stick to
Popperian methods of judgement coupled with quantitative techniques, rather
than subjective assessments of similarity of skull shape and the like, we come up
with a best estimate that there were multiple pre-modern species involved.

The new discoveries in Flores raise questions about pre-modern SE Asians in
terms of humanity, intellectual capabilities and material culture, as well as
change in settlement pattern, migration, etc. If these creatures did indeed under-
take significant water crossings, they may well have had attributes of language
and material culture previously unexpected. It may be that pre-modern hominids
in SE Asia possessed intellectual capacities previously only attributed to their
successors.

Much remains unknown and certainly much more research is required. If
current trends in discoveries are continued, it may turn out that pre-modern early
hominids in the SE Asian region are older than previously thought. Similarly,
they may also have survived until much later than has previously been thought,
existing into the modern human period. In addition, pre-modern hominids may
be smarter than we were willing to acknowledge, and it is quite likely there were
more varieties of them.

References

Cope, D.A. 1993. Measures of Dental Variation as Indicators of Multiple Taxa in
    Samples of Sympatric *Cercopithecus* Species. In: W.H. Kimbel and L.B. Martin
    (Editors) *Species, Species Concepts and Primate Evolution*. Plenum Press, New
    York: 211-37.
Franzen, J.L. 1985. Asian Australopithecines? In: P.V. Tobias (Editor) *Hominid
    Evolution: Past, Present and Future*. Liss, New York: 255-63.
Gibbons, A. 1996. *Homo erectus* in Java: a 250,000-year anachronism. *Science,* 274:
    1841-2.
Gibbons, A. 1997. A new face for human ancestors. *Science,* 276: 1331-3.
Groves, C.P. 1989. *A Theory of Human and Primate Evolution*. Clarendon Press,
    Oxford: 384 pp.
Howells, W.W. 1980. *Homo erectus* – who, when and where: a survey. *Yearbook of
    Physical Anthropology,* 23: 1-23.

Jacob, T. 1973. Palaeoanthropological discoveries in Indonesia with special reference to the finds of the last two decades. *Journal of Human Evolution,* 2: 473-85.

Kelley, J. and Plavcan, J.M. 1998. A simulation test of hominoid species number at Lufeng, China: implications for the use of the coefficient of variation in paleo-taxonomy. *Journal of Human Evolution,* 35: 577-96.

Lovejoy, C.O. 1970. The taxonomic status of the 'Meganthropus' mandibular fragments from the Djetis beds of Java. *Man,* 5: 228-36.

Marks, P. 1953. Preliminary note on the discovery of a new jaw of Meganthropus von Koenigswald in the lower Middle Pleistocene of Sangiran, Central Java. *Indonesian Journal of Natural Science,* 109: 26-33.

Martin, L. 1991. Teeth, sex and species. *Nature,* 352: 111-2.

Martin, L.B. and Andrews, P. 1993. Species Recognition in Middle Miocene Hominoids. In: W.H. Kimbel and L.B. Martin (Editors) *Species, Species Concepts, and Primate Evolution.* Plenum Press, New York: 393-427.

Morwood, M.J., O'Sullivan, P.B., Aziz, F. and Raza, A. 1998. Fission-track ages of stone tools and fossils on the east Indonesian island of Flores. *Nature,* 392: 173-6.

Pope, G.G. 1983. Evidence on the age of the Asian Hominidae. *Proceedings of the National Academy of Sciences of the USA,* 80: 4988-92.

Pope, G.G. and Cronin, J.E 1984. The Asian Hominidae. *Journal of Human Evolution,* 13: 377-96.

Rightmire, G.P. 1990. *The Evolution of* Homo Erectus: *Comparative Anatomical Studies of an Extinct Human Species.* Cambridge University Press, Cambridge: 272 pp.

Sartono, S. 1961. *Notes on a New Find of a Pithecanthropus Mandible.* Publikasi Teknik Seri Paleontologi No.2, Direktorat Geologi, Bandung.

Schwartz, J.H., Long, V.T., Cuong, N.L., Kha, L.T. and Tattersall, I. 1994. A diverse hominoid fauna from the Late Middle Pleistocene breccia cave of Tham Khuyen, Socialist Republic of Vietnam. *Anthropological Papers of the American Museum of Natural History,* 73: 1-11.

Schwartz, J.H., Long, V.T., Cuong, N.L., Kha, L.T. and Tattersall, I. 1995. A review of the Pleistocene hominoid fauna of the Socialist Republic of Vietnam (excluding Hylobatidae). *Anthropological Papers of the American Museum of Natural History,* 76: 1-24.

Swisher, C.C.III, Curtis, G.H., Jacob, T., Getty, A.G., Suprijo, A. and Widiasmoro. 1994. Age of earliest known hominids in Java, Indonesia. *Science,* 263: 1118-21.

Swisher, C.C.III, Rink, W.J., Antons, S.C., Schwarcz, H.P, Curtis, G.H., Suprijo, A. and Widiasmoro. 1996. Latest *Homo erectus* of Java: potential contemporaniety with *Homo sapiens* in Southeast Asia. *Science,* 274: 1870-4.

Tattersall, I. 1986. Species recognition in human paleontology. *Journal of Human Evolution,* 15: 165-76.

Tattersall, I. 1993. Speciation and Morphological Differentiation in the Genus *Lemur.* In: W.H. Kimbel and L.B. Martin (Editors) *Species, Species Concepts, and Primate Evolution.* Plenum Press, New York: 163-76.

Thorne, A. and Wolpoff, M.H. 1981. Regional continuity in Australasian Pleistocene hominid evolution. *American Journal of Physical Anthropology,* 55: 337-49.

Thorne, A., Grun, R., Mortimer, G., Spooner, N.A., Simpson, J.J, McCulloch, M., Taylor, L. and Curnoe, D. 1999. Australia's oldest human remains: age of the Lake Mungo 3 skeleton. *Journal of Human Evolution,* 36: 591-612.

Tobias, P.V. and von Koenigswald, G.H.R. 1964. A comparison between the Olduvai hominines and those of Java and some implications for hominid phylogeny. *Nature,* 204(4958): 515-18.

Tyler, D.E. 1991. A taxonomy of Javan hominid mandibles. *Human Evolution,* 6: 401-20.

Tyler, D.E. 1996. The taxonomic status of the 'Meganthropus' cranium Sangiran 31, and the 'Meganthropus' occipital fragment III. *Bulletin of the Indo-Pacific Prehistory Association,* 15: 235-41.

von Koenigswald, G.H.R. 1973. *Australopithecus, Meganthropus* and *Ramapithecus. Journal of Human Evolution,* 2: 487-91.

Walters, I. 1996. *Meganthropus* and the hominid taxa of Java. *Bulletin of the Indo-Pacific Prehistory Association,* 15: 229-34.

Walters, I. 1998. *Size and Species: Implications of Vietnam Hominoids.* Paper presented at the Indo-Pacific Prehistory Association 16th Congress, 1-7 July, 1998, Melaka, Malaysia.

Walters, I., Long, V.T., Cuong, N.L. and Oxenham, M. 1998. Pleistocene hominoid dental variation in Vietnam. *Bulletin of the Indo-Pacific Prehistory Association,* 17: 93-103.

Weidenreich, F. 1945. Giant early man from Java and South China. *Anthropological Papers of the American Museum of Natural History,* 40: 1-134.

Wolpoff, M.H., Wu, X.Z. and Thorne, A. 1984. Modern *Homo Sapiens* Origins: A General Theory of Hominid Evolution Involving the Fossil Evidence from East Asia. In: F. Smith and F. Spencer (Editors) *The Origin of Modern Humans: A World Survey of the Fossil Evidence.* Liss, New York: 411-83.

Asia and the Peopling of Polynesia:
Understanding Sequential Migration

Nicola van Dijk and Alan Thorne

It is thought that modern *Homo sapiens*, along with most other placental mammals, did not occupy the Pacific region east of Wallace's Line in the Middle Pleistocene. Earlier forms of *Homo sapiens* (or *Homo erectus* for some scholars) were present in China and Indonesia before this time, as far east as Flores (Morwood *et al.* 1999). In the Late Pleistocene evidence of human colonisation is recorded from New Guinea, the Bismarck Archipelago and Australia. Occupation of Remote Oceania (east of the Solomons) did not begin until well after the end of the Pleistocene, 10,000 years BP.

Recently re-dated to around 60,000 years ago, Lake Mungo in SE Australia now contains the earliest sign of humans in Greater Australia (Thorne *et al.* 1999; Grun *et al.* 2000; but see Bowler and Magee 2000; Gillespie and Roberts 2000 for differing views). As Lake Mungo (western New South Wales) is some 4500 km from the NW corner of the expanded landmass that then included New Guinea and the Australian continent, and irrespective of coastal or inland approach, it seems safe to regard 70,000 years as a minimum earliest arrival time if the re-dating reliably dates the Lake Mungo 3 burial.

The morphology of the two well-known skeletal examples from the earliest times, the female Lake Mungo 1 and the male Lake Mungo 3 individuals, are characterised by an overall gracility (Thorne and Curnoe 2000). The skeletally contrasting robust morphology seen at such sites as Cohuna, Kow Swamp and the Willandra Lakes (the WLH50 individual) are all dated to less than 20,000 years ago, at least 25,000 and likely more than 40,000 years later than the gracile earliest morphologies known (depending on which dating one subscribes to). As Thorne and Curnoe (2000) point out, it is unlikely that gracility gives rise to robusticity in Australia as this would be a southern exception to the global trend over the last 200,000 years or so for increasing gracilisation. This suggests that the robust morphology is a relatively late arrival in Australia. This is supported culturally by the appearance of edge-ground tools in northern Australia (and in Japan) 30-20,000 years BP, shortly before robusticity appears in southern Australia (Thorne and Curnoe 2000). Both these morphologies, gracile and robust, have their origins in SE and/or east Asia (Thorne 1980).

More recently in Australia there is the appearance in the mid Holocene of microlithic tools, which probably signifies a new migration from somewhere in Asia, one that reflected the then state of Asian morphology. Within the last 1000 years there is clear evidence of further migrations to Australia from Indonesia (the Macassans) and New Guinea (the Torres Strait Islanders) which produce

ISBN 3-923381-47-6
© 2002 by CATENA VERLAG, 35447 Reiskirchen

new morphological and genetic changes to Aboriginal Australian variation (Kirk 1981).

The paucity of prehistoric skeletal remains from New Guinea and Island Melanesia prevents a direct comparison with Australia, but it is unlikely that New Guinea lacked at least some substantial populations at the same time that they were first appearing in Australia. Van Dijk (1998, in press) demonstrates links cranially between recent Papua New Guinea and Australian populations. What this says about skeletal history is hard to read without a substantial Pleistocene record, but at the least it indicates that there are firm and widespread physical links between the two landmasses. The lowland physical differences between southern New Guinea and northern Australia, at the time of European contact, may represent some morphological migration to New Guinea that was not represented in Australia, but once again a better skeletal record is required before conclusions can be drawn.

However, what Australia and New Guinea do say clearly is that this huge region has seen sequential migration of physically distinct peoples as the principal source of Pleistocene skeletal and genetic variation. For the last 100 years and more this has been the view of the great bulk of anatomists and physical anthropologists studying skeletal remains (see Thorne and Curnoe 2000 for a list). This is of course no surprise, given that Africa, Europe, east and west Asia and the Americas have all seen similar fusion of physically distinct migrating populations. What we should expect, given the 40,000 to 70,000 years of occupation that we know of (depending on which chronology one accepts), is that Australo-Melanesia reflects the changing morphologies in Asia over this long period. Just as Australian and New Guinean populations today reflect ancient migrations, they also reflect ancient migrations within Asia.

SE Asia over this long time span is thus not only the homeland for the original inhabitants of Australia, New Guinea and Island Melanesia tens of thousands of years ago, but it is also generally thought to be the origin point for the initial colonisers of Polynesia and Micronesia. This view has not been passed without some opposition from both an archaeological (Allen and White 1989; Terrell 1986) and biological (Houghton 1989a, 1990, 1991, 1996; Kelly 1996) standpoint. The latter argument is that the Polynesians are derived from populations which have been resident in Melanesia for 40,000 years. Houghton (1989a: 229) states that 'the dominant settlement of Polynesia was from Island Melanesia, by a group that was part of the human population of that geographic region; that is, they were not in transit through Melanesia from more distant parts'.

There is some osteological evidence to support these claims (Houghton 1989a, 1989b), but the majority is strongly against (Brace and Hinton 1981; Brace *et al.* 1990; Hanihara 1992; Howells 1973; Ishida 1993; Katayama 1988; Pietrusewsky 1990a, 1990b, 1996; Turner 1989). Most would agree with Howells' (1973: 228) frequently quoted statement that 'the Polynesians – and the Fijians by and large are near them in body form – are simply too different from anything in Melanesia to be derived therefrom in a few thousand years'.

The genetic evidence is no more unified than the skeletal. Some researchers are adamant that there is little evidence for a solely Melanesian origin for Polynesians: 'the extreme view ... that Polynesians evolved within Melanesia from a population resident there for at least 30,000 years is untenable in the light of the genetic evidence' (Serjeantson and Hill 1989: 287). This view is based on data both from Human Leucocyte Antigen (HLA) loci (Serjeantson 1989; Serjeantson and Gao 1995) and some mitochondrial DNA (mtDNA) analyses (Hertzberg *et al.* 1989; Stoneking *et al.* 1989). Many (Lum *et al.* 1994; O'Shaughnessy *et al.* 1990), argue that while Polynesians did not evolve from populations resident in Melanesia, Melanesians have made a significant contribution to the Polynesian gene pool. Serjeantson and Gao (1995: 165) for example support 'an ultimate East Asian origin for Polynesians but ... some degree of past gene flow from Island Melanesian populations'.

Hagelberg and Clegg (1993) have a slightly different view again. On the evidence of ancient mtDNA extracted from prehistoric human skeletal material from both Melanesia and Polynesia, they argue that Melanesians were in fact the first to arrive in the Central Pacific, followed, rather than preceded by, the SE Asians:

> it appears that the earliest inhabitants of the central Pacific ... may have originated in Melanesia ... [this] implies that the Lapita culture was carried from its Melanesian homeland into the central Pacific by indigenous inhabitants of Island Melanesia rather than by Austronesian-speaking migrants from Southeast Asia ... our results give little credence to the traditional view that the Lapita people were essentially Polynesians. (Hagelberg and Clegg 1993: 168).

The basis for their argument is that a SE Asian mtDNA marker (the 9 base pair deletion), found in almost 100 percent of modern Polynesians, is only present in very low frequencies from the Melanesian islands of New Britain and Vanuatu, and absent from the early samples from the islands of Fiji, Tonga and Samoa. Hagelberg and Clegg (1993) consequently believe that Polynesians have a recent SE Asian ancestry. Austronesian speakers carrying the marker settled Island Melanesia, coastal New Guinea, and finally Polynesia, where its high frequency and the lack of genetic variability in modern Polynesians can be explained by population bottle-necking. As the authors themselves note, the extremely small samples of Lapita material available for testing, as well as possible contamination problems render this conclusion a tenuous one.

It would probably be fair to say that the consensus biological view (Houghton excepted) is of a present-day Polynesian population of largely SE Asian phenotype, with at least some degree of admixture with Melanesians, before and possibly after their voyage through Island Melanesia on their way into Polynesia.

As with the section above, which dealt with the relationship between SE Asia and Australia, we see evidence that the Pacific has again thwarted conventional evolutionary explanations of phenotypic change. Paralleling Australia, the picture here is not of a robust colonising population evolving into a more gracile form over time, but rather the opposite. Whilst both SE Asians and Lapita people are generally agreed to be gracile (Katayama 1990, no date; Pietrusewksy 1989a; but see Houghton 1989a for an opposing view), Polynesians are amongst the world's most robust people. Where this story differs to that of the Australian one is in its time span. Evolution in Australia has taken place over a minimum of 40,000-60,000 years, in Polynesia however a maximum of 3000 years is available, in most places considerably less. If, as we are led to believe from the archaeological evidence, the initial colonisers of Polynesia were the Lapita people, and if, as the evidence suggests, there were considerable morphological differences between the latter and their supposed descendants (Katayama 1990; Pietrusewsky 1989a; Visser 1995), by what mechanism(s) has this phenotypic change taken place? While it is easy to invoke natural selection as a primary impetus for change when 40,000-60,000 years are available, it is difficult to imagine the extremes of circumstance which would promote such a rate of differential survival (the mechanism for natural selection) when there are less than 3000.

Some have tried to see natural selection as an explanation for the unusual size and robusticity of the Polynesians. The most notable of these is a theory put forward by Houghton (1990, 1991, 1996) which argues that the large size of the Polynesians is a consequence of selection for a more robust phenotype in order to survive the extreme conditions of oceanic voyaging. The problems of this theory have been outlined in detail elsewhere (van Dijk 1991, 1993, 1999) not the least of which is that from Lapita sites in Polynesia it has been found that the seafaring Lapita people living in this region were still small. If the robust phenotype had evolved in order to cope with the difficult conditions of voyaging surely this would have been evident by 2000 years BP; indeed one would assume, from the moment of arrival. The fact that it is not makes such an adaptation both unnecessary and untenable.

Below we review some of the skeletal and genetic evidence for links between SE Asia and Polynesia, and propose that the morphological changes evident between SE Asian and Polynesian populations, including the progression from gracile to robust, do not require complex theories of natural selection on a grand scale, but simply a combination of migration, founder effect and physiological responses to a new environment.

The skeletal evidence

So what skeletal evidence exists for affinities between the inhabitants of Polynesia and SE Asia? Difficulties in assessing this question include the paucity of early skeletal material in Polynesia and Lapita sites elsewhere, and at present little comparative work completed on anything other than recent museum samples from both regions.

In recent years Pietrusewsky (1992a, 1992b, 1994) has included much SE Asian material in his multivariate analyses. A multivariate craniometric study of Pacific-Asian relationships (Pietrusewsky 1994) used approximately 2500 crania from Oceania and Asia, including prehistoric samples from Bronze Age China (Anyang), Bronze Age Thailand, and some early Japanese material (including Jomon, Yayoi, Kofun, Kamakura and Endo periods). As was found in previous studies, two major biological groups – one Australo-Melanesian and the other Asian (including Polynesia and Micronesia) – emerged. Polynesians were found to be closest to SE Asians, particularly from island SE Asia (for example, Lesser Sundas, Sulawesi, and Sulu). However, no support was found for a Jomon link. No links between Pacific and Jomon people were found by Ishida and Dodo (1993) either (for a contrasting view see Brace and Hunt 1990; Brace *et al.* 1990; Katayama 1990). On the basis of craniofacial measurements from populations worldwide, Brace *et al.* (1990: 345) argue that contrary to expectation they found 'no basis for deriving the inhabitants of the small islands in the Remote Pacific from Southeast Asia ... [while] in Japan, the Early Jomon ... already displayed the same constellation of features now demonstrable throughout Micronesia and Polynesia'. However, some aspects of the methodology, including the use of many more measurements than individuals (see van Dijk 1998) and some of the results obtained (such as a closer association between Polynesians, Eskimos and Amerindians than with either Jomon or SE Asian populations), make these results subject to debate.

Pietrusewsky (1994) found a connection between the Bronze Age Chinese sample, modern Chinese from Taiwan and the Hainan Islands, and the Austronesian speaking Atayal from Taiwan. This, he argues, parallels the linguistic reconstructions (Sagart 1993) linking Old Chinese with Proto-Austronesian. Other Asiatic samples linked together include SE Asian, Japanese (not Jomon or Ainu) and an eastern and northern Asian branch. The main differentiation within the Asiatic populations is between northern (east Asia) and southern (SE Asia) groups. The Bronze Age Thai sample, although somewhat isolated, groups with island SE Asia. There are also similarities between samples from Vietnam, island SE Asia and the Atayal. Other studies (Pietrusewsky 1992a) reiterate the relative isolation of Jomon and Ainu populations within the Asian cluster, while modern Japanese are more closely related to mainland and island SE Asian samples than with Chinese or Mongolians. The latter is the most isolated and well-differentiated Asiatic group. Pietrusewsky (1992a: 48) concludes from these results that 'Southeast Asia may have served as the ultimate homeland of both Polynesians and modern Japanese'.

Taiwan (and adjacent mainland China) is a region frequently cited in the literature (e.g. Bellwood 1989, 1995) as a possible origin point for the Lapita people and the Polynesians, due to its reputation as the home of the Austronesian language family and the close correlation between the archaeological and linguistic records along the Lapita trail. In comparison to other Austronesian speakers, the aboriginal peoples of Taiwan are thought to be the most internally differentiated linguistically, culturally and biologically (Pietrusewsky 1992b).

Aboriginal Taiwanese conform to a generalised 'Mongoloid' phenotype: dark brown/black straight coarse hair, brown eyes, medium brown skin colour, epicanthic eye fold, concave nasal bridge, short stature and brachycephalic head form. Craniometric data presented by Pietrusewsky (1992b) however does not substantiate the claim for a Taiwanese origin for Polynesians. The Atayal, Taiwan Chinese, Hainan Chinese and Anyang form a separate cluster within a larger mainland and offshore Asiatic grouping. Japanese, northern Chinese from Manchuria, and Korean form a separate cluster which attaches to a homogeneous Chinese sub-branch. These results are similar to those obtained by Howells (1989, 1990) which also contained the Atayal sample. Pietrusewsky argues that these results also reinforce Bowles' (1977, 1984) north-south division based on anthropometrics and also to a lesser extent Turner's distinction between Sinodont and Sundadont dental morphology (Turner 1987, 1990). On the basis of this, Pietrusewsky concludes that given the present-day distribution of Austronesian speakers:

the Atayal sample is closest to crania from Viet Nam, the Philippines and Indonesia, areas occupied by speakers of the Western Austronesian division. There is no evidence in the cranial data to connect Atayal directly with Polynesians or the Oceanic division of this language Family. These biological data broadly parallel the linguistic classification of this language Family which has characterised the Austronesian languages of Taiwan as the most isolated and closest to the western division. (Pietrusewsky 1992a: 8).

The skeletal evidence therefore does little to suggest a direct link between the Taiwanese and Polynesians, although admixture subsequent to departure and the long time period available for colonisation and expansion must be taken into account.

Lapita remains

So what of the Lapita skeletal remains and their affinities between SE Asia on the one hand, and Polynesia on the other? The fragmentary skeletal material found in Lapita-age sites has done little to solve the question of their biological affinities. Skeletons representing approximately 17 individuals have been recovered from Lapita-associated archaeological sites, surprisingly few when one considers the large overall number of sites and the amount of pottery and other artefacts found within them. Sites have been located in the Bismarck Archipelago, New Caledonia, Fiji, and Tonga. All are extremely fragmentary, and in most cases their dating has been complex and subject to debate. Most dates are considerably later than the 3600 years BP cited as the beginning of the Lapita period, and long after the initial colonisation of Polynesia, so it is questionable whether any colonising 'Lapita people' have in fact been recovered. The majority of analyses on these remains have been undertaken by Pietrusewsky (1989a, 1989b, 1991, 1996; Pietrusewsky *et al*. 1997) and it is mostly his work that will be cited here.

The largest collection of Lapita-associated skeletal remains (eight individuals) is from a site on Watom Island off the northern tip of New Britain, dated to around 2500 years BP. No complete skulls were found, and most of the information recovered was from the four mandibles. While the remains were considered by Houghton (1989a) to be Polynesian on the basis of their moderately robust and bowed limb bones, costo-clavicular sulcus, tall statures, moderately shovel-shaped incisors and rocker mandibles, Pietrusewsky (1989a: 241), amongst others, argues that 'none of these features are exclusive to Polynesians but are generally typical of most Pacific populations'. In addition, while some features of the Watom sample are similar to those of Polynesians, there are also marked differences, including broad divergent rami and short mandibular bodies, gracile lower limb bones and small tooth crown diameters. These features, Pietrusewsky (1989a: 241) argues, are 'without parallel among the inhabitants of the Pacific'. The only reasonable explanation for this phenomenon relies on a combination of admixture and founder effect. Whoever the Lapita people were, they must have their origins somewhere in the wider Pacific and there is insufficient time or impetus for these features to be the result of an evolutionary adaptation. Houghton (1989a: 223) is alone in arguing that 'the morphology of this group displays the large-bodied characteristics of ocean voyagers and small-island dwellers' and that this 'places the Watom people firmly with other large-bodied ocean voyagers ... whose evident living representatives are the Polynesians' (Houghton 1989a: 230).

Results of most multivariate comparisons on the Lapita mandibles show them to be unique, and generally isolated from the main Pacific groups (i.e. Polynesia, Melanesia etc.). In a study based on five mandibular measurements, Houghton (1989a: 230) found that the closest groups to Watom were Hawaii and New Zealand, followed by Tonga, Namu and New Britain. He concludes that 'these analyses support the thesis of affinities both with groups now regarded as Polynesian or having Polynesian associations (Namu), and with recent Island Melanesian groups'. Pietrusewsky (1991) disagrees, noting that large samples generally indicate SE Asian affinities, making a Melanesian origin for Polynesians unlikely.

Turner's (1989) comparative analysis of the dentition from Watom Islands with other Pacific populations found the Watom teeth to be more like those from New Britain than Thailand or Hawaii, the latter two populations being more similar to each other. This evidence leads Turner to conclude that:

> the Watom dentition suggests that the Lapita wares found on this small island were associated with Melanesians. Given the nearness of Watom to New Britain, this finding is reassuring in the sense that we would expect Watom teeth to be most like those of the larger and more heavily populated New Britain. (Turner 1989: 296).

This appears to be a more realistic conclusion, owing not only to the geographic proximity of Watom Island to New Britain, but also considering that the remains are dated to at least 1000 years after the initial colonisation of Polynesia. There is the possibility that the ancestral phenotype of these individuals had been subject to over 1000 years of gene flow and drift in New Britain itself. In contrast, those examining the dental remains from Mussau Island in the Bismarck archipelago argued that 'an extremely tentative affinity assessment would be that the Mussau Lapita people were slightly more like Indonesians than like Melanesians on the basis of dental morphology' (Kirch *et al.* 1989: 76)

That different methods and different regions of the skeleton produce different results seems to be the rule with the Lapita remains, and indeed a disturbing trend in the Pacific in general (see Green and van Dijk in prep; Pietrusewsky 1984). The Natunuku skeleton, for example, appears to have Polynesian affinities on the basis of univariate comparisons of morphological details of the skeleton and associated teeth, whilst other aspects of morphology, limb proportions and leg bone shape suggest a Melanesian link. Multivariate analyses of mandibular measurements further suggest Melanesian affinities, while similar analyses of non-metric traits suggest Polynesian relationships (Pietrusewsky 1989b). Why the discrepancies in results are so large in these cases is unclear. It may have something to do with small sample sizes (less than five in most cases) and statistical errors arising from this, or possibly one type of data (e.g. metrics) is more subject to environmental influence than the other.

Visser (1995), using dental information from a collection of post-Lapita remains from Sigatoka, Fiji, dated to around 2000 years BP, argued that while the Sigatoka people showed strong affinities both with Lapita populations and

Polynesian populations to the east, there were significant skeletal differences between them and more recent Fijians. He concluded that the differences between the Sigatoka and more recent Fijian material is due to subsequent admixture in Fiji from the west, probably Vanuatu.

Recent research using cranial non-metric traits (van Dijk 1998) looked at the biological relationship of Island Melanesian populations to others within the wider Pacific. It was found that Polynesians were readily distinguishable from Island Melanesians, New Guineans and Australians. The Island Melanesians in fact appeared to be a surprisingly homogeneous group, with not even the archaeological/temporal division into Near and Remote Oceania causing significant separation between the populations. While this may merely be a reflection of more recent interactions, it seems unlikely that the Polynesians evolved solely from a resident Island Melanesian population.

The genetic evidence

Genetic data has also been somewhat inconsistent in pinpointing the origins of the Polynesians in SE Asia. Ranford (1989: 190-1) notes that the complement gene distributions for coastal Melanesians and Polynesians are almost identical, suggesting either that Polynesians were derived from a coastal Melanesian popu-lation or that there has been extensive gene flow between them. As to why the Polynesians passed through Melanesia without leaving any noticeable genetic input is a matter of some debate. One possibility is that the continued influx of New Guinean markers has diluted the original gene pool. Another possibility is that present-day Polynesians are derived from two distinct Austronesian migrant groups, one which interbred with Island Melanesians, while the other by-passed the Solomons and Vanuatu before arriving in Fiji (Hill *et al.* 1989).

The Melanesian island of New Guinea has been postulated by some to be a 'major staging ground for the ensuing settlement of Polynesia' (Melton *et al.* 1995: 22). Redd *et al.* (1995) and Melton *et al.* (1995) argue that the Asian marker, the 9 base pair deletion (9 b.p.), originated in Asia at about 60,000 years BP. This mutation occurs at almost 100% frequency in Polynesia. Various other substitutions followed and became widespread throughout SE Asia, Taiwan and by 6000 years BP, the Philippines and Indonesia. The final mutation probably occurred somewhere in east Indonesia and rose to a high frequency in coastal New Guinea. The trio of nucleotide changes occurring with the 9 b.p. deletion is present in 79.2% of Samoans, 73.9% of coastal New Guineans, and 20% of east Indonesians (Melton *et al.* 1995).   Control Region (CR) data from the same region show that the sequences from coastal New Guinea were almost as homogeneous as the Samoan CR sequences (Redd *et al.* 1995). Phylogenetic analysis of these data links Samoans and New Guineans with east Indonesians, Taiwanese and Filipinos (although the latter branches of the trees are rather short and therefore are possibly not statistically significant). The highest diversity for the Polynesian mtDNA types is in Taiwan, and thus genetic (along with linguistic) evidence indicates that this may be the source of proto-Austronesian expansion. The fact that the 9 b.p. deletion is nearly fixed in Polynesia indicates a probable population bottleneck or marked founder effect occurring in that region, or a series of them, and consequently a reduction in genetic diversity. As

Redd *et al.* (1995: 604) note, there is a marked increase in the frequency of the 9 b.p. deletion from west to east: 'the frequency ranges from 3-18% from Japan to mainland Asia to the Malaysian peninsula; from 8-42% in Island Asia …; and from 77-100% in Polynesia'. Sykes *et al.* (1995) also argue for the origin of one of the more minor Polynesian mtDNA lineages in New Guinea.

Mitochondrial DNA evidence from 1178 individuals in Polynesia (Sykes *et al.* 1995) indicates that the prehistoric settlement of Polynesia was from the west involving two, possibly three distinct populations. The predominant lineage, which shares a specific 9 b.p. deletion, accounts for 94% of Polynesian mtDNA and has related lineages in Taiwan, Indonesia and the Philippines. The authors conclude that this indicates a 'relatively recent major eastward expansion into Polynesia, perhaps originating from Taiwan, in agreement with the archaeological and linguistic evidence' (Sykes *et al.* 1995: 305). The second lineage group, which accounts for 3.5% of Polynesian mtDNA, in fact indicates affinities with Vanuatu and Papua New Guinea. Low overall diversity of these lineages indicates severe genetic bottlenecking during the colonisation of Remote Oceania. The third lineage group shows shared variants between Polynesia and the Philippines.

Similar results have been found by Lum *et al.* (1994: 567) who argue that 'Polynesians, traditionally considered to be a single cohesive linguistic and cultural unit, contain at least three distinct mitochondrial DNA (mtDNA) groups which probably shared a common maternal ancestor over 85,000 years ago'. Lum *et al.* (1994) conclude that during the colonisation of the Pacific, admixture took place between Melanesians and SE Asians in Near Oceania, and this complex assortment of genes was carried into the islands of Polynesia.

Hill *et al.* (1989: 275) state that there is clear evidence of both Melanesian and 'Mongoloid' ancestry in Polynesians:

> These data argue very strongly that ancestors of the present-day Polynesians must have, at some time, been present in island Melanesia. [This] fits very well with the now generally accepted route for the pre-Polynesians through island Melanesia. [The alternative] view – of a wholly Melanesian origin for the Polynesians – [is] untenable. The data argue rather for a basically Mongoloid population which has interbred to some considerable extent with Melanesians. (Hill *et al.* 1989: 275).

This seems the most likely scenario, although the rather remarkable comment by Ranford (1989: 191) that 'there is little evidence of Mongoloid markers penetrating into the Melanesian and Polynesian islands of the Pacific' must also be noted.

Turning SE Asians into Polynesians

Despite the plethora of so-called 'objective' biological methods of looking at Polynesian origins, such as non-metric and metric skeletal measurements or genetic analyses, it is sometimes useful to go back a step, look at the people living in these regions today and observe whom they actually resemble, just as the early explorers and missionaries did. This quickly and decisively defeats any theories of a solely Melanesian origin. Quotes abound as to the thoughts of early explorers when they first saw the Polynesians and Micronesians, but all appear to have had the same idea – the Polynesians came from the west, they came from SE Asia. Even Cook couldn't fail to see the strong resemblance between the

> inhabitants of the Sandwich Islands [which] are undoubtably of the same race with those of New Zealand ... and the Marquesas ... they bear strong marks of affinity to some of the Indian tribes, that inhabit the Landrones [Mariana] and Caroline Islands; and the same affinity may be again traced amongst the Battas [Bataks of northern Sumatra] and the Malays. (Cook and King 1785: 124-5).

Early researchers also noted the so-called 'Mongoloid' features of these groups. Coon (1965: 11), in a rather racist description of the 'Mongoloid' phenotype, notes that the hair is black and straight, they have high cheekbones and 'slant' eyes [epicanthic eye folds]. The teeth are large, with shovel-shaped incisors whilst the body has a long trunk and short limbs, with particularly short forearms and lower legs. Later researchers (e.g. Akazawa 1996: 5) describe the 'Mongoloids' as having 'a relatively extensive distribution of subcutaneous fat ... [and] a generally flat face'. It must be noted that there is great diversity within this group and, although not described in such terms, all the above features can be found in descriptions of the Polynesian phenotype (which Micronesians also closely resemble) (Houghton 1996; Howells 1973). Melanesians and Australians do not have these traits, so for Polynesians to have evolved out of a population resident in Melanesia for 40,000 years would require a complete re-evolution of this suite of features upon arrival. Given the lack of selection pressure and small time frame available for such radical evolutionary changes, the idea seems unfeasible. Melanesians do not have shovel-shaped incisors, an epicanthic eye fold, a high frequency of the 9 b.p. deletion or even straight black hair. All this seems very obvious, and yet there is still a persistence on the part of some researchers to equate evidence of the evolution of pottery styles or language with the evolution of people.

Having said this, Polynesians are not simply tall Asians. There are many features of the phenotype in which they diverge. In the populations of west Polynesia and Fiji there is clear evidence of Melanesian influence, both in terms of outward physical appearance and genetics. Few would now deny, given the genetic evidence presented above, a certain and possibly considerable amount of admixture between the two groups. This is clearly one reason why determining a precise 'origin' for the Polynesians from skeletal material has produced such complex and contradictory results. There is simply no single, precise homeland to find. One crucial component of their phenotypic divergence from their ancestral population is clearly admixture, both during the voyages of colonisation and upon arrival in their new homeland.

Features which differ between SE Asians and Polynesians include the latters´ very large size and robusticity, and possibly the pentagonal head form (although we have seen examples which could be at the very least described as a 'proto-pentagonal' head form amongst some early cranial series from Thailand and elsewhere in SE Asia). Other features which are commonly noted as being distinctive to Polynesians generally are not, including the famed 'rocker jaw' form which occurs in frequencies of at least 50% in SE and east Asia, and over 90% in some parts of Papua New Guinea (Pietrusewsky 1984). This must therefore be ignored as a diagnostic taxonomic feature. It is interesting that

Bulbeck (1981: 399) in his study of SE Asian evolution since the Late Pleistocene notes a decrease in 'tooth size, jaw size and prognathism' as well as a trend towards brachycephalisation. Polynesians too have small teeth, a vertical facial profile and brachycephalic head form, all of which is in direct contrast to most Melanesian populations.

Another significant component in the creation of the Polynesian phenotype is likely to have been the small founding size of the population and limited gene flow after colonisation. Evidence for this has been found from genetic studies which suggest that the founding population in Polynesia is likely to have numbered fewer than 50 individuals (Hertzberg *et. al.* 1989). Any minor bias towards a particular feature present in these individuals (such as large body size) will have been considerably magnified after a few generations. A numerically small, already admixtured population (from the journey through Island Melanesia) arriving in a previously uninhabited region, is likely over time to look considerably different to the founding population in its ancestral homeland.

A final, and possibly considerable influence upon a changing phenotype, is physiological adaptation to novel environments. As discussed in some depth elsewhere (van Dijk 1991, 1993), Polynesian diets are likely to have been very high in protein (largely from seafoods) and for the most part food in general would have been abundant, allowing the population to reach its full growth potential. That this had at least some influence on the phenotype can be seen in the accounts of early voyagers who thought that the discrepancies seen between the physiques of commoners and high status individuals meant that they were different races. Ellis notes:

> the chiefs and persons of hereditary rank and influence ... are ... as much superior to the peasantry or common people ... in physical strength, as they are in rank and circumstances ... Their limbs are generally well-formed, and the whole figure proportioned to their height; which renders the difference between the rulers and their subjects so striking, that some have supposed they were a distinct race [the difference being attributable to] 'differential treatment in infancy, superior food and distinct habits of life'. (Oliver 1974: 782-3, citing Ellis)

Differential access to food resources is also likely to be the source of physical differences between peoples on the coast and inland in Island Melanesia. Friedlaender (1975: 28) writes of a bush/beach dichotomy in regions where coastal peoples with access to a rich protein diet are significantly larger in physique than their inland counterparts. While Houghton (1991, 1996) ascribes this difference to a cold adapted phenotype on the part of the coastal peoples, this seems highly unlikely. Natural selection requires either differential survival rates (i.e. large numbers of people dying on these fishing voyages) or differential fecundity. Irwin's (1992) book on the voyages of colonisation effectively defeats the idea of large numbers of people dying at sea, even on long voyages of exploration and colonisation, and it seems highly unlikely that during short fishing trips between islands in relatively close proximity to one another there was a significant mortality rate.

Large size (including obesity) was also a status symbol and deliberately encouraged through *Ha apori* or fattening rituals (van Dijk 1991) and there are accounts of high status women deliberately choosing to reproduce with men of impressive physique in order to ensure that their offspring would also be similarly blessed (Oliver 1974: 353).

In addition to an abundance of food there were also (until the arrival of the Europeans) fewer debilitating diseases in Polynesia than in Melanesia, the most devastating of these being malaria. That the pre-Polynesians had at some point passed through, or been in contact with people from malarial regions can be seen from the presence of the alpha thalassemia gene in Polynesians (Clark and Kelly 1993), which confers some resistance to malaria. Malaria can have devastating effects upon growth and nutrition, particularly during childhood and may impact on the overall size of a population (Black 1956: 136). The combination of abundant food and lack of disease may have produced a secular increase in size similar to that seen in Japan after the second World War (Greulich 1976), or in Europeans since the last century (Houghton 1996). Such increases are too rapid to be attributable to natural selection (Roberts 1981: 124-5).

From this brief review of osteological and genetic studies in SE Asia and Oceania some summary statements can be made. Firstly, Polynesians have SE Asian genes and physical characteristics which link them conclusively to SE Asia. Since it is highly unlikely that this combination of features re-evolved in Polynesians after they evolved out of populations resident in Melanesia, nor is there any evidence for substantial SE Asian contact subsequent to the colonisation of Polynesia, it is possible to conclude that Polynesians must ultimately be SE Asian in origin. Given the substantial number of genes of Melanesian origin in Polynesians, it must also be concluded that contact with Melanesians was significant at some point in prehistory. It is not necessary to invoke natural selection and people dying at sea in large numbers (e.g. Houghton 1990) to create the Polynesian phenotype. The combination of an essentially SE Asian phenotype, admixture with Melanesian populations, genetic bottlenecking and the not insignificant impact of physiological adaptation to a new environment, is sufficient. While phenotypic change in human evolution has traditionally been from a robust to a more gracile morphology, what we in fact see in this part of the world is a gradual increase in size, from the slight SE Asians to the taller, but still gracile Lapita people and finally the big, robust, rugby-playing Polynesians. Similarly in Australia, the gracile people of Mungo precede the robust inhabitants of Kow Swamp. This situation cannot conveniently be explained away as one of morphology gradually evolving over time. Consequently, our ubiquitous search for the origins of these populations is made even more interesting and challenging.

References

Allen, J., and White, J.P. 1989. The Lapita Homeland: some new data and an interpretation. *Journal of the Polynesian Society,* 98: 129-46.
Ballinger, S.W., Schurr, T.G., Torroni, A., Gan, Y.Y., Hodge, J.A., Hassan, K., Chen, K.H. and Wallace, D.C. 1992. Southeast Asian mitochondrial DNA analysis reveals genetic continuity of ancient Mongoloid migrations. *Genetics* 130: 139-52.
Bellwood, P. 1978. *Man's Conquest of the Pacific*. Collins University Press, Auckland: 462 pp.
Bellwood, P. 1989. The Colonisation of the Pacific: some current hypotheses. In: A.V.S. Hill and S. Serjeantson (Editors) *The Colonisation of the Pacific: A Genetic*

Trail. Clarendon Press, Oxford: 1-59.

Bellwood, P. 1993. Crossing the Wallace Line – with style. In: M. Spriggs, D.E. Yen, W.Ambrose, R. Jones, A.Thorne and A. Andrews (Editors) *A Community of Culture: the People and Prehistory of the Pacific*. Occasional papers in Prehistory No. 21. Department of Prehistory, Research School of Pacific Studies, Australian National University, Canberra: 152-63.

Bellwood, P. 1995. Austronesian Prehistory in Southeast Asia: Homeland, Expansion and Transformation. In: P. Bellwood, J.J. Fox and D. Tryon (Editors) *The Austronesians: Historical and Comparative Perspectives*. Department of Anthropology, Australian National University, Canberra: 96-111.

Black, R.H. 1956. The Epidemiology of Malaria in the South-west Pacific: changes associated with increasing European contact. *Oceania* 27: 136-42.

Bowler, J.M. and Magee, J.W. 2000. Redating Australia's oldest human remains: a sceptic's view. *Journal of Human Evolution* 38: 719-26.

Bowles, G.T. 1977. *The People of Asia*. Charles Scribner's Sons, New York: 414 pp.

Bowles, G.T. 1984. China, Mongolia, Korea. In I. Schwidetzky (Editor) *Rassengeschichte der Menschheit*. R. Oldenbourg Verlag, Munchen: 39-105.

Brace, C.L. and Hinton, R.J. 1981. Oceanic tooth size variation as a reflection of biological and cultural mixing. *Current Anthropology* 225: 549-69.

Brace, C.L., Brace, M.L., Dodo, Y., Hunt, K.D., Leonard, W.R., Yongyi, L., Sangvichien, S., Xiang-qing, S. and Zhenbiao, Z. 1990. Micronesians, Asians, Thais and Relations: A craniofacial and odontometric perspective. *Micronesica Supplement* 2: 323-48.

Brace, C.L. and Hunt, K.D. 1990. A non-racial craniofacial perspective on human variation: A (ustralia) to Z (uni). *American Journal of Physical Anthropology, 82*: 341-60.

Brace, C.L. and Tracer, D.P. 1992. Craniofacial continuity and change: a comparison of late Pleistocene and recent Europe and Asia. In: T. Akazawa, K. Aoki and T. Kimura (Editors) *The Evolution and Dispersal of modern Humans in Asia*. Hokusensha, Tokyo: 439-71.

Bulbeck, F.D. 1981. *Continuities in Southeast Asian Evolution Since the Late Pleistocene*. Unpublished M.A Thesis, Australian National University, Canberra.

Clark, J.T. and Kelly, K.M. 1993. Human Genetics, Paleoenvironments, and Malaria: Relationships and Implications for the settlement of Oceania. *American Anthropologist* 95(3): 612-30.

Cook, J. and King, J. 1785. *A Voyage to the Pacific Ocean*. M Hughs, London: 3 Vols.

Coon, C.S. 1965. *The Living Races of Man*. Knopf, New York: 344 pp.

Ellis, W. 1827. *A Narrative of a Tour through Hawaii*. Fisher and Jackson, London: 480 pp.

Friedlaender, J.S. 1975. *Patterns of Human Variation: The Demography, Genetics, and Phenetics of Bougainville Islanders*. Harvard University Press, Cambridge: 252 pp.

Gillespie, R. and Roberts, R.G. 2000. On the reliability of age estimates for human remains at Lake Mungo. *Journal of Human Evolution* 38: 727-32.

Green, R.C. 1991. Near and Remote Oceania: Disestablishing 'Melanesia' in Culture History. In A. Pawley (Editor) *Man and a Half: Essays in Pacific Anthropology and Ethnobiology in Honour of Ralph Bulmer*. The Polynesian Society, Auckland: 491-502.

Green, M.K. 1990. *Prehistoric Cranial Variation in Papua New Guinea*. Unpublished PhD thesis, Australian National University, Canberra.

Green, M.K. and van Dijk, N. In preparation. *Prehistoric Cranial Variation in Eastern New Guinea: Metric and Non-Metric Analyses of Highland, Lowland and Coastal Populations*.

Greulich, W.W. 1976. Some secular changes in the growth of American-born and native

Japanese children. *American Journal of Physical Anthropology,* 45: 553-68.

Grun, R., Spooner, N., Thorne, A., Mortimer, G., Simpson, J., McCulloch, M, Taylor, L. and Curnoe, D. 2000. Age of the Lake Mungo 3 skeleton, reply to Bowler and Magee and to Gillespie and Roberts. *Journal of Human Evolution* 38: 733-41.

Hagelberg, E. and Clegg, J.B. 1993. Genetic Polymorphisms in Prehistoric Pacific Islanders determined by analysis of ancient bone DNA. *Proceedings of the Royal Society of London,* B252: 163-70.

Hanihara, T., 1992. Dental and Cranial Evidence on the Affinities of the East Asian and Pacific Populations. In K. Hanihara (Editor) *Japanese as a member of the Asian and Pacific Populations.* International Symposium No. 4. International Research Centre for Japanese Studies, Kyoto: 119-37.

Hertzberg, M., Mickleson, K.N.P., Serjeantson, S.W., Prior, J.F. and Trent, R.J. 1989. An Asian-specific 9-bp deletion of mitochondrial DNA is frequently found in Polynesians. *American Journal of Human Genetics* 44: 504-10.

Hill, A.V.S., O'Shaughnessy, D.F. and Clegg, J.B. 1989. Haemoglobin and globin variants in the Pacific. In: A.V.S. Hill and S. Serjeantson (Editors) *The Colonisation of the Pacific: A Genetic Trail.* Clarendon Press, Oxford: 246-85.

Houghton, P. 1989a. Watom: the People. *Records of the Australian Museum,* 41: 223-33.

Houghton, P. 1989b. The Lapita-associated Human Material from Labeka, Fiji. *Records of the Australian Museum,* 41: 327-29.

Houghton, P. 1990. The adaptive significance of the Polynesian Body form. *Annals of Human Biology,* 17(1): 19-32.

Houghton, P. 1991. The Early Human Biology of the Pacific: some considerations. *Journal of the Polynesian Society,* 84: 325-36.

Houghton, P. 1996. *People of the Great Ocean: Aspects of Human Biology of the Early Pacific.* Cambridge University Press, Cambridge: 292 pp.

Howells, W.W. 1973. *The Pacific Islanders.* Reed, Sydney: 299 pp.

Howells, W.W. 1989. *Skull Shapes and the Map.* Papers of the Peabody Museum of Archaeology and Ethnology No. 79. Harvard University Press, Cambridge: 200 pp.

Howells, W.W. 1990. Micronesia to Macromongolia: Micro-Polynesian Craniometrics and the Mongoloid Population Complex. *Micronesica Supplement,* 2:363-72.

Irwin, G. 1992. *The Prehistoric Exploration and Colonisation of the Pacific.* Cambridge University Press, Cambridge: 240 pp.

Ishida, H. 1993. Cranial non-metric variation of Circum-Pacific populations with special reference to the Pacific peoples. *Japan Review,* 4: 27-43.

Ishida, H. and Dodo, Y. 1993. Non-metric cranial variation and the populational affinities of the Pacific peoples. *American Journal of Physical Anthropology,* 90: 49-57.

Katayama, K. 1988. A comparison of the incidences of non-metric cranial variants in several Polynesian populations. *Journal of the Anthropological Society of Nippon,* 96: 357-69.

Katayama, K. 1990. A scenario on prehistoric Mongoloid dispersals into the South Pacific, with special reference to hypothetic Proto-Oceanic connection. *Man and Culture in Oceania,* 6: 151-9.

Katayama, K. no date. *Were the Polynesians Hypermorphic Asiatics? A Scenario on Prehistoric Mongoloid Dispersals into Oceania.* Paper presented at the Prehistoric Mongoloid Dispersals Symposium, Tokyo, 21 November 1992.

Kelly, K.M. 1996. The end of the trail: the genetic basis for deriving the Polynesian peoples from Austronesian speaking palaeopopulations of Melanesian Near Oceania. In: J. Davidson, G. Irwin, F. Leach, A. Pawley, and D. Brown (Editors) *Oceanic Culture History: Essays in Honour of Roger Green.* New Zealand Journal of Archaeology Special Publication, Dunedin: 355-64.

Kirch, P.V., Swindler, D.R. and Turner, C.G. II. 1989. Human Skeletal and Dental

Remains from Lapita Sites (1600-500 B.C.) in the Mussau Islands, Melanesia. *American Journal of Physical Anthropology,* 79: 63-76.

Kirk, R.L. 1981. *Aboriginal Man Adapting.* Clarendon Press, Oxford: 229 pp.

Lum, J.K., Rickards, O., Ching, C. and Cann, R.L. 1994. Polynesian mitochondrial DNA's reveal three deep maternal lineage clusters. *Human Biology,* 66(4): 567-90.

Melton, T., Peterson, R., Redd, A., Sofro, S.M., Martinson, J. and Stoneking, M. 1995. Polynesian genetic affinities with southeast Asian populations as identified by mitochondrial DNA analysis. *American Journal of Human Genetics,* 57: 12-56.

Morwood, M.J., Aziz, F., O'Sullivan, P., Nasruddin, Hobbs, D.R. and Raza, A. 1999. Archaeological and palaeontological research in central Flores, east Indonesia: Results of fieldwork 1997-8. *Antiquity,* 73(280): 273-86.

Oliver, D. 1974. *Ancient Tahitian Society.* University of Hawaii Press, Hawaii: 1419 pp.

O'Shaughnessy, D.F., Hill, A.V.S., Bowden, D.K., Weatherall, D.J. and Clegg, J.B. 1990. Globin Genes in Micronesia: Origins and Affinities of Pacific Island Peoples. *American Journal of Human Genetics,* 46: 144-55.

Pietrusewsky, M. 1984. Metric and Non-metric Cranial Variation in Australian Aboriginal Populations Compared with Populations from the Pacific and Asia. *Occasional Papers in Human Biology,* 3: 1-113.

Pietrusewsky, M. 1989a. A Study of Skeletal and Dental Remains from Watom Island and Comparisons with other Lapita People. *Records of the Australian Museum,* 41: 235-92.

Pietrusewsky, M.1989b. A Lapita-associated Skeleton from Natunuku, Fiji. *Records of the Australian Museum,* 41: 297-325.

Pietrusewsky, M. 1990a. Craniometric Variation in Micronesia and the Pacific: A Multivariate Study. *Micronesica Supplement,* 2: 373-402.

Pietrusewsky, M. 1990b. Lapita-Associated Skeletons from Watom Island, Papua New Guinea, and the Origins of the Polynesians. *Asian Perspectives,* 28(1): 83-9.

Pietrusewsky, M., 1991. Lapita People and the Origins of the Polynesians: An Osteological Assessment. *Newsletter of Chinese Ethnology,* 28: 1-18.

Pietrusewsky, M. 1992a. Japan, Asia and the Pacific: A Multivariate Craniometric Investigation. In: K. Hanihara (Editor) *Japanese as a Member of the Asian and Pacific Populations.* International Symposium No. 4. International Research Centre for Japanese Studies, Kyoto: 9-52.

Pietrusewsky, M. 1992b. *Taiwan Aboriginals, Asians and Pacific Islanders: A Multivariate Investigation of Skulls.* Paper presented at the International Symposium on Austronesian Studies relating to Taiwan. Academia Sinica. Dec. 28-30.

Pietrusewsky, M. 1994. Pacific-Asian Relationships: A Physical Anthropological Perspective. *Oceanic Linguistics,* 33(2): 407-30.

Pietrusewsky, M. 1996. The Physical Anthropology of Polynesia: A review of some cranial and skeletal studies. In: J. Davidson, G. Irwin, F. Leach, A. Pawley and D. Brown (Editors) *Oceanic Culture History: Essays in Honour of Roger Green.* New Zealand Journal of Archaeology Special Publication, Dunedin: 343-53.

Pietrusewsky, M., Hunt, T.L. and Ikehara-Quebral, R.M. 1997. A new Lapita-associated skeleton from Fiji. *Journal of the Polynesian Society,* 106(3): 284-95.

Ranford, P.R. 1989. Genetic variants of the serum complement components. In: A.V.S. Hill and S. Serjeantson (Editors) *The Colonisation of the Pacific: A Genetic Trail.* Clarendon Press, Oxford: 174-93.

Redd, A.J., Takezaki, N., Sherry, S.T., McGarvey, S.T., Sofro, A.S.M. and Stoneking, M. 1995. Evolutionary history of the COII/tRNA^{Lys} intergenic 9 base pair deletion in human mitochondrial DNA's from the Pacific. *Molecular Biology and Evolution,* 12(4):604-25.

Roberts, D.F. 1981. Selection and Body Size. In D.N. Walcher and N. Kretcher

(Editors) *Food, Nutrition and Evolution: Food as an Environmental Factor in the Genesis of Human Variability.* Masson Publishing, New York: 121-33.

Serjeantson, S.W. and Hill, A.V.S. 1989. The Colonisation of the Pacific: the Genetic Evidence. In: A.V.S. Hill and S. Serjeantson (Editors). *The Colonisation of the Pacific: A Genetic Trail.* Clarendon Press, Oxford: 120-73.

Serjeantson, S. and Gao, X. 1995. Homo Sapiens is an Evolving Species: Origins of the Austronesians. In P. Bellwood, J.J. Fox and D. Tryon (Editors) *The Austronesians: Historical and Comparative Perspectives.* Department of Anthropology, Research School of Pacific and Asian Studies, Australian National University, Canberra: 165-80.

Spriggs, M. 1997. *The Island Melanesians.* Blackwell Publishers, London: 326 pp.

Stoneking, M., Sofro, A.S.M. and Wilson, A.C. 1989. *Implications of a Mitochondrial DNA Marker for the Colonisation of the Pacific.* Paper presented at Circum-Pacific Prehistory Conference, Seattle, Washington USA.

Sykes, B., Leiboff, A., Low-Beer, J., Tetzner, S. and Richards, M. 1995. The Origins of the Polynesians: An interpretation from mitochondrial lineage analysis. *American Journal of Human Genetics,* 39: 305-16.

Terrell, J. 1986. *Prehistory in the Pacific Islands.* Cambridge University Press, Cambridge: 299 pp.

Thorne, A. 1980. The arrival of man in Australia. In: A. Sherratt (Editor) *The Cambridge Encyclopedia of Archaeology.* Cambridge University Press, New York: 495 pp.

Thorne, A., Grun, R., Mortimer, G., Spooner, N., Simpson, J., McCulloch, M., Taylor, L. and Curnoe, D. 1999. Australia's oldest human remains: age of the Lake Mungo 3 skeleton. *Journal of Human Evolution,* 36: 591-612.

Thorne, A. and Curnoe, D. 2000. Sex and significance of Lake Mungo 3: reply to Brown "Australian Pleistocene variation and the sex of Lake Mungo 3". *Journal of Human Evolution,* 39: 587-600.

Turner, C.G. II. 1987. Late Prehistoric and Holocene population history of East Asia based on dental variation. *American Journal of Physical Anthropology,* 73: 305-21.

Turner, C.G. II. 1989. Dentition of Watom Island, Bismarck Archipelago, Melanesia. *Records of the Australian Museum,* 41: 293-6.

Turner, C.G. II. 1990. Origin and Affinity of the people of Guam: a dental anthropological assessment. *Micronesia Supplement,* 2: 403-16.

van Dijk, N. 1991. The Hansel and Gretel Syndrome: a critique of Houghton's cold adaptation hypothesis and an alternative model. *New Zealand Journal of Archaeology,* 13: 65-89.

van Dijk, N. 1993. *The Evolution of the Polynesian Phenotype: An Analysis of Skeletal Remains from Site To-At-36, Tongatapu, Tonga.* Unpublished M.A. Thesis, University of Auckland, Auckland.

van Dijk, N. 1998. *The Melanesians: an osteological study of their biological relationships within the Pacific.* Unpublished PhD thesis. Department of Archaeology and Natural History, Research School of Pacific and Asian Studies, Australian National University, Canberra.

van Dijk, N. 1999. Who are these People? Human Skeletal Remains from the Pacific Region. In: J.C. Galipaud and I. Lilley (Editors) *The Western Pacific, 5000-2000BP: Colonisation and Transformations.* Institut de Recherche pout le Developpement, Paris: 201-9.

van Dijk, N. in press. Biological relationships within New Guinea and between New Guinea and Australia: the skeletal evidence. *Proceedings of the Papuan Pasts Conference,* December 2000.

Visser, E.P. 1995. *The People of Sigatoka.* Unpublished PhD Thesis. University of Otago, Dunedin.

From Savannah to Rainforest:
Changing Environments and Human Occupation at Liang Lembudu, Aru Islands, Maluku (Indonesia)

Sue O'Connor, Ken P. Aplin, Matthew Spriggs, Peter Veth and Linda K. Ayliffe

Introduction

The Aru Islands lie near the edge of the Australian continental shelf in the Arafura Sea, approximately 150 km south of the coast of Papua (formerly Irian Jaya). For at least the first 40,000 years of occupation of Sahul they formed part of a continuous land bridge linking Australia and New Guinea. During this time they would have been a dissected limestone plateau on the exposed Carpentarian Plain. About 14,000 years BP sea level rose and began to encircle the island group, separating it from Australia and by 11,500 years BP it was completely separated from New Guinea. The presence on Aru of numerous marsupials, the cassowary and Birds of Paradise attest to this shared history, a fact first recognised by Darwin's co-discoverer of the theory of evolution by natural selection, the naturalist Alfred Russel Wallace (Wallace 1869).

While the waters to the east of the Aru Islands are relatively shallow, reflecting the previous land bridge with Papua and NW Australia, the continental shelf to the west descends steeply. Along most of the northwest coastline of the Aru Islands the 100 m isobath is located as little as 10 km away (Figure 15.1). Recent research into sea level changes in the Australian region indicates that between 70,000 and 40,000 years BP sea level was between -50 and -90 m below its present position (Chappell 1994). This means that along most of the north and northwest coast of Australia the earliest evidence for colonisation will now be submerged and the earliest date for colonisation probably unknowable. Sites along the greater part of the present coastline may have been located up to 200 km inland at the time of their initial occupation. The steeply shelving offshore profile on the west coast of Aru makes it one of the few entry points into Sahul where evidence of early settlement is likely to be preserved.

The narrow water corridor which separated the islands of Wallacea to the west from the Aru Islands throughout the last 50,000 years, also make this one of the most likely routes of colonisation into Sahul (Birdsell 1977; Irwin 1992). Two of the key routes proposed by Birdsell pass through or close to the Aru Islands (Figure 15.2). The first passes through Maluku via Buru, Seram and finally the Kei Islands with a landfall directly on the Aru uplands. The second route passes along the Lesser Sundas to Timor then via Maluku through Wetar, Babar and Tanimbar with landfall on the Pleistocene coastline in the general

ISBN 3-923381-47-6

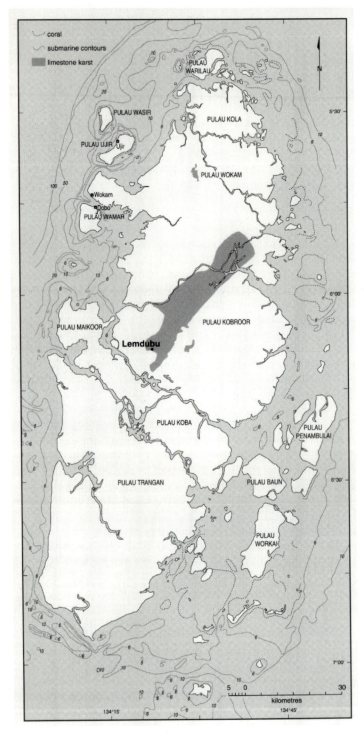

Fig. 15.1 Map of
Aru Islands
showing location of
Liang Lemdubu.
Submarine
contours and
distribution of
karst limestone
indicated.

region of the Aru group. Such proximity would also have facilitated two-way voyaging, both before and after Aru was separated from Sahul approximately 12,000 years BP.

The islands are predominantly uplifted marine sedimentary rocks of late Tertiary to Quaternary age, and have a combined area of 7770 km^2 and a maximum elevation of 241m (Szalay 1995). The various major islands are divided by narrow sea channels (sungai). The present-day vegetation of the Aru Islands is a mosaic of moist rainforest/vineforest and drier savannah communities (Monk *et al.* 1997: 198). Rainfall records are not available but an annual rainfall in excess of 2000 mm is probable.

Initial reconnaissance in the Aru Islands began in 1995 and was focussed on the northwest islands of Wamar, Wokam, Ujir and Wasir as these lay closest to the edge of the Continental Shelf (Figure 15.1). Ground survey of the west coast, however, indicated that suitable cave formations would not be found in this region, hence survey shifted to the karst region of Kobroor and Wokam Islands (Figure 15.1). This strategy resulted in the location of several caves including Liang Lemdubu, the largest of the karst caves recorded during the 1995 field season. Excavation of Liang Lemdubu began the following field season in November 1996 and confirmed our expectation that the limestone caves would provide an excellent environment for faunal and floral preservation.

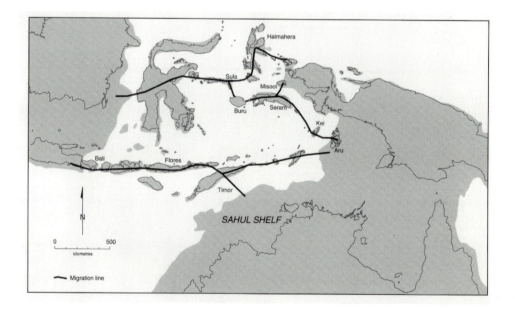

O/C SAHUL SHELF - Cartography RSPAS 98006 ih

Fig. 15.2 Key migration routes into Australasia (after Birdsell 1977). Australasia and Wallacean islands with -200 m contour.

Peter Kershaw, Bruno David, Nigel Tapper, Dan Penny and Jonathan Brown (Eds): Bridging Wallace's Line

Liang Lemdubu

Lemdubu Cave is located in the western interior of Pulau Kobroor (Figure 15.1) and may be approached by boating to the upper reaches of Sungai Papakula (the very sungai where Wallace spent six weeks collecting in 1857, at the hamlet he called 'Wanumbai'), followed by a two hour walk through rainforest. It is a large, double-entranced cave formed from an ancient subterranean river passage cut through the limestone.

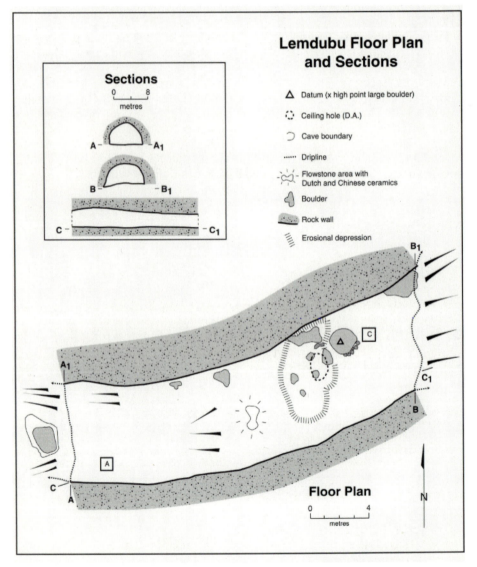

Fig. 15.3 Plan of Liang Lemdubu showing location features discussed in the text and locations of Test Pits A and C.

Fig. 15.4 Lemdubu Cave looking west. The flowstone platform where Dutch and Chinese ceramic bowls have been placed to collect water dripping through the cave roof can been seen mid centre. The water is believed to have sacred properties.

While no more than about 25 m above sea level, the cave represents a high point in the local landscape. It is surrounded by evergreen rainforest which has a fairly simple structure (Monk *et al.* 1997: 198, 203). Around the cave the rainforest is uncleared but within a few hundred metres to the east it is interrupted by sago swamps and small garden clearings. Lemdubu runs in length for 30 metres, is up to eight metres wide and has an average height of three metres (Figures 15.3 and 15.4). There is a small hole in the central portion of the roof through which water and limited sediments enter. This minor collapse appears quite recent. Stalactites and stalagmites occur in the central third of the cave; beneath the former pieces of Dutch and Chinese porcelain have been placed to collect the water which is believed to have sacred properties. These historical artefacts are now cemented to a patch of flowstone on the cave floor.

The densest concentrations of cultural materials on the extensive sediment floor of the cave are located near both of the driplines. Shellfish, (*Geloina* sp., *Anadara* sp. and *Terebralia* sp.), pottery, and animal bone were noted on the surface along with pieces of matting and bamboo indicating the contemporary use of the cave by villagers while hunting. Our initial judgment was that the deposits immediately inside the driplines at both ends of the cave were likely to be the deepest and the least disturbed by water action, roots and major roof fall events. These areas were also more likely to have been the focus of daily occupation than the darker reaches in the centre of the cave. Two test pits were excavated at either end of the cave.

Peter Kershaw, Bruno David, Nigel Tapper, Dan Penny and Jonathan Brown (Eds): Bridging Wallace´s Line

Initially a 1 x 1 m test-pit (Test-pit A) was dug at the west end of the cave (Figure 15.3). Our method was to excavate in five centimetre spits, wet-sieving all materials through fine mesh (1.5 mm), record volumes of recovered materials and to sort all cultural residues. Culturally sterile deposits, apparently weathered bedrock, were reached at approximately 50 cm below the surface. Test-pit A revealed a homogeneous, loose grey-brown sediment which changed to an orange-brown mottled clay immediately above the sterile, basal deposit. The upper part of the grey-brown sediment contained charcoal, terrestrial fauna, earthenware pottery, marine shellfish, a fragment of metal and stone artefacts. Below this the shellfish disappeared but terrestrial fauna and stone artefacts continued while the orange-mottled clays yielded a sparse assemblage of fauna. From this first test-pit it became apparent that the deposits of Lemdubu contain a phenomenal quantity of terrestrial fauna.

Test-pit C was located near a massive in situ boulder at the eastern end of the shelter (Figure 15.3). This 1 x 1 m pit revealed a similar basic sequence to Test-pit A (Figure 15.5) with sterile deposits reached at approximately 160 cm below the surface.

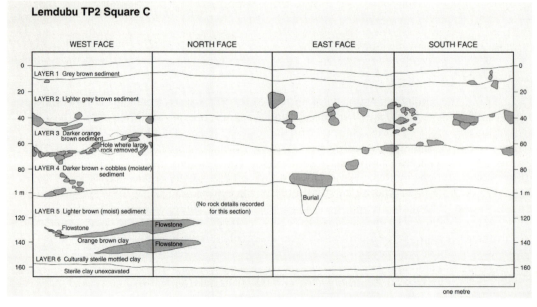

Fig. 15.5 Sections of Test Pit C.

The Lemdubu cultural assemblage

Several features were evident during the on-site sorting of material. Faunal material was abundant and a major marsupial component was seen through the presence of small to medium-sized wallabies (macropodids), cuscus (phalangerids), bandicoots (peroryctids and peramelids) and native cats (dasyurids). Of these, remains of small to medium-sized wallabies appeared most abundant. Introduced species such as pig and dog appeared to be confined to the very top few spits.

Marine/estuarine shellfish indicating post-transgressive sea levels commonly occur in Spits 1-4 but decline below Spit 4 where several species of land snail (as yet unidentified) occur in abundance. The best represented marine/estuarine species are the mangrove clam *Geloina coaxans* and the gastropod *Terebralia* sp. In Spits 5 and 6 some marine/estuarine shell occurs but it makes up a small proportion of the total shell weight. For example, of the 226 g of shell in Spit 6, only 12.5 g are from marine/estuarine species, the rest being from land snail. Fragments of *Geloina* occur occasionally down to Spit 25. These may have been brought into the site for use as artefacts as has been documented also from deposits in northern Australia at far greater distances inland (O'Connor 1996). Final separation of the shell into species has not been undertaken and the weights presented are the combined weights for marine and terrestrial shell. In general, most shell above Spit 5 is of marine/estuarine species and most shell from below Spit 5 is of terrestrial species.

Pottery was present in very small quantities in the upper two spits of Test-pit C. While no analysis has been undertaken of the pottery, superficially it resembles the earthenware pottery currently manufactured in eastern Aru at Batu lei and traded today into this region.

Stone artefacts occurred throughout the excavation, generally in low numbers. The industry is essentially percussion flake-based, with a minor modified component represented by a few retouched/utilised flakes. There is, however, a change in dominant lithologies used through time, with silicified limestone dominating the upper five spits and cherts in larger proportions in the lower levels. One interesting and unusual feature of the stone artefact assemblage is the near absence of cores and small flakes resulting from flake manufacture or retouch, indicating that manufacturing took place off-site, i.e. the flakes in the assemblage were brought into the site for use. The stone artefacts are still under analysis and only artefact weights are presented here. Most artefacts occur between Spits 7 and 10 (the weight in Spit 1 is high because of a single large specimen).

Single bone bipoints were identified in Spits 2 and 15, with one other possible example in Spit 16. The two definite artefacts both appear to be manufactured on long-bone shaft fragments.

Seeds of *Celtis philippinensis* were found in Spits 1-10. It is uncertain whether these represent food remains or were brought into the site on branches destined for firewood. Charcoal was sparse below Spit 3. In view of the large quantity of burnt bone and shell, the small quantity of charcoal is presumed to be due to differential preservation with depth.

Human bones recovered from Spits 19-25 appear to be part of a burial. This material is currently under analysis and will be published in detail elsewhere.

Weights for all cultural material are presented in Table 15.1. The weights are somewhat preliminary as some bone has calcium carbonate adhering and both shell and bone have some sediment adhering that will tend to inflate the weights slightly.

Peter Kershaw, Bruno David, Nigel Tapper, Dan Penny and Jonathan Brown (Eds): Bridging Wallace's Line

Table 15.1. Liang Lemdubu, weight in grams of cultural materials by spit.

Spit	Bone	Shell	Stone artefact	*Celtis* seed	Charcoal	Cassowary Eggshell
LC1	1178.7	2585.2	58.0	11.29	57.6	4.35
LC2	430.5	792.0	12.3	7.4	5.5	2.25
LC3	361.0	594.9	23.8	5.25	0.8	0.16
LC4	475.0	346.2	3.2	4.57	1.0	2.64
LC5	617.4	178.0	6.4	3.85		0.32
LC6	546.1	226.5	5.7	1.0	0.2	
LC7	1085.2	160.6	166.6	0.84		
LC8	1162.9	29.7	89.4	0.08		
LC9	2062.1	33.2	54.0	0.16	0.8	
LC10	3482.7	15.4	217.9	0.07	0.01	
LC11	1352.6	6.3	11.3	0.09		
LC12	1241.7	missing	53.7			
LC13	1168.8	4.2	4.7			
LC14	1591.8	5.3	26.1			
LC15	1488.2	4.9	32.7			
LC16	1658.4	4.3	3.2			
LC17	1566.3	11.6	7.9			0.08
LC18	1421.3	3.1	6.0			0.07
LC19	2112.9	2.1	6.3			1.65
LC20	2400.6	0.7	.5			0.19
LC21	2433.8	1.7	0.0			
LC22	1567.6	2.0	0.0			
LC23	1531.9	5.0	0.0			
LC24	3356.4	7.1	2.5			
LC25	5665.8	9.8	1.6			
LC26	1912.0	12.1	0.0			
LC27	1004.2	11.4	0.0			
LC28	640.9	13.3	0.0			
LC29	207.9	2.2	0.1			
LC30	91.7	0.0	0.0			
LC31	59.3	2.8	0.0			

The Lemdubu fauna

This paper focuses on the faunal data in tandem with the chronology as an indicator of changing human occupation of the Sahul Shelf in the context of palaeo-environmental change both prior to and following insularisation. The faunal analysis was carried out by Aplin at the Western Australian Museum using additional reference material from the Australian Museum, Sydney.

The contemporary fauna of Aru Islands

The contemporary fauna of the Aru Island group has not been systematically surveyed. Most records date from material collected last century, much of it by Alfred Wallace himself. Recent visits by Pat Woolley (personal communication) and a W.A. Museum team (no date) have added several species to the list compiled by Szalay (1995) from historic sources. An updated mammal species list for Aru contains nine marsupials, six native rodents and 13 bats, but it is almost certainly incomplete. Six introduced mammals are recorded, including pig, dog and deer. The contemporary native fauna is very similar in composition to that recorded from the Trans-Fly region of mainland New Guinea, and contains a mixture of typically Australian (*Sminthopsis virginae*) and New Guinean elements (e.g. *Murexia longicauda, Myoictis melas, Thylogale brunii, Melomys platyops, M. lorentzii*). All species recorded from the Aru Islands are found also in the Trans-Fly area or in adjacent regions of Papua or Papua New Guinea. Eight of the 15 terrestrial mammal species occurring on the Aru Islands are common to both New Guinea and northern Australia (see Table 15.2).

The archaeological fauna

The bone is moderately well preserved throughout the sequence, with little indication of differential degradation. In the lower levels, especially around Spit 25, the bone is often coated with a cemented matrix, sometimes forming a weak 'bindstone' of bone fragments and sediment.

The relative paucity of small mammal remains through the sequence, and preponderance of medium- to large-bodied macropodids and possums, indicates that the Lemdubu Cave fauna is predominantly derived from human activity, rather than being the product of non-human carnivore or raptor activity. This inference applies to all spits down to Spits 28/29. The basal Spits 30 and 31 contain a higher proportion of smaller birds and mammals and may be derived wholly or in part from non-human activity. The fact that the lowest stone artefact was recovered from Spit 29 supports this view. Material from the upper levels of the sequence (above about Spit 10) is more thoroughly burnt and appears more highly fragmented than that from the lower levels. The degree of burning and fragmentation may be causally related to some extent. However, material from the 'bindstone' layers (Spits 25-28) is exceptionally complete and preserves many examples of partial articulation. The articulation of these elements indicates that little effort was being made to extract all available protein from these specimens, in turn suggesting an abundance of animal foods.

Peter Kershaw, Bruno David, Nigel Tapper, Dan Penny and Jonathan Brown (Eds): Bridging Wallace´s Line

Table 15.2. List of mammal species recorded from the Aru Islands. The modern fauna is contrasted with the archaeological fauna identified from Lemdubu Cave. The modern fauna is compiled from the following sources: Flannery (1995b); Kitchener (n.d.); Woolley (pers.comm.); and O'Connor (pers. comm.).

Taxa	Modern	Lembudu
Tachyglossidae (Echidnas)		
Tachyglossus aculeatus (Short-beaked Echidna)	-	+
Dasyuridae (Marsupial Mice and Native Cats)		
Dasyurus sp. cf *D. spartacus*[1] (Bronze Quoll)	-	+
Murexia longicauda (Short-furred Dasyure)	+	+?
Myoictis melas (Three-striped Dasyure)	+	+
Sminthopsis virginae (Red-cheeked Dunnart)	+	-
Peroryctidae (New Guinean Bandicoots)		
Echymipera sp. cf. *E. kalubu*[2] (Common Echymipera)	-	+?
Echymipera rufescens (Long-nosed Echymipera)	+	+
Peramelidae (Australian Bandicoots)		
Isoodon macrourus (Northern Brown Bandicoot)	-	+
Phalangeridae (Cuscuses)		
Phalanger gymnotis (Ground Cuscus)	+	+
Phalanger intercastellanus (Southern Common Cuscus)	+	+
Spilocuscus maculatus[3] (Common Spotted Cuscus)	+	+
Petauridae (Gliders and Striped Possums)		
Dactylopsila trivirgata (Striped Possum)	+	-
Macropodidae (Kangaroos and wallabies)		
Macropus agilis (Agile Wallaby)	-	+
Thylogale brunii[4] (Dusky Pademelon)	+	+
Thylogale stigmatica (Red-legged Pademelon)	-	+
Thylogale? sp.[5] (Pademelon species)	-	+
Large macropodid[6] (Kangaroo species)	-	+
Muridae (Rats and Mice)		
Hydromys chrysogaster (Common Water Rat)	+	+
Melomys rufescens (Black-tailed Melomys)	+	-
Melomys sp. cf. *M. lutillus*[7] (Grassland Melomys)	-	+
Paramelomys lorentzi (Lorentz's Melomys)	+	-
Paramelomys platyops (Lowland Melomys)	+	-
Pogonomys sp.[8] (Tree-mouse)	-	+
Rattus leucopus (Cape York Rat)	+	+
Rattus rattus (Black Rat)	+	-
Uromys caudimaculatus (Mottled-talied Giant Rat)	+	+
Soricidae (Shrews)		
Crocidura maxi	+	-
Suncus murinus (House Shrew)	+	-
Viverridae (Civets etc.)		
Paradoxurus hermaphroditus (Palm Civet)	+	-

Suidae (Pigs)		
Sus scrofa (Domestic Pig)	+	+
Cervidae (Deer)		
Cervus timorensis[9] (Rusa Deer)	+	-
Pteropodidae (Flying Foxes and Fruit Bats)		
Dobsonia molluccensis[10] (Bare-backed Fruit-bat)	+	+
Macroglossus minimus (Northern Blossum-bat)	+	-
Nyctimene albiventer (Common Tube-nosed Bat)	+	-
Pteropus macrotis (Big-eared Flying Fox)	+	-
Pteropus melanopogon (Black-bearded Flying Fox)	+	-
Syconycteris australis (Common Blossum-bat)	+	-
Hipposideridae (Horseshoe-bats)		
Hipposideros ater (Dusky Horseshoe-bat)	+	-
Hipposideros cervinus (Fawn Horseshoe-bat)	+	-
Rhinolophidae (Horseshoe-bats)		
Rhinolophus euryotis (New Guinea Horseshoe-bat)	+	-
Vespertilionidae (Insectivorous Bats)		
Miniopterus australis (Little Bentwing-bat)	+	-
Pipistrellus javanicus (Javan Pipistrelle)	+.	-
Pipistrellus papuanus (Papuan Pipistrelle)	+	-

Taxonomic notes (K. Aplin)

1. Two partial lower jaws and a proximal femur from the lower levels of the deposit represent a species of *Dasyurus*. They have been compared directly with specimens of *D. albopunctatus* from the Idenberg River, Papua, and to published accounts only of *D. spartacus* (Van Dyck 1988). They are tentatively referred to the latter species on account of the moderately large molar dimensions, relatively deep mandibular ramus, and relatively unreduced metaconids and uncrowded premolars. Direct comparisons are required to confirm the identification. *Dasyurus spartacus* has been definitely recorded only from the vicinity of Morehead in Papua New Guinea, although there are unconfirmed reports from Papua.

2. *Echymipera kalubu* is tentatively recorded from a single specimen (an incomplete lower jaw lacking teeth) from Spit 3. This specimen is not referrable to either *Isoodon macrourus* or *Echymipera rufescens*, both of which are well-represented in the deposit. It is generally consistent with specimens of *E. kalubu* but should not be considered as more than suggestive evidence for this species, which is otherwise unrecorded from the Aru Group (although present in the Trans-Fly region of New Guinea and generally found in lowland forest habitats). The material referred to *E. rufescens* could also include examples of *E. echinista*, a rare species known only from the Western Province of Papua New Guinea (Menzies 1990). However, the more complete upper teeth among the archaeological sample are consistent with *E. rufescens* and lack the enlarged hypocones of *E. echinista*.

3. *Spilocuscus maculatus* is represented by excellent material. The sample is characterised by the small size of the teeth and jaws, in which regard they are consistent with Australian populations of the species.

4-5. Three small macropodids are represented in the archaeological fauna. Two are abundant and represented by excellent diagnostic material. The third is uncommon and represented by less complete material. However, one very well preserved

Peter Kershaw, Bruno David, Nigel Tapper, Dan Penny and Jonathan Brown (Eds): Bridging Wallace´s Line

maxilla with an adult toothrow demonstrates the distinctness of this taxon. Uncertainty regarding the taxonomic identity of these taxa is due to a paucity of modern comparative material. Dr T. Flannery kindly compared a sample of each taxon with material in the Australian Museum. He was able to confirm that one of two common forms is *Thylogale brunii,* recorded in the modern Aru fauna. The second common species compared most closely (except for being a little smaller) with north Queensland specimens of *Thylogale stigmatica,* a species which also occurs in the Trans-Fly region of mainland New Guinea. Unfortunately, no material of the New Guinean race, *T. s. oriomo,* is available in Australia for direct comparisons. Tate and Archbold (1935) comment that *oriomo* is comparable in size to north Queensland *stigmatica.*The third species also appears to be referrable to the genus *Thylogale* rather than the closely related genus *Petrogale.* Based on the limited sample available to him, Flannery expressed the opinion that the third group might be the same taxon as the second, i.e., *T. stigmatica,* although he acknowledged a significant size difference between the samples. Having seen the full range of materials, I remain convinced that the third taxon is distinct. Perhaps both taxa are indeed *Thylogale stigmatica* – one representing the race *oriomo* and the other typical *stigmatica,* the two 'races' possibly coming into contact during periods of exposure of the Carpentarian Plain.

6. A large macropodid is represented by two pedal phalanges, one from each of Spits 20 and 21. Initial suspicion that these specimens were referrable to a cassowary was eliminated by direct comparisons with material in the Australian Museum (W.E. Boles, pers.comm.). Both specimens represent the second phalanx from the fourth toe, and they clearly come from different individuals, one being slightly larger than the other (possibly an expression of sexual dimorphism). Comparisons with the limited comparative materials in the W.A. Museum indicated a fairly close resemblance to the second pedal phalanx of *Macropus robustus* and *M. antilopinus,* although the archaeological specimens are considerably larger than either of these species. The specimens were subsequently examined by Flannery at the Australian Museum, where specimens of a New Guinean *Protemnodon* species and other *Macropus* species were available for comparison. He reported that the second phalanx in the New Guinean *Protemnodon* species is 'much shorter and more robust' that the Lemdubu specimens. He also confirmed the general similarity to phalanges from the extant large kangaroos, especially members of the *Osphranter* group. For the present it may be concluded that a very large species of kangaroo, different to any of those present in Northern Australia today and perhaps larger than any of the extant species, was present in the vicinity of the Aru Plateau during late Pleistocene times. It appears to have overlapped in time with early human populations; whether it qualifies as an element of the Pleistocene 'megafauna' is largely a matter of semantics.

7. Two specimens (both mandibles with molar teeth) from the lower levels of the site represent a small species of *Melomys.* The first lower molar has three roots on the lingual side, a feature seen in species of the genus *Melomys* sensu stricta (Menzies 1996). It is considerably smaller than *M. rufescens,* the only species of the group recorded in the modern fauna (Kitchener *et al.* 1995) and is probably referrable to a species of the *M. lutillus* group. Musser and Carleton (1993) have synonymised *M. lutillus* and several other, closely related taxa with *M. burtoni,* a species of the northern Australian grasslands.

8. This species is probably the same taxon recorded in north Queensland and variously identified as *Pogonomys mollipillosus* or *P. loriae.* The systematics of the larger, white-bellied *Pogonomys* species is in need of revision. For the present, it is best referred to simply as *Pogonomys* sp.

9.   Deer are not listed by Flannery (1995b) as part of the modern Aru fauna. O'Connor (pers.comm.) reports hunting of deer in open grassland habitats on Kobroor. The common deer in eastern Indonesia is *Cervus timorensis*, the Rusa Deer.

10. *Dobsonia mollucensis* was recorded by Anderson (1912) for the Aru Islands, based on specimens in the British Museum and in Leiden. Bergmans and Sarbini (1985) questioned the origin of these specimens and questioned the occurrence of the genus in the Aru Group. This doubt is maintained by Flannery (1995a) who does not record any *Dobsonia* species from Aru. The Western Australian Museum expeditions to the Kei and Aru Islands obtained *Dobsonia* in both island groups. These have been tentatively identified as *D. mollucensis,* with more detailed study pending.

In the light of these initial impressions, it was decided to focus on the pattern of distribution of the various species through the stratigraphic column and the quantification of changes in abundance among the major species. The critical requirements were therefore: 1) to determine the range of species present; 2) to establish the pattern of distribution of each species through the sequence; and 3) to estimate the relative abundance of the major species, both within each unit (interspecific) and through the sequence (intraspecific).

Prior to analysis, all samples were washed in a fine-mesh sieve and air dried. Given the large quantity of material recovered and the overwhelming predominance of macropodid remains, it was considered impractical to identify every potentially diagnostic element. It was therefore decided to concentrate on two areas of the macropodid skeleton – the cranium and the foot (tarsus and metatarsus) – which would provide best species discrimination. For other mammals, all recognisable body elements were identified at least to family, as was all bone referrable to fish, reptiles and birds. Fragments of crab chelae, eggshell and several small bone points were also separated from the remaining unallocated bone material, the great bulk of which is thus of macropodid origin (non-head, non-foot). To date, the various identified faunal classes have not been weighed or otherwise fully quantified. As a preliminary measure the number of bone fragments referrable to each taxon was simply counted. For the various macropodid species, separate counts were made of: teeth or tooth-bearing elements; other cranial pieces; calcanea and astragali. Pythons (Boidae) were counted in two classes: vertebrae and cranial, including tooth bearing elements. Because the abundance of the animal groups has been calculated in slightly different ways, it is not possible at present to directly equate the abundance data to any measure of economic importance. However, the values are consistent through the sequence and so an increase or decrease in any particular taxon is significant, but not necessarily equally scaled from group to group. Full taphonomic and economic analysis of the fauna is underway.

The archaeological fauna includes several species not previously recorded from the islands, but also lacks several species which are present today. The major additions are the macropodids *Macropus agilis*, *Thylogale stigmatica* and *Thylogale* sp., the bandicoots *Isoodon macrourus* and (tentatively identified)? *Echymipera kalubu*, and the Short-Nosed Echidna *Tachyglossus aculeatus*. Other additions are a Native Cat (*Dasyurus* sp. *cf. D. spartacus*), and several rodents (*Pogonomys* sp., *Melomys* sp. *cf. M. lutillus*). A full list of species is given in Table 15.2. Species which have not been recorded previously from the Aru Islands are highlighted.

A very large macropodid which is not clearly referrable to any modern species found in New Guinea or northern Australia is represented by two pedal phalanges from two different individuals. These have been compared to a variety of both living and extinct taxa including both Australian and New Guinea *Protemnodon* spp. The closest comparison is with species of *Macropus (Ophranter)*, the group containing the modern wallaroos; however the fossil specimens are considerably larger than any of the modern species. This taxon appears to be confined to the lower part of the sequence.

Macropus agilis is a true grassland/savannah wallaby and it is rare or absent from closed, moist forest types. The species has been intensively hunted in coastal Papua both during prehistoric times and historically (Allen 1977), and its numbers have declined in some areas as a consequence (Flannery 1995a). Although it has not been previously recorded from the Aru Islands, suitable habitats are present in the southeast and it may still occur there.

The ecology of the various New Guinean *Thylogale* species is unfortunately not well known. In Australia, *T. stigmatica* occurs mainly in rainforests but it also utilises wet sclerophyll and drier vine forest communities (Johnson and Vernes 1995). Tate and Archbold (1935) described this species' habitat on the Oriomo Plateau of Papua New Guinea as 'mixed grasslands and gallery woods', while Waithman (1979) reports it to be uncommon in low mixed savannah and woodland near swamps in the Morehead area. The endemic New Guinean *T. brunii* is reported to occur in dense monsoonal rainforest in the Morehead area (Waithman 1979), however in the recent past it was apparently present in grassland/savannah habitats around Post Moresby (Flannery 1995a). Nothing can be inferred regarding the ecology of the unidentified, third small macropodid, which is listed as ?*Thylogale* sp.

Isoodon macrourus is found in a wide variety of habitats across Northern Australia but in New Guinea it appears to be broadly associated with major tracts of savannah/grassland (Flannery 1995a). In New Guinea *Echymipera rufescens* is primarily a rainforest inhabitant (Flannery 1995a). On Cape York Peninsula its distribution is centred on patches of closed forest but it also occurs in 'eucalypt grassy woodland and closed coastal heath near closed forest and ... low layered open forest' (Gordon 1995: 192). The endemic New Guinean *Echymipera kalubu* is found in low to mid-montane rainforests but also inhabits anthropogenic grassland and woody regrowth (Flannery 1995a).

Dasyurus spartacus is known only from the Morehead region of the Fly River Plains of Papua New Guinea, a region of low mixed savannah. Flannery (1995a) notes unconfirmed reports of the species from similar habitats in Wassur National Park, Papua. A second species, *D. albopunctatus*, is present in lowland rainforest habitats throughout New Guinea. The Lemdubu specimens are as yet only tentatively referred to *D. spartacus*.

The Spotted Cuscus (*Spilocuscus maculatus*) is generally associated with lowland to mid-montane rainforest habitats, but in Australia it has been observed 'in freshwater and saline mangroves, in large paperbarks in thin riparian forest strips and in open forest up to half a kilometre from the nearest rainforest' (Winter and Leung 1995a: 266). The Grey Cuscus (*Phalanger intercastellanus*) is also widespread in lower altitude rainforests in southern New Guinea and is largely confined to rainforests on Cape York Peninsula. Winter and Leung (1995b: 269) notes that the species 'penetrates the acacia fringes of rainforest'. The endemic New Guinean *P. gymnotis* has not been recorded outside of rainforest habitats.

Among the smaller mammals, *Pogonomys* sp. and *Myoictis melas* are both indicative of rainforest habitats. The small *Melomys* species is tentatively identified as *M. lutillus*. This species, along with its close relative *M. burtoni* of northern Australia, is associated with grassland habitats, including anthropogenic grassland patches within forested regions.

The major omission from the archaeological fauna is the Striped Possum, *Dactylopsila trivirgata*, which occurs today on both Wokam and Kobroor Islands. However, the contemporary fauna also includes three *Melomys* species, none of which is definitely recorded in the Lemdubu Cave fauna. Elsewhere in New Guinea, *Dactylopsila* spp. are generally regarded as desirable food items and they are also quite well-represented in a number of archaeological faunas. The absence of this species from the archaeological remains is thus not due to unpalatability, and must either reflect strong cultural preference or the very recent introduction of the species to the islands (see Flannery and White 1991 for other examples of human translocation of species).

Other than pig and dog, no introduced mammals are recorded in the archaeological fauna. From this we can perhaps infer that Deer, Palm Civets and the two Shrews found in the Aru Islands today are relatively recent introductions. While deer hunting is now a major occupation, it is possible that deer is a very recent introduction. In this context it is relevant to note that Wallace does not mention deer. The species is likely to be *Cervus timorensis*, the Rusa Deer, which is common in eastern Indonesia.

Changes through the sequence

The changing proportions of various major classes of animals through the deposit are shown in Figure 15.6. Major changes in species representation take place in the uppermost part of the sequence (Spits 1-3) with the appearance of pig and dog, and a marked increase in the quantity of fruitbat and fish bone. Changes in the relative abundance of the different macropodids occur just below this at about Spit 4 and are even more informative. As shown in Figure 15.7, the relative abundance of *M. agilis* increases somewhat from the basal layers then almost disappears above Spit 4. Only a few fragments of *M. agilis* were found above Spit 5 and these are most likely a result of reworking of bone from lower levels.

Bandicoots contribute a small proportion of the identified material, but rise in abundance between Spits 17-25. Through most of the sequence, the savannah-dwelling *Isoodon macrourus* is more abundant than the rainforest species *Echymipera rufescens*. However the reverse is true in Spits 1-3, paralleling the apparent recent decline in *Macropus agilis*. *Echymipera kalubu* is tentatively identified from a single isolated tooth from Spit 3.

Peter Kershaw, Bruno David, Nigel Tapper, Dan Penny and Jonathan Brown (Eds): Bridging Wallace's Line

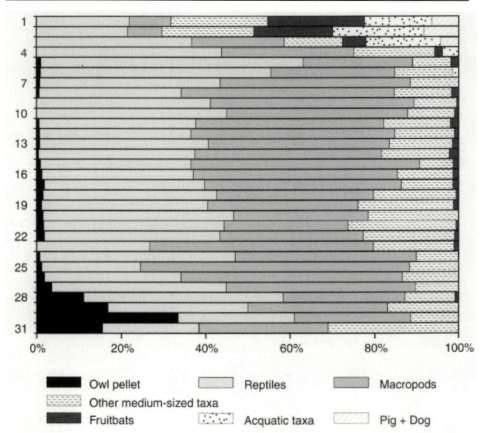

Fig. 15.6 Distribution of major faunal categories through the Lemdubu sequence.

Other than for these very conspicuous changes, the overall faunal composition shows little systematic change. Minor points of interest are the relative abundance of small murid remains in the lowermost levels of the deposit and the sustained peak in bandicoot abundances between Spits 17 and 25. The large kangaroo, tentatively identified as a species of *Macropus (Osphranter)*, is represented only in Spits 20 and 21.

Cuscuses comprise between 6 and 16% of the identified material, with no obvious trend through the sequence. Three species (*Phalanger intercastellanus*, *P. gymnotis* and *Spilocuscus maculatus*) are present throughout the sequence; the same species are recorded in the modern fauna. *Phalanger gymnotis* is an uncommon element at all levels in the deposit, while *Spilocuscus maculatus* is more abundant than *P. intercastellanus* in all levels. Nothing is known regarding the relative abundance of these species in the contemporary fauna.

Boid snake vertebrae are present in large numbers throughout the sequence save for in the lowermost levels. If numbers of tooth-bearing elements only are examined, boid snakes and agamid lizards appear to be relatively most abundant in the uppermost levels, while varanid lizards are more evenly distributed throughout.

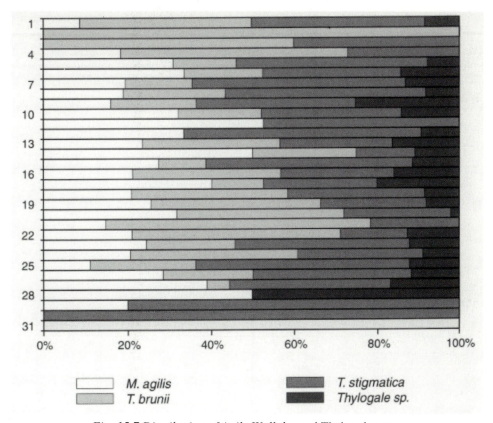

Fig. 15.7 Distribution of Agile Wallaby and Thylogale spp.

The fish rcmains are derived mostly from small- to medium-sized fish and are badly fragmented. Other than for some fragmentary head shields of an ariid catfish (freshwater to estuarine), very few potentially diagnostic elements are represented. Small quantities of freshwater turtle carapace in several levels also attest to use of riverine or wetland resources. Very small quantities of frog and bird bone, concentrated in the upper and lower levels of the site, complete the assemblage.

Dating the sequence

Dating of the Lemdubu sequence is still underway. Currently four radiocarbon dates and two Uranium Thorium dates have been obtained (Tables 15.3a and 15.3b). These samples date critical horizons in the Lemdubu occupation sequence. Both the shell dates are on *Geloina coaxans* valves (Table 15.3a). No marine reservoir correction has been applied to the shell dates as the extent to which this is appropriate for northern Australia/Papua New Guinea is presently unknown.

Peter Kershaw, Bruno David, Nigel Tapper, Dan Penny and Jonathan Brown (Eds): Bridging Wallace´s Line

The date from Spit 2 of 1830±60 years BP (ANU-10782) is consistent with the appearance of pottery and domestic animals in this level. Pottery and domestic animals make their first appearance in northern Moluccan sites about 3500 years ago (Bellwood 1997: 227). On this basis Spit 3 which contains pig and dog is likely to have a maximum age of 3500 years BP and is probably more recent.

The presence of large quantities of marine/estuarine shellfish only in Spit 4 and above indicated that the post-transgression period was recorded in the upper 25 cm of the deposit. It was therefore anticipated that Spit 5, the lowest spit with any quantity of marine shellfish would date to the period of sea level stabilisation approximately 6000 years ago. The date from LC5 of 11,700±130 years BP (ANU-10792) is therefore considerably older than anticipated.

*Table 15.3a. Carbon isotope composition and radiocarbon age estimates of samples from Liang Lemdubu, Aru Island. Radiocarbon years were converted into calender years using the calibration curves of Stuvier and Reimer (1993). Errors given are ± 1 s. * estimated values.*

Laboratory code	Site code	Material	Approximate depth below surface (cm)	d^{13} C (per mil, PDB)	±	Radiocarbon Years BP	±	Calendar Years	±
ANU 10782	LC 2	charcoal	5 - 10	-24.0	2.0	1830	60	1730	90
ANU 10792	LC 5	*Geliona* shell	25 - 30	-10.7	0.2	11,700	130	13,21	150
OZC 776	LC 19	*Geliona* shell	95 - 100	0.0	2.0*	17,750	450	20,62	660
OZC 777	LC 19	charcaol	130 - 136	-24.0	2.0*	13,300	300	15,89	440

Table 15.3b. Isotopic ratios for U and Th (activity) and calculated ages of speleothems from Liang Lemdubu, Aru Island. Analytical methods follow that of Ayliffe et al., 1998. 2σ errors are calculated by error propagation. Decay constants: $\lambda^{230}Th=9.1954x10^{-6}\pm7.19x10^{-8}$; $\lambda^{234}U=2.8349x10^{-6}\pm5.7x10^{-9}$; $\lambda^{238}U=1.55125x10^{-10}\pm1.66x10^{-13}$

Site code	Material	Approximate depth below surface (cm)	Uranium content (ppm)	$(^{230}Th/^{238}U)$	±	$(^{234}U/^{238}U)_o$
LC 26	flowstone	130 - 135	0.44	0.206	0.003	0.996
LC 28	flowstone	140 - 145	0.91	0.215	0.001	0.995

Table 15.3b. Continuation, right part

Site code	±	$(^{234}U/^{238}U)_t$	±	$(^{230}Th/^{232}Th)$	Calender Years	±
LC26	0.003	0.997	0.002	42	25,700	460
LC28	0.002	0.995	0.002	41	27,020	290

Spit 19 has a date on marine shellfish of 17,750±450 years BP (OZD-776). Two dates were obtained for LC26, one on a tiny fragment of what appeared to be charcoal (OZD-777), and the other on a fragment of a flowstone which formed a partial seal over the deposit at *c*.130 cm below surface. A further U/Th date was obtained from a lower flowstone LC28, at approximately 145 cm below surface. The date returned for the charcoal sample from LC26 is obviously at odds with the U/Th dates on the flowstones (one from the same unit), as well as the marine shell date LC19. In view of the extremely small size of the charcoal fragment from LC26, and the fact that the two U/Th flowstone dates and shell dates are in agreement, we believe the latter should be regarded as more reliable.

Faunal change and palaeoenvironment

The archaeological fauna is more diverse than the modern recorded fauna, with higher numbers of both savannah (most notably *Macropus agilis* and *I. macrourus*), and rainforest inhabitants (eg, *Thylogale stigmatica*, *Echymipera kalubu*, *Pogonomys* sp.). These differences may in part reflect the incomplete knowledge of the contemporary fauna. However, the fact that *M. agilis* and *I. macrourus* have not been detected in the modern fauna, points to their being either absent on this island entirely or at least severely restricted in distribution. As mentioned above, only a single fragment of *M. agilis* was found above Spit 4. This is most likely to result from limited reworking of bone from lower levels. Alternatively, it is possible that *M. agilis* was brought to the site from some distance away as the lowland evergreen rainforest environment around the site today does not provide a suitable habitat for either *M. agilis* or *I. macrourus*.

The fauna from the lower levels upwards to Spit 4 span the late Pleistocene and are indicative of a mosaic of closed and relatively open vegetation types. Extensive tracts of savannah are indicated by the abundance of *M. agilis* and *Isoodon macrourus*, while the presence of the various *Thylogale* species, *Echymipera rufescens* and the cuscuses points to patches of dense, moister vegetation, most likely located in sheltered contexts around the margins of the Aru plateaux and along watercourses (including the major channels). There is little evidence for faunal change throughout this part of the sequence suggesting an essentially stable environment from *c*.25,000 years BP to *c*.12,000 years BP The peak in bandicoot abundance between Spits 17 and 25 corresponds with the time when the most open conditions would be anticipated.

Nix and Kalma (1972) postulated the presence on the exposed Carpentarian Plain of extensive, continuous savannah habitats, with a large but shallow, brackish lake in the present Gulf of Carpentaria. These theoretical predictions were confirmed by sedimentological and palynological analyses of cores spanning ~35,000 years BP from the Gulf of Carpentaria (Torgersen *et al.* 1988). Their study confirms the presence of an extensive, shallow, fresh-to-brackish lake in the present Gulf, through to about 12,000 years BP when a sea connection was established to the west. A return to fully marine conditions around 8000 years BP marks the time of easterly connection to the Coral Sea and establishment of modern circulation patterns. The pollen spectra from *c*.30,000 years BP to the marine transgression remain fairly constant and resemble those generated by vegetation of the black soil plains currently distributed around the southern margins of the Gulf of Carpentaria, with sedge communities in lowlying areas and woodland on the higher, well-drained ground. The pollen record from the continental shelf adjacent to the wet tropics shows a similar trend (Kershaw 1995: 664).

Inundation of the Carpentarian Plain presumably brought wetter conditions to the newly formed coastal regions of southern New Guinea including the Aru Islands. This should have resulted in the relatively rapid expansion of lowland rainforest communities across large parts of the newly formed Aru Islands, especially on the western side of the island group. The upward migration or reconstitution of montane communities in New Guinea indicates a gradual increase in mean annual temperature from the height of the last glacial until *c.*12,000 years BP although De Decker *et al.* (1991) suggest that a peak in aeolian dust in the Lake Carpentarian record between *c.*11,375 and 10,430 years BP indicates drier conditions at this time. Some support for drier conditions around this time comes from other studies (e.g. Edwards *et al.* 1993, van der Kaars 1995). Amelioration to conditions approximating those of today appears to have occurred between 10,500 and 8500 years BP (Haberle 1994; van der Kaars 1995). The archaeological fauna tracks the expected changes in vegetation well, with the virtual disappearance of open savannah species between Spits 4 and 5. The date of 11,700 years BP appears slightly early for the expansion of wetter forest but it should be kept in mind that this is a single date obtained from a 5 cm slice of deposit which potentially spans several thousand years.

Summary of the sequence

Overall, the faunal sequence indicates that over much of the Aru Islands prior to 12,000 years BP rainforest was dispersed without large and continuous blocks such as are found today. Savannah communities were evidently extensive and supported a diverse savannah and forest edge fauna including several significant game species. The sequence provides positive evidence for the expansion of wetter communities after insularisation and evidence of almost immediate faunal loss of savannah-dependent species. While this finding was not unexpected, the fact that there is little or no evidence of occupation of Lemdubu following the expansion of the wet forests until the appearance of people with the pig, dog and presumably a fully developed horticultural economy, is surprising indeed. Although no dates are yet available for Spits 3 and 4, the date of 1800 years BP in Spit 2 and the other dates for the arrival of the SE Asian Neolithic in Maluku (Bellwood 1997) would suggest a likely maximum age of 3500 years BP for Spit 3. On this basis it seems likely that Spit 4 will represent a mixing horizon containing material from the terminal Pleistocene and late Holocene levels. Although further dating is clearly needed, the available evidence suggests that Lemdubu was either abandoned or visited only very occasionally during the first two thirds of the Holocene.

While it is premature to reflect on the extent to which this pattern might reflect regional abandonment as against disuse of this particular cave, it is worth considering in the context of other regions where rainforest expands at the beginning of the Holocene.

Aru in the context of the rainforest debate: Greater Australia and Island Melanesia

The extent to which hunter-gatherers can make a living in rainforest and the origin and antiquity of rainforest adaptations has been a subject of considerable recent debate (Bailey et al. 1989; Bailey and Headland 1991; Roosevelt et al. 1996). Bailey et al. (1989) confined their argument to tropical rainforest. Important evidence for this debate, however, comes from the Greater Australian region from a diversity of rainforest environments which have not been considered previously, including cool temperate and montane forests.

The two areas in Australia where archaeological research has contributed evidence pertinent to this debate are western Tasmania and NE Queensland. Today Western Tasmania is cloaked in dense closed-canopy cool temperate forest. The first Europeans to enter the region found the rainforest to be un-occupied, showing no traces of fires or any evidence of visitation (Jones 1995: 429). Their findings led to the view that western Tasmania was a true 'wilder-ness', never having been subject to human use or manipulation. Research over the last decade, however, has demonstrated that the region was occupied, and intensively so prior to the advance of rainforest communities in the terminal Pleistocene/early Holocene. Rockshelters and caves in SW Tasmania record extremely high artefact discard rates prior to the Last Glacial Maximum between c.30,000 and 26,000 years BP and again in some sites from 16,000 to 14,000 years BP Rainforest must have been present at this time but restricted to small pockets in the valley bottoms. At around 14,000 years BP the rainforest began to expand and it appears that some higher altitude sites cease to be used by c.15,000 years BP and most went out of use by 13,000 years BP (Porch and Allen 1995: 725). As Porch and Allen conclude, the best explanation for aban-donment of this region is still that of Kiernan et al. (1983) 'who suggested that recolonization of the region by forest tree species reduced the preferred habitats of the game species; game became scarce and humans also left' (Porch and Allen 1995: 725).

Further to the north, the most extensive tract of tropical rainforest in Australia occurs in NE Queensland, between the towns of Ingham and Cook-town. Pollen records indicate that the rainforest here expanded out from refugia to reach its current extent approximately 9000 years ago (Horsfall 1996: 177). The earliest evidence of human occupation comes from Jiyer Cave, about 50 km south of Cairns where the first occupation levels date to c.5000 years BP. This early occupation does not appear to be marked by distinctive rainforest adapta-tions. Almost all nut shells from toxic species which require processing derive from upper deposits dated to less than 1000 years BP. Although Horsfall suggests decay may be responsible for this pattern, pecked and battered basal stones which may have been used as anvils for grinding nuts and other materials are confined to the top 25 cm of deposit covering the last 2000 years. An open site, Mulgrave River 2, located 30 km south of Cairns, has a basal date of 2690 years BP At this site, nuts of toxic species were found in deposits nearly 2000 years old, however, specialised nut cracking stones were confined to the surface of the deposit (Horsfall 1996: 187). Horsfall (1996: 188) concludes that in her excavations 'the vertical distribution of stone artefacts possibly associated with complex processing appears to indicate that intensive use of toxic plants is relatively recent (ie. within the last millennium)'.

The main island of New Guinea provides considerable evidence for utilisation of rainforest habitats during the late Pleistocene and early Holocene but this evidence is mostly not from lowland tropical rainforest but rather from the high altitude cool montane forests. The earliest recorded archaeological sites on raised coral terraces of the Huon Peninsula dated to c.40,000 years BP have yielded numerous waisted axes, including some which would have been hafted, which have been viewed as circumstantial evidence for a great antiquity for rainforest use (Groube et al. 1986). Groube (1989) argued that these tools were used in forest clearance to promote conditions for growth of useful plants, implying considerable investment of energy into the management of forest resources.

The open site of Kosipe (White et al. 1970) at 2000 m contains waisted axes but no direct evidence of economic activities. When first occupied at c.25,000 years BP it was positioned close to the upper limit of forest growth and may have been strategically placed for exploitation of subalpine game resources and/or Pandanus fruits from the adjacent swamp. Other sites in Papua New Guinea which record human activities in forested environments during the late Pleistocene include two open sites, NFX at 1550 m in the Eastern Highland Province and Wañelek at 1680 m in the Madang Province, and the rockshelter Yuku at 1280 m in the Hagen Range (Hope and Golson 1995: 823). Three additional sites record human occupation between the LGM and the end of the Pleistocene, namely Kafiavana at 1350 m in the Eastern Highlands Province and Kiowa at 1530 m in the Simbu Province and Manim Cave, Western Highlands Province (Hope and Golson 1995: 821). All of these sites would have been situated in the montane forest zone during the LGM and several contain good economic evidence for the exploitation of forest fauna (Mountain 1991). Other evidence of forest utilisation comes from various pollen and charcoal records which suggest anthropogenic burning and forest clearance as early as 32,000 years BP at 1580 m in the Bailem Valley of Papua (Haberle et al.1991; Swadling and Hope 1992) and elsewhere in highland Papua New Guinea in the late Pleistocene (Hope and Golson 1995: 823)

By far the best faunal record of rainforest exploitation comes from the Nombe Rockshelter at 1720 m in the Simbu Province (Mountain 1991). Nombe contains evidence of sporadic human visitation between 25,000 years BP and c.14,000 years BP The faunal remains found between these levels suggest a variety of available environments were being exploited including open alpine/shrub/grassland. However the high proportion of forest species such as Dorcopsulus vanheurni, Dendrolagus goodfellowi and Phalanger carmelitae demonstrate the availability of primary mid-montane forest within easy reach of the cave. The fauna from the levels covering the period 14,000-10,000 years BP registers the presence of thicker forest as the tree line presumably rose with ameliorating climate. By 10,000 years BP the fauna indicates that mid-montane forest was dominant around the cave and grassland out of reach of the human occupants of the site (Mountain 1991: 840).

This geographically and temporally widespread evidence of forest utilisation in Papua New Guinea points to a long-established tradition of montane rainforest utilisation in New Guinea back to earliest occupation. Evidence of human activity in lowland tropical rainforest is far less compelling, however, far less archaeological and palynological research has been undertaken in such areas.

In the Birds Head region of Papua at an altitude of 325 m asl there is evidence of continuous occupation and utilisation of lowland rainforest fauna back to at least 8000 years BP at Kria Cave (Pasveer 1998, Pasveer and Aplin in press). At Toe Cave at the same altitude, a montane fauna precedes the rainforest fauna indicating an earlier phase of occupation than that reflected at Kria during colder drier conditions. While undated, this is presumed to reflect the climatic changes of the Last Glacial Maximum (Jelsma 1998, Pasveer and Aplin in press). Again, the pollen evidence indicates burning of local forests at 780 m in the Cyclops Mountains by 11,500 years BP. (Hope and Tulip 1994).

Other possible evidence for rainforest exploitation in the Pleistocene in Melanesia comes from the West New Britain site of Yombon (Pavlides and Gosden 1994). Here stone artefacts sealed by dated tephras demonstrate occupation between 35,000 years BP and 14,000 years BP. It is concluded that while direct pollen evidence is lacking, 'there is every reason to believe that the area was covered by rainforest during the late Pleistocene. Its equatorial position, low altitude and high rainfall all favour the establishment of a rainforest zone' (Pavlides and Gosden 1994: 609). Unfortunately, the Yombon site lacks dietary remains or any indication of specialised technology which might inform on the nature of cultural adaptations allowing humans to use the rainforests.

In summary it would appear that in the Greater Australian region there were a diversity of responses to rainforest and the ability of humans to use them effectively. Good evidence exists in Papua New Guinea and Papua for exploitation of closed montane forest in the Pleistocene and Holocene. Significantly, in Papua New Guinea we have specialised tools for forest clearance dated to 40,000 years BP coupled with pollen evidence suggesting that anthropogenic firing and clearance was underway at a similarly early date. Aside from Kria Cave in Papua and possibly Yombon, little firm evidence for early lowland tropical forest exploitation exists at present. While the inhabitants of Liang Lemdubu were using rainforest in the Pleistocene, it was as a component of a strategy focussed on a mosaic of environments. It is possible that in lowland areas such as Aru and NE Queensland where the earliest occupants would have inhabited open environments such as savannah, grassland or woodland, the expansion of the lowland rainforests may have initially presented difficulties for human exploitation. Being untooled for, and unskilled in, specialist use of these environments, an entire new set of hunting and gathering approaches had to be acquired independently and therefore took some time. The fact that Kria Cave shows early exploitation of lowland rainforest may be because the inhabitants of this region had only to make the transition from montane to lowland rainforest. Many of the same exploitation strategies and technology may have been able to be applied and therefore no lag time would be expected.

Conclusion

The earliest dated level at Lemdubu is 27,000 years old. As the lowest artefact is found in Spit 29, and the fauna in Spits 30 and 31 is probably not humanly derived, it seems likely that this date is close to the real age for initial human occupation at the site.

The Lemdubu Cave faunal sequence beginning 27,000 years ago, documents the presence on the late Pleistocene Kobroor Plateau of open savannah with denser, lusher vegetation present in pockets along watercourses and in other sheltered areas. The mammal fauna present at that time was basically similar to that found today in the Trans-Fly region of New Guinea and in certain parts of Cape York Peninsula, but included more species than either of these areas has today. People using the site at this time focussed their efforts on the procurement of the large-bodied Agile Wallaby, *Macropus agilis*, and several smaller wallabies (*Thylogale* spp.), with more casual acquisition of various medium-sized animals including cuscuses, bandicoots and pythons. Game was evidently abundant such that animal carcasses were incompletely utilised for available protein. There is little faunal change from the time of initial occupation throughout the last major glacial maximum at 18,000 years BP until the terminal Pleistocene about 12,000 years BP. Scarce remains of an as yet unidentified large kangaroo points to a significant temporal overlap between humans and at least one 'megafaunal' element.

The terminal Pleistocene saw the inundation of the Carpentarian Plain and the change to insular conditions. The expected spread of wetter forests in response to higher temperatures and increased rainfall is reflected in the decreased faunal diversity and the loss of the open savannah element, most prominently the Agile Wallaby and two of three smaller wallaby species. Human occupation of the cave effectively ceased at this time.

The cave appears to be reoccupied only in the late Holocene when there is a shift in patterns of resource use and presumably site use. Specific, targeted hunting of wallabies declines, replaced by increased use of riverine resources and various lesser game items (cuscuses, fruitbats). Wallace's observation of the interior groups of this area that 'Now and then they get wild pig or kangaroo, but too rarely to form anything like a regular part of their diet, which is essentially vegetable ... e.g. plantains, yams, sweet potatoes and raw sago; sugar cane, betel nuts, gambir and tobacco' (Wallace 1869: 343) contrasts sharply with the faunal deposit from the Pleistocene levels of Lemdubu but sits more comfortably with the Holocene assemblage. It seems probable that when the site was reoccupied it was by people with a horticulturally-based economy, domestic pig and dog and access to widespread marine-oriented trade systems.

Acknowledgements

Funding for research in the Aru Islands was obtained from the Australian Research Council Small Grant 1995 and an ARC Large Grant 1996-98. The 1996 Lemdubu excavation team consisted of O'Connor, Veth and Spriggs, Husni Mohammad from Puslit Arkenas, Branch Menado and Widia Nayati, Gajah Mada University, Yogyakarta. The research was sponsored in Indonesia by Puslit Arkenas in Jakarta and Universitas Pattimura in Ambon. Dr Peter Hiscock, Australian National University, provided the stone weights. Dr Pat Woolley (La Trobe University) and Dr Ric How, W.A. Museum, provided unpublished information on the modern mammal fauna of the Aru Islands. Dr Tim Flannery and Mr Walter Boles, both from the Australian Museum, assisted with some of the taxonomic determinations. Drs Juliette Pasveer, University of Groningen, assisted with the faunal analysis. Dr Geoff Hope, Australian National University, provided useful advice on the vegetation surrounding Liang Lemdubu.

References

Allen, J. 1977. Fishing for Wallabies: trade as a mechanism for social interaction, integration and elaboration on the central Papua coast. In: J. Friedman and M.J. Rowlands (Editors) *The Evolution of Social Systems*. Duckworth, London: 419-55.

Ayliffe, L.K., Marianelli, P.C., Moriarty, K.C., Wells, R.T., McCulloch, M.T., Mortimer, G.E. and Hellstrom, J.C. 1998. 500 Ka precipitation record from southeastern Australia: evidence for interglacial relative aridity. *Geology* 26: 147-150.

Bailey, R.C., Head, G., Jenike, M., Owen, B., Rechtman, R. and Zechenter, E. 1989. Hunting and gathering in tropical rainforest: is it possible? *American Anthropologist,* 91(1): 59-82.

Bailey, R.C. and Headland, T.N. 1991. The tropical rainforest: is it a productive environment for human foragers? *Human Ecology,* 19(2): 261-85.

Bellwood, P. 1997. *Prehistory of the Indo-Malaysian Archipelago (2nd Edition)*. University of Hawaii Press, Honolulu: 384 pp.

Birdsell, J.B. 1977. The Recalibration of a Paradigm for the First Peopling of Greater Australia. In: J. Allen, J. Golson and R. Jones (Editors) *Sunda and Sahul: Prehistoric Studies in Southeast Asia, Melanesia and Australia*. Academic Press, London: 113-67.

Chappell, J.M.A. 1994. Upper Quaternary sea levels, coral terraces, oxygen isotopes and deep-sea temperatures. *Journal of Geography,* 103: 828-40.

De Deckker, P., Correge, T. and Head, J. 1991. Late Pleistocene record of cyclic aeolian activity from tropical Australia suggesting the Younger Dryas is not an unusual climatic event. *Geology,* 19: 602-5.

Edwards, R.L., Beck, J.W., Burr, G.S., Donahue, D.J., Chappell, J., Bloom, A.L., Druffel, E.R.M. and Taylor, F.W. 1993. A large drop in atmospheric ^{14}C /^{12}C and reduced melting in the Younger Dryas, documented with ^{230}Th ages of corals. *Science,* 260: 962-8.

Flannery, T.F. 1995(a). *Mammals of New Guinea (Revised Edition)*. Reed Books, Chatswood: 568 pp.

Flannery, T.F. 1995(b). *Mammals of the South-West Pacific and Moluccan Islands*. Reed Books, Chatswood: 464 pp.

Flannery, T.F. and White, J.P. 1991. Animal translocations. *National Geographic Research and Exploration,* 7(1): 96-113.

Gordon, G. 1995. Rufous Spiny Bandicoot, *Echymipera rufescens.* In: R. Strahan (Editor) *The Mammals of Australia.* Reed Books, Chatswood: 191-2.

Groube, L., Chappell, J., Muke, J. and Price, D. 1986. A 40,000 year-old occupation site at Huon Peninsula, Papua New Guinea. *Nature,* 324: 453-5.

Groube, L. 1989. The taming of the rainforests: a model for Late Pleistocene forest exploitation in New Guinea. In: D.R. Harris and G.C. Hillman (Editors) *Foraging and Farming. The Evolution of Plant Exploitation.* Unwin Hyman, London: 292-304.

Haberle, S., Hope, G.S. and De Fretes, Y. 1991. Environmental change in the Baliem Valley, montane Irian Jaya, Republic of Indonesia. *Journal of Biogeography,* 18: 25-40.

Haberle, S. 1994. Anthropogenic indicators in pollen diagrams: problems and prospects for late Quaternary palynology in New Guinea. In: J.G. Hather (Editor). *Tropical Archaeobotany: Applications and New Developments*: Routledge, London: 172-201.

Hope, G. and Golson, J. 1995. Late Quaternary Change in the mountains of New Guinea In: F.J. Allen and J.F. O'Connell (Editors) *Pleistocene to Holocene in Australia and Papua New Guinea (Antiquity Special Issue 265),* 69: 818-30.

Hope, G. and Tulip, J. 1994. A long vegetational history from lowland Irian Jaya, Indonesia. *Palaeogeography, Palaeoclimatology, Palaeoecology,* 109: 385-98.

Horsfall, N. 1996. Holocene occupation of the tropical rainforests of north Queensland. In: P. Veth and P. Hiscock (Editors) *Archaeology of Northern Australia Regional Perspectives (Tempus Vol. 4).* University of Queensland Press, Brisbane: 175-90.

Irwin, G. 1992. *The Prehistoric Exploration and Colonisation of the Pacific.* Cambridge University Press, Cambridge: 240 pp.

Jelsma, J. 1998. Room with a view. An excavation in Toe Cave, Ayamaru district, Bird's Head, Irian Jaya, Indonesia. *Modern Quaternary Research in Southeast Asia,* 15: 41-65.

Johnson, P.M and Vernes, K.A. 1995. Red-legged Pademelon, *Thylogale stigmatica.* In: R. Strahan (Editor) *The Mammals of Australia.* Reed Books, Chatswood: 397-9.

Jones, R. 1995. Tasmanian Archaeology: establishing the sequences. *Annual Review of Anthropology,* 24: 423-46.

Kershaw, A.P. 1995. Environmental change in Greater Australia. In: F.J. Allen and J.F. O'Connell (Editors) *Pleistocene to Holocene in Australia and Papua New Guinea (Antiquity Special Issue 265),* 69: 656-75.

Kiernan, K., Jones, R. and Ranson, D. 1983. New evidence from Fraser Cave for glacial age man in southwest Tasmania. *Nature,* 301: 28-32.

Kitchener, D.J. no date. *Report on Phase X of a Study of the Vertebrate Fauna of Nusa Tenggara and the Moluccas, Indonesia. Ambon, Banda, Aru and Kai Islands.* Unpublished Report.

Kitchener D.J., Cooper N. and Maryanto I. 1995. The Myotis adversus (Chiroptera: Vespertilionidae) species complex in eastern Indonesia, Australia, Papua New Guinea and the Solomon Islands. *Records of the Western Australian Museum,* 17: 191-212.

Monk, K.A., De Fretes, Y.and Reksodiharjo-Lilley, G. 1997. *The Ecology of Nusa Tenggara and Maluku.* Periplus, Hong Kong: 966 pp.

Mountain, M.J. 1991. *Highland New Guinea Hunter-Gatherers: The Evidence of Nombe Rockshelter, Simbu, with Emphasis on the Pleistocene.* Unpublished PhD thesis, Australian National University, Canberra.

Musser, G. and Carleton, M.D. 1993. Muridae. In: D.E. Wilson and D.M. Reeder (Editors) *Mammal Species of the World: A Taxonomic and Geographic Reference.* Smithsonian Institute Press, Washington D.C: 501-756.

Nix, H.A. and Kalma J.D. 1972. Climate as a dominant control in the biogeography of northern Australia and New Guinea. In: D. Walker (Editor) *Bridge and Barrier: The Natural and Cultural History of Torres Strait*. Department of Biogeography and Geomorphology, RSPAS, The Australian National University, Canberra: 61-91.

O'Connor, S. 1996. 30,000 years in the Kimberley: results of excavation of three rockshelters in the coastal west Kimberley, W.A. In: P. Veth and P. Hiscock (Editors) *Archaeology of Northern Australia Regional Perspectives (Tempus Vol. 4)*. University of Queensland Press, Brisbane: 26-49.

Pasveer, J.M. 1998. Kria cave: An 8000-year occupation sequence from the Bird's Head of Irian Jaya. *Modern Quaternary Research in Southeast Asia*, 15: 67-89.

Pasveer, J.M. and Aplin, K.P. in press. Late Pleistocene to modern vertebrate faunal succession and environmental change in lowland New Guinea: evidence from the Bird's Head of Irian Jaya, Indonesia. In: J. Miedema, C. Ode and M.A.C. Dam (Editors) *Perspectives on the Bird's Head Peninsula of Irian Jaya, Indonesia. Proceedings of an Interdisciplinary Conference*. Rodopi Publishers, Leiden.

Pavlides, C. and Gosden, C. 1994. 35,000-year-old sites in the rainforests of West New Britain, Papua New Guinea. *Antiquity,* 68(260): 604-10.

Porch, N. and Allen, J. 1995. Tasmania: archaeological and palaeo-ecological perspectives. *Antiquity,* 69(265): 714-32.

Roosevelt, A.C., Lima Da Costa, M., Lopes Machado, C., Michab, M., Mercier, N., Valladas, H., Feathers, J., Barnett, W., Imazo Da Silverira, M., Henderson, A., Silva, J., Chernoff, B., Reese, D.S., Holman,J.A., Toth, N. and Schick, K. 1996. Paleo-indian cave dwellers in the Amazon: the peopling of the Americas. *Science,* 272: 373-84.

Stuiver, M. and Reimer, P.J. 1993. *Radiocarbon,* 35: 215-230.

Swadling, P. and Hope, G.A.S. 1992. Environmental change in New Guinea since human settlement. In: J.R. Dodson (Editor) *The Naive Lands: Prehistory and Environmental Change in Southwest Pacific*. Longman Cheshire, Melbourne: 13-42

Szalay, A. 1995. Annotated faunal list for the South-West Pacific and Moluccan Islands. In: T. Flannery (Editor) *Mammals of the South-West Pacific and Moluccan Islands*. Reed Books, Chatswood: 409-23.

Tate, G.H.H. and Archbold, R. 1935. Results of the Archbold Expedition No. 4. An apparently new race of wallabies from southern New Guinea. *American Museum Novitates* No. 804: 2.

Torgersen, T., Luly, J., De Deckker, P., Jones, M.R., Searle, D.E., Chivas A.R. and Ullman, W.J. 1988. Late Quaternary environments of the Carpentaria Basin, Australia, *Palaeogeography, Palaeoclimatology, Palaeoecology,* 67: 245-61.

van der Kaars, S. 1995. Preliminary palynological results on the Pleistocene-Holocene transition, Seram Trench, offshore Irian Jaya, Indonesia. *Geologie en Mijnbouw,* 74: 285-6. (extended abstract)

van Dyck, S.M. 1988. The Bronze Quoll, *Dasyurus spartacus* (Marsupialia: Dasyuridae), a new species from the savannas of Papua New Guinea. *Australian Mammalogy,* 11(1-2): 145-156.

Waithman, J. 1979. A report on a collection of mammals from southwest Papua. 1972-1973. *Australian Zoologist,* 20: 313-26.

Wallace, A.R. 1869. *The Malay Archipelago, the Land of the Orang-Utan and the Bird of Paradise: a Narrative of Travel, with Studies of Man and Nature*. Macmillan, London: 653 pp.

White, J.P., Cook, K.A.W. and Buxton, B.P. 1970. Kosipe: a late Pleistocene site in the Papuan Highlands. *Proceedings of the Prehistory Society*, 36: 152-70.

Peter Kershaw, Bruno David, Nigel Tapper, Dan Penny and Jonathan Brown (Eds): Bridging Wallace´s Line

Winter, J.W. and Leung, L.K-P. 1995(a). Common Spotted Cuscus *Spilocuscus macuatus*. In: R. Strahan (Editor) *The Mammals of Australia*. Reed Books, Chatswood: 266-8.

Winter, J.W. and Leung, L.K-P. 1995(b). Southern Common Cuscus *Phalanger intercastellanus*. In: R. Strahan (Editor) *The Mammals of Australia*. Reed Books, Chatswood: 268-70.

Long-term Archaeological and Environmental Trends:
A Comparison from Late Pleistocene-Holocene Australia

Harry Lourandos and Bruno David

Introduction

The relatively isolated, island continent of Australia offers archaeologists the challenge of modelling long-term trends in hunter-gatherer settlement; a further example of processes more familiar on other land masses. Australia has been settled by Aboriginal people for at least 40,000 years and perhaps considerably longer. Between about 30,000 and 20,000 years BP most, if not all, major environments had been occupied. Long-term archaeological trends, however, have rarely been explored for the Australian continent as a whole, and have been difficult to measure, describe and compare. Substantial, regional data sets, of archaeological sites and radiocarbon dates, are now available for many parts of Australia, enabling us to model long-term archaeological trends, and compare these to contemporaneous long-term environmental trends. Comparisons of this kind allow us to assess the relationship between human behaviour and environment over long stretches of time. Further, it is argued here that common methods exist by which regional archaeological and environmental trends can be compared from contrasting environmental sectors of the continent. In these ways general models of land-use and socio-demographic patterns can be generated covering vast time periods – in this case the late Pleistocene-Holocene – and across diverse landscapes, on a continental scale.

Recently, Australian regional and continental patterns and trends have been investigated by employing the temporal distribution of radiocarbon dates (Bird and Frankel 1991a, 1991b; Smith and Sharp 1993; Allen and Holdaway 1995; Allen and O'Connell 1995; Holdaway and Porch 1995), as well as by more individual methods (O'Connor *et al.* 1993; Morwood and Hobbs 1995; also Lourandos 1993). For example, O'Connell and Allen (1995) have recently noted close associations between archaeological and environmental patterns during the terminal Pleistocene-early Holocene period in Australia. They argued that regions which have witnessed population growth and increases in regional land use during particular episodes of prehistory 'should be marked archaeologically by diachronic increases in site numbers and rates of refuse discard across most habitats. Departures from a monotonic pattern in either parameter should correlate with local fluctuations in rainfall or other measures of general resource availability' (O'Connell and Allen 1995: 857).

ISBN 3-923381-47-6

This general approach has been extended by David and Lourandos (1997) who demonstrated both general correlations as well as divergences between long-term archaeological and environmental trends covering the last 37,000 years in SE Cape York Peninsula, NE Australia. That is, change needs to be considered in relation to both environmental shifts and cultural-historical trajectories. The way people react to their environment is situated within particular historical and socio-cultural contexts (Ingold 1980, 1988; Ellen 1982, 1988; Lourandos 1987, 1988, 1993, 1997; David 1994). Here we extend the latter investigations further still by using common methods to compare long-term archaeological and environmental trends from diverse Australian environmental zones – tropical, arid and temperate – for the lengthy late Pleistocene-Holocene period[1]. In this way, regional and continental trends are compared, as are various epochs of the late Pleistocene and Holocene. We can therefore address questions concerning periods of socio-cultural change, and their relationship to environmental shifts; each a key issue in Australia and hunter-gatherer studies generally.

The method or index used here in constructing the regional archaeological trends is the temporal distribution of radiocarbon dates from rockshelter (and cave) sites. Rockshelters were chosen in preference to open sites (including shell middens) and sites on more unstable landforms, as they offered the potential for long, relatively well stratified sequences in long-standing features of the landscape. Eleven study areas have been investigated here: three in the tropical north, two in subtropical regions, two in arid and semiarid Australia, and four more in the temperate southeast (Figure 16.1). Long-term regional palaeo-environmental sequences have been obtained from studies on lake levels and pollen analyses. Essentially we argue that: a) from at least the last glacial maximum until about the mid Holocene, archaeological trends, *at this scale of analysis*, appear to follow closely upon environmental trends; and b) archaeological and environmental trends diverge significantly during the last 4000 years in all study areas. We conclude by suggesting that changing hunter-gatherer socio-demographic patterns are linked to these trends.

Environment

Mainland Australia lies roughly between tropical southern latitudes of 10 degrees and southern temperate latitudes of about 43.5 degrees. Today, tropical Australia (including arid and semiarid areas) receives summer rainfall with dry winters, while the temperate zone (including its more arid regions) has winter rainfall and drier summers. Subtropical regions experience a less seasonal rainfall. Forests are most widespread in the humid belt, with more open vegetation in areas of lower rainfall. During the Pleistocene the continent of Greater Australia was significantly more extensive and also incorporated the present-day islands

---

[1] While this paper stands alone, it is also part of a series that explores temporal trends in Australian prehistory at various temporal and geographical scales, using a variety of methods and interpretative perspectives (e.g. David and Lourandos 1997, in press a, in press b; Lourandos 1996).

of New Guinea and Tasmania, situated respectively to the north and south. At the time of the last glacial maximum (18,000 years BP), for example, the arid zone extended beyond the present semiarid belt (see below; also Figure 16.1).

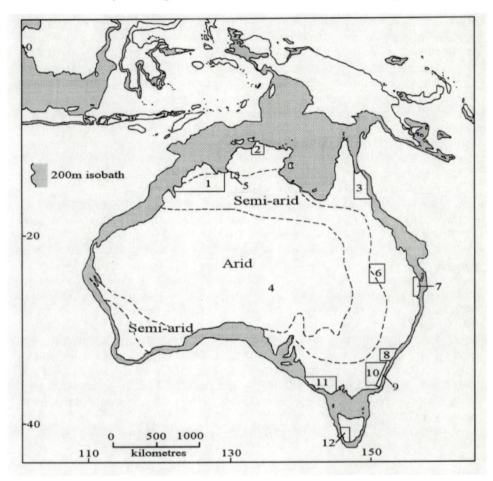

Fig. 16.1 Map of Greater Australia, including mainland Australia, Papua New Guinea and Tasmania, together with the 11 study areas (plus one extra region) mentioned in the text. A more humid region surrounds the semiarid zone. Study areas include: 1) Kimberley; 2) Kakadu; 3) SE Cape York Peninsula; 4) arid zone; 5) Wardaman; 6) Central Queensland Highlands; 7) SE Queensland; 8) Sydney and Blue Mountains; 9) South Coast New South Wales; 10) Southern Uplands; 11) SW Victoria-SE South Australia; and 12) SW Tasmania. The Pleistocene coastline is indicated by the 200 m submarine contour.

Long-term trends

We investigate long-term archaeological trends across 11 regions of Australia by considering all radiocarbon dates associated with cultural deposits as constituting, in themselves, an archaeological data base worthy of further exploration (Rick 1987). Our analyses are based on a modified version of Rick's (1987) method, whereby the frequency of radiocarbon dates for any given region is plotted on a graph in sliding 1000 year intervals measured every 500 years. Rick's method has been employed widely in recent Australian studies of regional occupational trends (Bird and Frankel 1991a, 1991b; Allen and Holdaway 1995; Holdaway and Porch 1995; Ulm 1995; David and Lourandos 1997). In our version, the temporal scale is divided into 1000 year time periods, and each date is plotted three times in order to even-out minor fluctuations in the curve. For example, a radiocarbon date of 7370 ± 60 years BP would be plotted in the 7000-7500 years BP time 'spike', as well as in the preceding and proceeding 500 year 'spikes'. The only exception is the most recent time spike, which receives its own radiocarbon dates plus those from the preceding 500 year period, but of course not from the subsequent one (as this does not yet exist). Therefore, the most recent 1000 year period should show an approximate 33% artificial *downplay* of the trend as a result of the method itself.

As Rick (1987) argued, this method treats large regional samples of radiocarbon dates as potentially indicative of occupational trends for any given region of study. It was designed principally to investigate general trends *between* regions, and not minor fluctuations evident at finer-grained levels. Patterns in the regional, temporal distribution of radiocarbon dates, including differences between regions, lead us to ask why certain periods of time are represented by numerous radiocarbon dates, while others possess relatively few. Are these differences due to occupational trends, or to biases in the preservation, retrieval and/or selection of appropriate materials for dating (see David and Lourandos 1997 for further discussion). Therefore, one should be cautious when interpreting the resulting graphs, which in turn should ideally be based on large data bases in order to minimise the effects of site-specific selective biases. Nevertheless, we argue that general patterns in the frequency of dates obtained through time are potentially meaningful archaeological indicators. Given these limitations, we view the approach as a *first step* in generating archaeological models, to establish broad, general outlines only. Perhaps the emerging trends can be adequately accounted for by reference to taphonomic (including post-depositional) factors. But if not, they may then be taken as useful indicators of changing regional occupational patterns through time in hunter-gatherer societies and, by implication, also of regional land use. The reader is directed to Rick's (1987) original paper and to Allen and Holdaway (1995) for further discussions of the usefulness and limitations of Rick's method to this type of research.

An extensive, but not exhaustive, coverage was undertaken of all available information on radiocarbon dates from the 11 study areas, including published and unpublished data. In all, 696 radiocarbon dates from 189 rockshelters and caves were used, averaging 3.7 dates per site (Table 16.1; Figure 16.2; Appendix 16.1)[2].

---

[2] The radiocarbon dates from the 11 Australian regions analysed in this paper were compiled by considering all dates accepted as reliable by the original researchers and reported in the following journals: *Archaeology and Physical Anthropology in*

Table 16.1: Number of sites and dates used in this paper.

Region	# of Radiocarbon-Dated Sites	# of Radiocarbon Dates	Mean # of Dates/Site
Kimberley	13	32	2.5
Kakadu	15	61	4.1
SE Cape York Peninsula	28	152	5.4
Central Queensland Highlands	11	47	4.3
SE Queensland	7	25	3.6
Arid Zone	43	139	3.2
Wardaman Country	8	26	3.3
Sydney and the Blue Mountains	27	82	3.0
S Coast of New South Wales	13	40	3.1
Southern Uplands	10	40	4.0
SW Victoria and SE S Aust.	14	52	3.7
Total	**189**	**696**	**3.7**

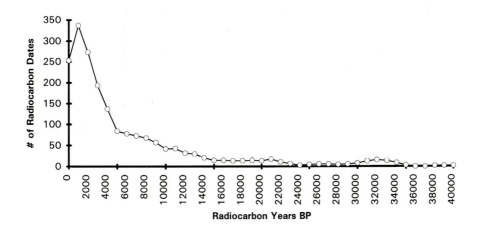

Fig. 16.2 Radiocarbon date curve for the 696 dates from Australia considered in this paper.

Oceania, Archaeology in Oceania, The Artefact, Australian Aboriginal Studies, Australian Archaeology, Mankind, Newsletter of the Australian Institute of Aboriginal Studies, Queensland Archaeological Research, Records of the Australian Museum and *Rock Art Research.* In addition, the radiocarbon dates from all other sites known by us to have been excavated in these regions were also included by consulting the original site reports and all other known published reports. This necessitated some selective perusal of *Antiquity, Asian Perspectives, Current Anthropology, Proceeding of the Prehistoric Society, Radiocarbon,* other journals, books, monographs and unpublished theses (see Appendix 16.1).

Results

Given the broad-based nature of the data, in some cases of relatively small sample sizes, only general trends are considered here. Our data sets nevertheless include some of the most studied regions of Australia, and from these more radiocarbon dates have been obtained than elsewhere.

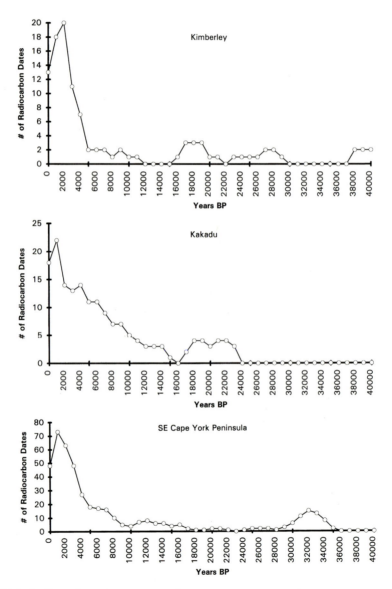

Fig. 16.3 Radiocarbon date curves for regions in tropical Australia: Kimberley; Kakadu; SE Cape York Peninsula.

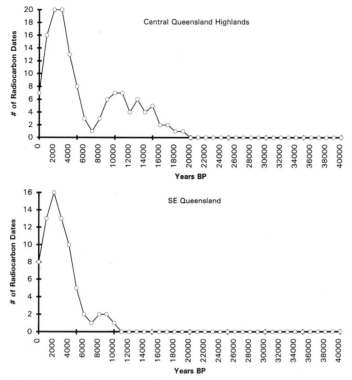

Fig. 16.4 Radiocarbon date curves for regions in sub-tropical Australia: Central Queensland Highlands; SE Queensland.

Tropical North

In SE Cape York Peninsula (Figure 16.3) the spread of radiocarbon dates covers over 37,000[3] years and indicates a cluster of dates prior to about 30,000 years BP with decreasing numbers until after the glacial maximum. From the terminal Pleistocene an increasing trend in numbers of dates is indicated which continues into the early Holocene; increasing more than threefold from around 4000 years BP and continuing so throughout the late Holocene until recent times (also David and Lourandos 1997). This general trend is reflected also in the two other tropical study areas, Kakadu and Kimberley (Figure 16.3), with some minor differences. For example, in the latter areas numbers of dates cluster also during the glacial maximum around 18,000 years BP, unlike in SE Cape York Peninsula. The general tropical Australian trend is reflected also in both the subtropical Central Queensland Highlands and SE Queensland (Figure 16.4), beginning from around 20,000 years BP in the former and 11,000 years BP in the latter. Minor fluctuations in these two curves, such as a suggested decline in numbers of dates in the mid Holocene, may be real or a result of small sample sizes.

---

[3] The single radiocarbon date of >37,170 years BP from Ngarrabullgan Cave is a minimum date, and could therefore not be plotted on the graph.

Peter Kershaw, Bruno David, Nigel Tapper, Dan Penny and Jonathan Brown (Eds): Bridging Wallace's Line

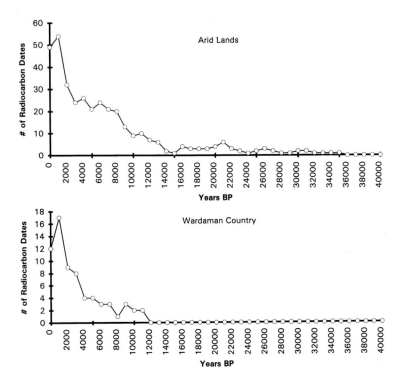

Fig. 16.5 Radiocarbon date curves for regions in arid and semiarid Australia: arid zone; Wardaman.

Arid and semiarid Australia

From arid Australia (Figure 16.5) the spread of dates indicates low numbers from around 35,000 years BP until about 14,000 years BP after which numbers increase and continue to do so until recent times, the steepest increase beginning in the last 2000 years. The Wardaman sample (Figure 16.5), from semiarid terrain, indicates similar results beginning around 12,000 years BP.

Temperate SE Australia

Similar general trends in distribution of radiocarbon dates are indicated in temperate SE Australia, beginning around 24,000 years BP and 22,000 years BP respectively in the Sydney-Blue Mountains region and South Coast of New South Wales (Figure 16.6). In neither region are dates recorded for the last glacial maximum (*c*.18,000 years BP), whereas numbers of dates begin to increase during the terminal Pleistocene and peak in the late Holocene after 4000 years BP. For the Southern Uplands and SW Victoria-SE South Australia (Figure 16.6), the trends begin around 23,000 years BP and 12,000 years BP respectively, both peaking markedly in the late Holocene. Small decreases are apparent around 5000 years BP.

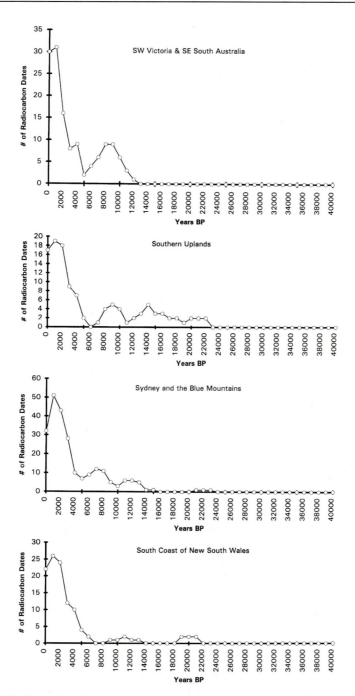

Fig. 16.6 Radiocarbon date curves for regions in temperate Australia: Sydney and Blue Mountains; South Coast New South Wales; Southern Uplands; SW Victoria-SE South Australia.

Peter Kershaw, Bruno David, Nigel Tapper, Dan Penny and Jonathan Brown (Eds): Bridging Wallace´s Line

The apparent downturn in the curves during the last 1000 years can be explained by three main factors: a) the upper levels of most archaeological sites have not been dated; b) the recent European contact phase of Australian history has often been disturbed or obliterated in upper levels of sites, or has witnessed an abandonment of sites due to European intrusions in all regions. Furthermore, where European goods are found, radiocarbon dates are simply not obtained; and c) our methods of pooling dates from three periods of time except for the last 1000 years where dates are only pooled from two periods (see above). The only regions where such factors need to be assessed more closely are those of the Central Queensland Highlands and SE Queensland where the recent 'downturn' in the curve covers a slightly longer period and thereby may represent occupational trends. That these two regions neighbour each other suggests that cultural-historical factors may be operating here. Downturns in curves during the mid Holocene in both subtropical and temperate regions also need to be investigated more closely to distinguish genuine trends from biased results due, for example, to small sample sizes. The most marked 'downturns' during the period are found in some of the smaller-sized samples (Central Queensland Highlands, Southern Uplands, SW Victoria-SE South Australia) (but see Hiscock 1986).

Can the general trends revealed in each of the 11 regions studied be explained without recourse to *socio-cultural* explanations? Are they a taphonomic illusion? Are the radiocarbon dates a function of archaeologists preferentially selecting mid and late Holocene samples from longer sequences for dating? Conversely, is the absence or relative paucity of dates from the glacial maximum due to biases in the preservation or retrieval of charcoal? We argue against the above alternatives, as there is a far greater emphasis by Australian archaeologists on the search for, and dating of, *Pleistocene* sites. Considerably less emphasis is placed on dating the surface of sites or the most recent levels of occupation. Each of these factors skews our results towards the older periods of time; and yet the patterns observed in the graphs repeatedly show major increases during relatively recent times (in particular the late Holocene). Are we, then, perhaps also observing a decay curve, one illustrating loss of data through time? Or could it be that the major increases in the incidence of radiocarbon dates after *c.*4000 years BP is due to the proximity of sites to fluctuating coastlines, which themselves only stabilised during the mid to late Holocene. We have addressed these two possibilities by considering the following factors. First, by using only rockshelter sites, open sites in more vulnerable, less geomorphologically stable locations have been eliminated. Many of the rockshelters have culturally sterile sediments underlying their archaeological deposits, which indicates that occupation of the sites post-dated by some time the existence of the shelters themselves. Second, in order to minimise or eliminate the direct influence of changing sea levels on the trends (and especially the post-glacial stabilisation of sea levels in the mid Holocene), we have plotted only the radiocarbon dates (N=329) from those rockshelters (N=84) which are located more than 100 km from the coast (Figure 16.7). The curve obtained in this way again corresponds with the general curves discussed above. And third, so as to eliminate the possibility that rockshelters which only formed recently (that is, during the late

Holocene) have been included (thus biasing our overall sample in favour of recent natural site formation processes), in Figure 16.8 we have plotted only the radiocarbon dates from *Pleistocene* rockshelters located more than 100 km inland. It can be seen that the latter group of 177 radiocarbon dates from 23 inland Pleistocene sites also produces a curve similar to the general curve presented above; indicating that the most significant increases in radiocarbon dates occur in the post-4000 years BP period of the late Holocene.

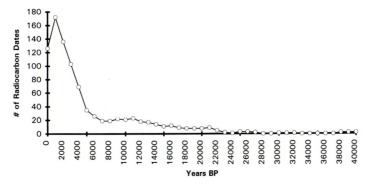

Fig. 16.7 Radiocarbon date curve for sites greater than 100 km inland.

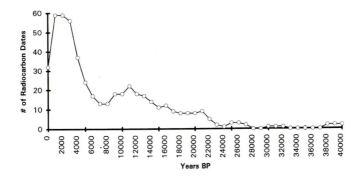

Fig. 16.8 Radiocarbon date curve for Pleistocene sites greater than 100km inland. The only sites included here are those where occupation begins during the Pleistocene; the downturn in radiocarbon dates immediately following 10,000 years BP can thus be explained by this absence of newly established, Holocene sites from this point onwards.

   In study areas such as SE Cape York and the South Coast of New South Wales the date curves generally have been correlated positively with other indices (David and Lourandos 1997, figures 10-11; Hughes and Lampert 1982: 21, 6-7). These include curves based on a) the number of occupied rockshelters through time and b) the initial occupation of sites. These correlations, therefore, lend considerable support to our results and to the cultural significance of the general curves themselves. We now compare these archaeological results to the palaeoenvironmental data and trends.

Palaeoenvironmental trends

General palaeoenvironmental information and long-term temporal trends arc discussed from three key Australian study regions where the data are most plentiful
and, to some extent, resolved. These include the three major environmental
sectors of the continent – tropical, arid and temperate (Figure 16.9). Information
comes mainly from lake levels reconstructed from a variety of sedimentological
and palaeoecological lines of evidence which provide information on water
availability, and dry land pollen records which provide evidence on past vegetation and inferred climatic conditions (Harrison and Dodson 1993; Kershaw
1995). Throughout the continent moisture availability was relatively high prior
to about 30,000 years BP, with increasingly drier conditions from this time to the
height of the glacial period, *c.*24,000 to 15,000 years BP; maximum dryness
occurred at some time between 15,000 and 11,000 (Kershaw and Nanson 1993).
Effective precipitation then increased and peaked at some time during the mid
Holocene. Maximum effective precipitation and highest lake levels were greatest
between about 6500 and 3500 years BP in NE Australia, and between 7000 and
5000 years BP in SE Australia (Hiscock and Kershaw 1992; Kershaw 1995).
The late Holocene, from about 5000 years BP onwards, is characterised by climatic variability (McGlone *et al.* 1992), with lowest moisture levels centred
around 2000 years BP in northern Australia and about 3500 years BP in SE
Australia.

Temperatures were lower than today until the Holocene, with minimum
values – between 4° and 8°C lower than present in southern Australia and probably much less in northern Australia – during the last glacial period. Maximum
temperatures, about 10°C higher than today, may have been achieved in the early
Holocene in southern Australia (Lloyd and Kershaw 1997), but not until 5000 to
3500 years BP in NE Australia (Bowler 1981; Bowler and Wasson 1984;
Dodson 1974, 1992; Kershaw 1995; Kershaw and Nix 1989; Ross *et al.* 1992;
Woodroffe *et al.* 1988) (Figures 16.9 and 16.10).

Natural levels of bioproduction have been correlated generally with long-
term environmental trends in other regions of the world, including the European
Upper Palaeolithic (Mellars 1985; Soffer 1985; Gamble 1986). In a similar vein,
the Australian data (based on indices such as precipitation and temperature;
Figures 16.9 and 16.10) would suggest that natural levels of bioproduction were
being reduced as climate became more stressful (drier and cooler) after about
24,000 years BP, reaching their lowest levels sometime during the last 8000
years of the Pleistocene (timing depending on location). Following this, bioproduction began to increase slowly, although this may not have happened in
some parts of tropical Australia until 13,000-10,000 years BP. This increasingly
upward trend continued throughout the terminal Pleistocene and into the early
Holocene, peaking towards the mid Holocene. Relatively lower levels of bioproduction appear to have characterised the late Holocene (after 5000 years BP)
when climate was generally drier, and driest between *c.*3000-2000 years BP.

A comparison of the archaeological and palaeoenvironmental trends indicates
the following. In all cases, a general correlation of both trends is apparent
between the terminal Pleistocene (17,000-11,000 years BP) until about the mid
Holocene. In some regions (Kimberley, Kakadu), there is evidence of this also

before the glacial maximum; and in others (SE Cape York Peninsula) back even further to around 35,000 years BP and beyond. After 4000 years BP, however, the archaeological and environmental patterns diverge significantly in all sampled regions. In all cases, the general pattern is essentially the same, although the meaning of minor fluctuations is, at this stage, uncertain. This is most evident in regions with small databases, indicating that they may be a function of sample sizes. This cannot be argued for the general pattern, however, nor does it appear to be explained by methodological bias or data loss, as discussed above. Further supporting evidence of results comparable to our own have been obtained recently from SW Tasmanian rockshelters covering the Pleistocene period *c*.35,000-10,000 years BP, by using similar methods (Rick's method) (Holdaway and Porch 1995). In this case also, archaeological trends appear to closely follow environmental oscillations, and thereby again reinforce the methods employed to define the archaeological trends. We can, therefore, now consider possible interpretations of all this evidence.

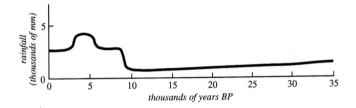

Fig. 16.9 General rainfall levels, based on pollen curves from the Atherton Tablelands, NE Queensland (based on Ross et al. 1992: 91, figure 5.12).

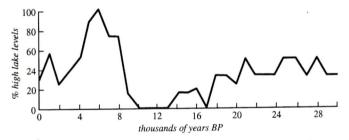

Fig. 16.10 General high lake levels from near-coastal South Australian and Victorian lakes (based on Ross et al. 1992: 80, figure 5.3).

Summary of trends

While information varies regionally, the following general sequence can be proposed from the above archaeological trends for the key Australian environmental sectors studied here, tropical, arid and temperate.

Prior to about 25,000 years BP archaeological data are relatively sparse on present evidence.

Between about 22,000-17,000 years BP, the last glacial maximum, there is a decline in site use in some drier areas (e.g. SE Cape York Peninsula), while in other regions, including the arid zone, this trend is not apparent. In contrast, in some more humid tropical areas (Kakadu, Kimberley) increased site use occurs. Changing patterns in site use also appear in SW Tasmania at this time.

From the terminal Pleistocene to the early Holocene, as climate ameliorated in most regions, there are indications of increased site establishment and use in all study areas and environmental zones (tropical, arid and temperate, including SW Tasmania). Between 10,000-*c*.5000 years BP, the period of maximum climate amelioration and highest levels of bioproduction, the trend in site establishment and use commenced during the terminal Pleistocene continued or increased in most regions studied.

After 4000 years BP a steep acceleration of the trend in site use is evident throughout the drier, more stressful late Holocene period, in all sampled regions of mainland Australia.

Scale, trends, trajectories

To evaluate long-term trends in both archaeological and environmental information both data and interpretations need to be scaled at the same level to achieve comparable results (Marquardt 1985; David and Chant 1995; Lourandos 1996). In this case we emphasise long-term chronological trends and general interpretations, set within general theoretical frameworks. Finer-grained analyses – pitched at middle range levels, for example – would require a different set of analytical tools. The long-term trends themselves are statistical phenomena, represented here by curves and graphs, and need to be distinguished from long-term historical trajectories – the behavioural dimension of the data. Trends and trajectories, therefore, describe the statistical and historical properties of the data, respectively.

Socio-demography

Failing to adequately account for the above trends by other means, we argue that they represent long-term *occupational* trends of sites and regions, effectively reflecting socio-demographic patterns within hunter-gatherer societies. We argue that increasing levels of 'closure' among hunter-gatherer populations led to more dynamic relations between people and the natural environment (for further details see Lourandos 1997; David and Lourandos 1997). More 'open', flexible and fluid social formations have been documented for ethnographic populations of lower density and reduced patterns of aggregation, such as those in more arid regions of Australia. Pardoe (e.g. 1990) has coined such social systems as tending towards 'inclusiveness'. In contrast, more 'closed' formations are generally found among populations of higher density and increased levels of aggregation, as among Australian Aboriginal populations in richer coastal regions. These are deemed tending towards 'exclusiveness' (Pardoe 1990). In effect, a continuum existed between these two categories, with all Australian populations possessing both principles of 'openness' and 'closure'. We argue that the nature and balance between the two could be reflected through time and be identified archaeologically.

More 'open ' systems are characterised by more homogeneous cultural traits, whereas in more 'closed' systems there is a greater emphasis upon territoriality, social hierarchy and ritual. Among the latter, for example, 'boundary' maintenance is often expressed through territorial marking (e.g. rock-art), ritual (e.g. population aggregation for ceremonies) and exchange (Wobst 1976; Bender 1981, 1985; Jochim 1983; McBryde 1984; Gamble 1986; Lourandos 1993, 1997). 'Closure' can be accentuated by increasing population size and density. It can also be associated with 'opening' up of relations at other levels (through ceremony and exchange) as an attempt to overcome the territorial and social constrictions imposed by 'closure' (Lourandos 1993, 1997). In these ways more dynamic social formations are created, and examples are to be found in SE Australia (McBryde 1984; Lourandos 1983, 1985, 1997; Pardoe 1988) and NE Queensland (David and Cole 1990; David 1994). We argue, however, that these conditions are not soley the product of environmental change or increasing population sizes, but also of particular historical conditions, and potentially can occur in any environment.

Interpretations

A general model incorporating environmental (biogeographical, ecological) and socio-demographic factors and encompassing all the Australian regions studied here can be proposed to explain the observed regional trends (see also David and Lourandos 1997; Lourandos 1997).

Prior to about 30,000 years BP, the Australian continent contained large bodies of water, lakes and swamps. Evidence of Aboriginal settlement at this time is patchy but includes parts of all major environments (tropical, arid and temperate, including Tasmania). Increasingly more arid conditions after *c.*25,000 years BP resulted in a contraction of Aboriginal settlement, with a concentration of population in the better-watered sectors of the continent at the glacial maximum. The latter regions include parts of tropical, temperate and arid Australia and the Tasmanian southwest. Following *c.*17,000 to 10,000 years BP – depending on the region under consideration – Aboriginal settlement began to expand and increase in all environments as climate ameliorated. This trend continued in mainland Australia throughout the early to mid Holocene, when climate was most humid and bioproduction highest, and is clearest in tropical, arid and semiarid areas. After about 4000 years BP, however, the expansion of Aboriginal settlement accelerated in all regions of the Australian mainland studied here; a trend that did not follow that of environment, which was becoming drier with reduced levels of bioproduction.

Therefore, during the terminal Pleistocene and early Holocene at least (*c.*25,000-*c.*6000 years BP), Aboriginal settlement more closely followed general environmental trends in many of the regions studied here. However, after about 4000 years BP settlement and environmental trends across mainland

Australia diverged significantly. The close relationship between archaeological and environmental trends prior to the mid Holocene may be the product of the coarse-grained nature of our data, and finer-grained analyses may reveal more variation between Aboriginal and environmental patterns, as well as between regions. Our ability to address finer-grained enquiries, however, will have to await considerably larger sample sizes. We stress here that cultural-historical conditions, as well as scales of analysis, have to be considered in conjunction with environmental shifts. It is only by situating the above trends in historical context that the archaeological trajectories can be characterised. The coarse-grained nature of the data, however, appears to be no barrier to indicating clear divergences in archaeological and environmental trends after 4000 years BP.

As an explanation of the general trend we suggest the following scenario (also David and Lourandos 1997; Lourandos 1997). The increasing archaeological trend in site and regional land use after the glacial maximum (17,000 years BP) in some regions, and after 13,000-10,000 years BP in others, can be generally associated with ameliorating climate which peaked in the early to mid Holocene. These events may have stimulated changing patterns of Aboriginal demography (including population dispersal, aggregation and increase), as well as increasing levels of 'closure'. The suggestion here is that increased 'closure' of social systems may be associated with an acceleration of change; the socio-demographic dynamic linking 'closure' to heightened structural formalisation and competition between and within Aboriginal groups. Therefore, Once set in motion in the early Holocene or earlier, the trend continued on throughout the late Holocene against the environmental trend. Had the archaeological trend followed that of environment, then a general decline in land and site use should be visible in the late Holocene.

Additional archaeological evidence

As well, qualitative archaeological evidence from our study areas lends further support and helps expand upon our interpretations. This has implications for Aboriginal land and resource use patterns, and aspects of territoriality and 'closure', and includes the following factors: a) rates of occupational discard in sites; b) plant species; c) particular environments; d) stone and other artefacts; e) rock-art, burial and exchange; f) offshore island use.

a) Increased rates in discard of artefacts (stone, bone, charcoal etc), for example, may indicate periods of increased site use (as noted above), and examples of this are found pre-25,000 years BP (O'Connor *et al.* 1993), at the glacial maximum (Hiscock 1984; Lamb 1993) and terminal Pleistocene (Cosgrove *et al.* 1990). This is also a characteristic widely reported for Australian sites of the late Holocene (Attenbrow 1982; Hughes and Lampert 1982; Lourandos 1983, 1985, 1987, 1993, 1997; Ross 1985; Lourandos and Ross 1994).

b) Processing of key plant species, more costly in terms of energy expenditure, appears in the late Holocene. At this time in tropical and subtropical Queensland toxic cycads were employed in the Central Queensland Highlands

(Beaton 1982), Cape York (Morwood and Hobbs 1995) and Whitsunday Islands (Barker 1991,1995) and toxic (rainforest) species in the rainforests south of Cape York (Horsfall 1987). The processing of grass seeds – mainstay of recent arid zone Aboriginal diets of the ethnographic period – appears throughout arid and semiarid Australia (Smith 1986), and also in Cape York (Morwood and Hobbs 1995), in the late Holocene.

c) Certain environments, used perhaps more peripherally in the past, appear also to have been exploited more widely in the late Holocene. These include rainforests in northern Queensland (Hiscock and Kershaw 1992), Mt. Mulligan, also in north Queensland (David *et al.*, in press), the Lamington Plateau, SE Queensland (Hall 1986; Hiscock and Hall 1988) and Cape Otway in SW Victoria (Lourandos 1983, 1985). Further examples are the extensive wetlands of temperate SE Australia where earth mounds, as habitation sites, first appear in the last 2500 years (Lourandos 1983, 1985; Williams 1987; Lourandos and Ross 1994; Lourandos 1997). The harshest sandy desert regions of central Australia had a late Holocene settlement (Veth 1989,1993) and apparently also the semiarid Mallee region of NW Victoria (Ross 1981). Offshore islands and their resources, in many parts of northern and eastern Australia, also fall into this category (Rowland 1982; Lourandos 1983, 1985; Barker 1991, 1995; Beaton 1985; O'Connor 1992).

d) Rationing of stone material appears during the glacial maximum in drier areas of tropical Queensland (Hiscock 1984; David 1991a; Lamb 1993), and in SW Tasmania during the terminal Pleistocene perhaps linked to more mobile settlement (McNiven 1994). Comparable changes in stone use have been reported from tropical Queensland during the early Holocene (Morwood and Hobbs 1995). The most extensive changes in stone artefacts (e.g. new lithic types etc.) occurs in the late Holocene along with more widespread changes (Hiscock and Veth 1991; McNiven 1994; Morwood 1981; Mulvaney 1975) and have been linked to risk management strategies, part of expanding settlement of the continent and its environments (Hiscock 1994). As well, new fishing equipment, some with Melanesian parallels, appears also in the last thousand years or so along the east coast of Australia (Lampert 1971; Bowdler 1976; Rowland 1981; Barker 1989, 1991, 1995).

e) Studies of rock-art throw further light on changes in inter-regional relations during the mid to late Holocene. Evidence of more regionalised rock-art traditions has been studied in Cape York by David (David 1991a, 1994; David and Cole 1990; David and Chant 1995) who argued that prior to about 3500 years BP rock-art was more homogeneous; suggesting the development of more 'closed' territorial relations. Morwood and Hobbs (1995) came to similar conclusions, and suggested that some aspects of 'closure' were evident from the terminal Pleistocene (also from the Central Queensland Highlands) (Morwood 1980, 1984). Comparable studies of late Holocene regionalisation in rock-art have come from tropical Arnhem land (Taçon 1993; Lewis 1988) and the more temperate Sydney region (McDonald 1994). Taçon and Chippindale (1994) also have argued that changes in depictions of fighting in Arnhem Land rock-art, beginning in the mid and accelerating in the late Holocene, may indicate increasing levels of 'social complexity and the development of the highly complicated kinship relations' of ethnographic times (but see Welch 1997). Extensive exchange patterns, including greenstone axes, occur in temperate SE Australia in relatively recent contexts, although dating of these is poor at present (McBryde 1984). Cemeteries first appear in the temperate SE Murray River corridor, which lies immediately outside our study areas, in the terminal Pleistocene about 13,000 years BP, but are more numerous and complex in the region in the last few thousand years (Pardoe 1988, 1990). Pardoe argued that this signalled the appearance of territorially corporate Aboriginal groups.

f) As already noted above, O'Connor (1992) has argued that systematic use of distant offshore islands in northern Australia commenced during the mid to late Holocene. Barker (1991, 1995) has similarly argued that in the Whitsunday Islands of NE Australia, higher degrees of territorial boundedness emerged after around 3000 years BP. The control and distribution of resources from specific islands, such as stone artefacts from the South Molle Island quarry, became more circumscribed, and access to individual islands more restricted politically (i.e. 'closed').

As can be seen from this evidence, while socio-cultural changes are apparent from the terminal Pleistocene and before, most occur in the late Holocene period in many of our study areas, and thereby coincide with the period of maximum use of Australian sites and landscapes as determined from the general archaeological trends above. This late Holocene archaeological pattern has been widely discussed (Attenbrow 1982; Bird and Frankel 1991a; Hughes and Lampert 1982; Lourandos 1983, 1985, 1987, 1993, 1997; Beaton 1985; Ross 1985; Hiscock 1994; Lourandos and Ross 1994), and we suggest that it is somewhat clarified here by situating it historically in relation to preceding trends and patterns of the early Holocene and terminal Pleistocene periods in particular.

Conclusions

By employing common methods of analysis based upon the temporal spread of radiocarbon dates from 11 study areas across mainland Australia, comparable general long-term archaeological trends have been observed in tropical, arid and temperate zones. Archaeological trends appear to follow more closely upon environmental trends from at least the terminal Pleistocene, after *c*.17,000 years BP (and in some areas back to *c*.37,000-30,000 years BP), until the mid Holocene. In the more recent, post-4000 years BP period, however, archaeological and environmental trends diverge significantly in all areas studied. We suggest that the somewhat linear trajectories outlined here are valid at the coarse-grained analytical scale by which we have viewed the data. More varied results, we suggest, may be apparent when finer-grained methods are employed. Linearity is also evident in the long-term environmental trends used here, which are scaled at similar general levels.

We have argued also that explanations of these archaeological trends can be viewed from both ecological and socio-demographic perspectives. While increasing archaeological trends in the terminal Pleistocene-mid Holocene might be closely associated with amelioration in climate (including rising levels of bioproduction), we suggest also that competition between and within Aboriginal groups, and more 'closed' social formations, produced more dynamic socio-demographic contexts. Such processes could potentially occur at any point in time, and upon finer-grained scrutiny may be evident in earlier contexts in Australian prehistory. Here, our coarse-grained analysis has indicated that these more dynamic socio-demographic trends appear during the terminal Pleistocene, and are most obvious during the late Holocene (post-4000 years BP) on the Australian mainland.

Our results lend support to the use of dynamic models to explain *long-term regional trends* in Australian prehistory; those which emphasise complex relationships between people, their demography and the natural environment (Lourandos 1985, 1987, 1993, 1997; David 1994; David and Lourandos 1997; Williams 1987; Veth 1989, 1993; Hiscock and Veth 1991; Hiscock 1994; Ross *et al.* 1992; Bowdler 1981,1990). We have argued also that population increase (or decrease, for that matter) is not an independent factor in social change, for it is only part of a broader socio-demographic process (Beaton 1985, 1990; Smith 1989; Smith and Sharp 1993; Morwood and Hobbs 1995). Our results also contrast with models which emphasise long-term stability between people, their numbers and environment and minimise the effects of change (Birdsell 1953, 1957, 1977; Jones 1990; Allen 1989; Cosgrove *et al.* 1990; Cosgrove 1995; Bird and Frankel 1991a, 1991b). Indeed, interpretations of Australian prehistory can be viewed as largely still divided between these two general approaches (Lourandos 1985, 1997). We suggest that resolution of these issues in part lies in rescaling methods and interpretative frameworks when evaluating archaeological trends and the interpretation of likely cultural-historical trajectories (Lourandos 1997; David and Lourandos 1997). More accurate comparison can only take place between comparable scales of analysis; like needs to be compared with like. Our ability to refine interpretations of these data, however, will always be subject to the limitations of particular data sets, for example sample size (another aspect of scale).

These issues – of scale, trends and trajectories – also have broader implications for modelling change in other parts of the world. For example, during the European Palaeolithic of the late Pleistocene there is a clear correlation between settlement-subsistence patterns (including aspects of 'intensification') and climatic episodes both before and after the glacial maximum (Mellars 1985; Soffer 1985; Gamble 1986). Also, terminal- and post-Pleistocene shifts in resource use towards more complex patterns (including 'intensification', and in some cases 'agriculture') are viewed principally as the result of two key processes: either a) changing ecological resource structures (Henry 1985); or b) changing socio-economic relations among hunter-gatherer populations (Bender 1978, 1981, 1985; Lourandos 1983, 1985, 1988). The Australian evidence presented here does not contradict either proposition; but strongly suggests that the timing and pattern of these events is based on both regional ecology and socio-cultural trends and historical trajectories. In this way, the apparent separation in time in the Australian example, between changing climates of the terminal Pleistocene-early Holocene period and socio-demographic patterns of the late Holocene, can be viewed as related to individual regional historical trajectories. The timing of the Australian pattern can be generally equated with those of nearby Papua New Guinea where agriculture is clearly evident only after 6,000-5,000 years BP, and its intensification in the last 2,000 years. Indeed, the Australian historical trajectory fits well with a regional Australasian-Pacific 'signature' – where increasing patterns of resource intensification (and agriculture) are largely late Holocene phenomena; also linked broadly to patterns in Southeast Asia.

Other regions of the world, also peripheral to agricultural 'heartlands', where 'time lags' in broadly similar processes occur, include the North American Archaic and Eurasian Mesolithic. During the Archaic, archaeological changes post-date environmental changes (Brown 1985), and socio-cultural patterns on the ground do not necessarily correspond with those of environment (Bender 1985). Likewise, considerable time was involved before the adoption of agriculture in Mesolithic northern Europe (Zvelebil 1986). In all these world examples, therefore, including Australia, both environmental trends and historical trajectories need to be individually considered.

Appendix 16.1

Rockshelters and caves whose radiocarbon dates have been used in this paper.

Sites	References
SOUTHEAST CAPE YORK PENINSULA	
Walkunder Arch Cave, Pillar Cave, Echidna's Rest, Fern Cave, Racecourse Site, Initiation Cave, Lookout Shelter, Ngarrabullgan Cave, Courtyard Rock, Mitchell River Cave, Hearth Cave, Mordor Cave, Early Man Rockshelter, Platform Gallery, Hann River 1, Sandy Creek 1, Sandy Creek 2, Yam Camp, Magnificent Gallery, Giant Horse, Red Horse, Red Bluff 1, Mushroom Rock, Green Ant, Echidna Shelter, Endaen Rockshelter, Walaemini Rockshelter, Alkaline Hill	Beaton 1985, Campbell and Mardaga-Campbell 1993, David 1991a, David 1991b, David 1993, David unpublished, David and Chant 1995, David and Dagg 1993, David et al. 1995a, David and Lourandos in press b, David et al. 1997, Flood and Horsfall 1986, Mardaga-Campbell 1986, Morwood 1995a, Morwood 1995b, Morwood and Dagg 1995, Morwood and Jung 1995, Morwood and L'Oste-Brown 1995a, Morwood and L'Oste-Brown 1995b, Morwood et al. 1995a, Morwood et al. 1995b, Polach et al. 1967, Rosenfeld et al. 1981, Watchman and Cole 1993, Wright 1971
KAKADU	
Malakunanja 2, Nauwalabila 1, Nawulandja, Nangalawurr, Leichhardt Site, Ngarradj Warde Djobkeng, Anbangbang 1, Spirit Cave, Blue Painting Site, Yiboiog, Paribari, Malangangerr, Nawamoyn, Jimeri 1, Jimeri 2, Borngolo Shelter	Allen and Barton 1989, Jones and Negerevich 1985, Jones and Johnson 1985, Roberts et al. 1990, Roberts et al. 1994, Schrire 1972, 1982
KIMBERLEY	
Windgingarri 1, Windgingarri 2, Koolan 2, Carpenter's Gap Rockshelter, High Cliffy, Miriwun, Monsmont, Kununurra, Pincombe Rockshelter, Pilchowski Crossing, Canyon Rockshelter, Bangorono, Ngurini, Wundalal	Attenbrow et al. 1995, Bowdler and O'Connor 1991, Dortch 1977, O'Connor 1987, O'Connor 1989, O'Connor 1992, O'Connor 1995, Veth 1995

ARID ZONE	
Newman Rockshelter, Newman Orebody Rockshelter, Malia Rockshelter, Manganese Gorge 8 Cave, Packsaddle Ridge P4623, Packsaddle Ridge P4627, Packsaddle Ridge P5315, Packsaddle Ridge P5316, Puntutjarpa Main Cave, Puntutjarpa West Cave, Noala 1 Rockshelter, Noala 2 Rockshelter, Haynes Cave, Kwerlpe, James Range NW Rockshelter, Kweyunpe 1, Kweyunpe 2, Kweyunpe 6, Japilyari, Winakurijuna, Yulpul, Puritjarra, Tjungkupu 1, Ilarari 17, Rrewurlpmurlpme Kweke, Intirkekwerle (James Range East North Cave), Walga Rock, Clarke's Cave, Anadara Shelter, Not-So-Secret Shelter, Kuyunba 106, Kuyunba 107, Mandu Mandu Creek Rockshelter, Pilgonoman Creek Rockshelter, Yardie Well Rockshelter, Cuckadoo 1, Monkey Mia Rockshelter 1, Monkey Mia Rockshelter 2, Karlamilyi, Murderers Bore Rockshelter, Kyeenee 1 Rockshelter	Bordes *et al.* 1983, Bowdler 1995, Bradshaw 1995, Brown 1987, Davidson *et al.* 1993, Gould 1980, Gould 1977, Gould 1968, Maynard 1980, Morse 1993, Napton and Greathouse 1985, Robins 1995, Smith 1987, Smith 1988, Smith 1989, Veth 1993, Veth 1995
WARDAMAN COUNTRY	
Yiwarlarlay 1, Delamere 3, Jalijbang 2, Mennge-ya, Gordol-ya, Garnawala 1, Garnawala 2, Yingalarri 1 (Ingaladdi)	Attenbrow *et al.* 1995, Barker unpublished, Clarkson and David 1995, Cundy 1990, David *et al.* 1991, David *et al.* 1992, David *et al.* 1995b, David unpublished, McNiven *et al.* 1992,
CENTRAL QUEENSLAND HIGHLANDS	
Kenniff Cave, The Tombs, Wanderer's Cave, Cathedral Cave, Rainbow Cave Rockshelter, Rainbow Cave, Ken's Cave, Turtle Rock, Native Well 1, Native Well 2, Goat Rock 1 Rockshelter	Beaton 1991a, Beaton 1991b, Kelly 1982, Morwood 1981, Mulvaney 1975, Mulvaney and Joyce 1965
SOUTHEAST QUEENSLAND	
Bishop's Peak, Boonah Shelter, Brooyar Shelter, Bushranger's Cave, Gatton Rockshelter, Maidenwell Shelter, Platypus Rockshelter	Hall and Hiscock 1988, Kelly 1982, McNiven 1988, Morwood 1986, Morwood 1987, Ulm 1995
SW VICTORIA and SE SOUTH AUSTRALIA	
Glenaire Shelter 2, Moonlight Head Rockshelter, Black Range Shelter 2, Bridgewater Cave South, Camp of Emu's Foot B, Cave of Hands, Cave of Hands B, Drual, Glen Isla 1, Koongine Cave, Mount Burr, Mount Talbot 1, Narcurrer, Piccininnie Cave	Barker 1987, Bird 1995, Bird and Frankel 1991a, Coutts 1978, Coutts and Witter 1977, Frankel 1986, 1988, Head 1985, Lourandos 1980, 1983, Mulvaney 1962, Zobel *et al.* 1984
SYDNEY and BLUE MOUNTAINS	
Lapstone Creek, Capertee 3, Curracurrang 1, Curracurrang 2, Curracurrang 7, Wattamolla, Yowie Bay, Shaws Creek 2 Rockshelter, Bantry Bay Shelter, Gymea Bay, Walls Cave, Lyre Bird Dell (La), Lyre Bird Dell (Lb), Springwood Creek, Kings Table, Noola, Horseshoe Falls, Angophora Reserve Rockshelter, Great Mackeral Shelter, Upside-Down Man Shelter, Upside-Down Man Shelter Rock-art, Native Animals Rock-art, Yengo 1, Bindea Road, Henry Lawson Drive Rockshelter, Mill Creek 11, Barden's Creek 9, Bull Cave, Wolloby Gully, White	Attenbrow 1982, Attenbrow and Steele 1995, Bermingham 1966, Glover 1974, Johnson 1979, Kohen *et al.* 1984, McCarthy 1978, McDonald 1991, McDonald 1992, McDonald 1994, McDonald 1996, McDonald and Ross 1990, McDonald *et al.* 1990, Megaw 1966, Megaw 1968, Megaw and Roberts 1974, Moore 1970, Moore 1981, Poiner 1974, Polach *et al.* 1967, Ross and Specht 1976, Stockton 1977, Stockton and

Figure, Waterfall Cave, MR1, Daley's Point (Milligan's) Shelter, Big L, Black Hands, Dingo, Elongated Figure, Roo and Echidna, Dingo and Horned Anthropomorph, Loggers Shelter, Mussel Shelter, Bracken Shelter, Deep Creek Shelter, Rainforest 2 Shelter, Mt. Trefle Rockshelter, Dendrobium Shelter, Sandy Cave	Holland 1974, Tracey 1974
SOUTH COAST OF NSW	
Bomaderry Creek Rockshelter, Bob's Cave, Bulee Brook 2, Kangaroo Hill 4, Rock Pool Shelter, Gnatilia Creek 3, Burrill Lake 357, Bourkes Road 2, Durras North, Burrill Lake, Currarong 1, Currarong 2	Boot 1993, Hughes and Djohadze 1980, Hughes and Lampert 1982, Lampert 1966, Lampert 1971, Lampert and Steele 1993, Polach *et al.* 1967
SOUTHERN UPLANDS	
Sassafras 1, Sassafras 2, Nursery Swamp 2, Caddigat Shelter, Bogong Shelter, Yankee Hat 2, Hanging Rock, New Guinea 2 Cave, Birrigai, Clogg's Cave	Flood 1980, Flood *et al.* 1987, Ossa *et al.* 1995, Rosenfeld *et al.* 1983

References

Allen, H. and Barton, G. 1989. *Ngarradj Warde Djobkeng: White Cockatoo Dreaming and the Prehistory of Kakadu.* University of Sydney, Sydney: 150 pp.

Allen, J. 1989. When did humans first colonise Australia? *Search,* 20: 149-154.

Allen, J. and Holdaway, S. 1995. The contamination of Pleistocene radiocarbon determinations in Australia. *Antiquity,* 69: 101-112.

Allen, J. and O'Connell, J.F. (Editors) 1995. Pleistocene to Holocene in Australia and Papua New Guinea. *Antiquity,* 69, Special Issue No. 265: 360 pp.

Attenbrow, V. 1982. *The archaeology of Upper Mangrove Creek catchment: research in progress.* In S. Bowdler (Editor) Coastal Archaeology in Eastern Australia. Department Prehistory, Australian National University, Canberra: 67-79.

Attenbrow, V., David, B. and Flood, J. 1995. Mennge-ya and the origins of points: new insights into the appearance of points in the semi-arid zone of the Northern Territory. *Archaeology in Oceania,* 30: 105-20.

Attenbrow, V. and Steele, D. 1995. Fishing in Port Jackson, New South Wales - more than met the eye. *Antiquity,* 69: 47-60.

Barker, B.C. 1987. *A Faunal Analysis from Narcurrer Shelter, South-East South Australia.* Unpublished BA (Honours) thesis, University of Queensland, St. Lucia.

Barker, B.C. 1989. Nara Inlet 1: A Holocene sequence from the Whitsunday Islands, central Queensland coast. *Queensland Archaeological Research,* 6: 53-76.

Barker, B.C. 1991. Nara Inlet 1: Coastal resource use and the Holocene marine transgression in the Whitsunday Islands, Central Queensland. *Archaeology in Oceania,* 26:102-109.

Barker, B. 1995. *'The Sea People': Maritime Hunter-Gatherers on the Tropical Coast. A Late Holocene Maritime Specialisation in the Whitsunday Islands, Central Queensland.* Unpublished PhD thesis, Department of Anthropology and Sociology, University of Queensland, St Lucia.

Beaton, J.M. 1982. Fire and water: aspects of Australian Aboriginal management of cycads. *Archaeology in Oceania,* 17: 59-67.

Beaton, J.M. 1985. Evidence for a coastal occupation time-lag at Princess Charlotte Bay (North Queensland) and implications for coastal colonisation and population growth theories for Aboriginal Australia. *Archaeology in Oceania,* 20: 1-20.

Beaton, J.M. 1990. *The importance of past population for prehistory.* In: B. Meehan and N. White (Editors) Hunter-Gatherer Demography: Past and Present. Oceania Monograph No. 39: 23-40.

Beaton, J.M. 1991(a). Excavations at Rainbow Cave and Wanderer's Cave: two rockshelters in the Carnarvon Range, Queensland. *Queensland Archaeological Research,* 8: 3-32.

Beaton, J.M. 1991(b). Cathedral Cave: a rockshelter in Carnarvon Gorge, Queensland. *Queensland Archaeological Research,* 8: 33-84.

Bender, B. 1978. Gatherer-hunter to farmer: a social perspective. *World Archaeology,* 10: 204-22.

Bender, B. 1981. *Gatherer-hunter intensification.* In: A. Sheridan and G. Bailey (Editors) Economic Archaeology. British Archaeological Reports, International Series No. 96: 149-157.

Bender, B. 1985. *Prehistoric developments in the American midcontinent and in Brittany, northwest France.* In: T.D. Price and J.A. Brown (Editors) Prehistoric Hunter-gatherers: The Emergence of Cultural Complexity. Academic Press, Orlando: 21-57.

Bermingham, A. 1966. Victoria natural radiocarbon measurements 1. *Radiocarbon,* 8: 507-21.

Bird, C. and Frankel, D. 1991(a). Chronology and explanation in western Victoria and southeast South Australia. *Archaeology in Oceania,* 26: 1-16.

Bird, C. and Frankel, D. 1991(b). Problems in constructing a prehistoric regional sequence: Holocene south-east Australia. *World Archaeology,* 23(2): 179-192.

Bird, C. 1995. Mount Talbot 1: a rockshelter in the southern Wimmera, Victoria, Australia. *The Artefact,* 18: 12-21.

Birdsell, J.B. 1953. Some environmental and cultural factors influencing the structuring of Australian Aboriginal population. *American Naturalist,* 87: 171-207.

Birdsell, J.B. 1957. *Some population problems involving Pleistocene Man.* Cold Spring Harbor Symposia on Quantitative Biology, 22: 47-69.

Birdsell, J.B. 1977. *The recalibration of a paradigm for the first peopling of Australia.* In: J. Allen, J. Golson and R. Jones (Editors) Sunda and Sahul. Academic Press, London: 113-167.

Boot, P. 1993. Pleistocene date from archaeological excavations in the hinterlands of the New South Wales coast. *Australian Archaeology,* 37: 59.

Bordes, F., Dortch, C., Thibault, C., Raynal, J.P. and Bindon, P. 1983. Walga rock and Billabong spring: two archaeological sequences from the Murchison basin, Western Australia. *Australian Archaeology,* 17: 1-26.

Bowdler, S. 1976. Hook, line and dillybag: an interpretation of an Australian coastal shell midden. *Mankind,* 10: 248-258.

Bowdler, S. 1981. Hunters in the highlands: Aboriginal adaptations in the eastern Australian uplands. *Archaeology in Oceania,* 16: 99-111.

Bowdler, S. 1990. *Peopling Australasia: the `coastal colonisation' hypothesis considered.* In: P. Mellars and C. Stringer (Editors) The Human Revolution: Behavioural and Biological Perspectives on the Origins of Modern Humans (Volume 2). University of Edingburgh Press, Edinburgh: 327-343.

Bowdler, S. 1995. The excavation of two small rockshelters at Monkey Mia, Shark Bay, Western Australia. *Australian Archaeology,* 40: 1-13.

Bowdler, S. and O'Connor, S. 1991. The dating of the Australian Small Tool Tradition, with new evidence from the Kimberley, WA. *Australian Aboriginal Studies,* 1991(1): 53-62.

Bowler, J.M. 1981. Australian salt lakes, a palaeohydrologic approach. *Hydrobiologia,* 82: 431-444.

Bowler, J.M. and Wasson, R.J. 1984. *Glacial age environments of inland Australia*. In: J.C. Vogel (Editor) Late Cainozoic Palaeoclimates of the Southern Hemisphere. Balkema, Rotterdam: 183-208.

Bradshaw, E. 1995. Dates from archaeological excavations on the Pilbara coastline and islands of the Dampier Archipelago, Western Australia. *Australian Archaeology*, 41: 37-8.

Brown, J.A. 1985. *Long-term trends to sedentism and the emergence of complexity in the American midwest*. In: T.D. Price and J.A. Brown (Editors) Prehistoric Hunter-Gatherers: the Emergence of Cultural Complexity. Academic Press, Orlando: 201-231.

Brown, S. 1987. *Toward a Prehistory of the Hamersley Plateau, Northwest Australia*. Occasional Papers in Prehistory No. 6. Department of Prehistory, Australian National University, Canberra: 65 pp.

Campbell, J. and Mardaga-Campbell, M. 1993. *From macro- to nano-stratigraphy: linking vertical and horizontal dating of archaeological deposits with the direct dating of rock art at 'The Walkunders', Chillagoe (north Queensland, Australia)*. In: J. Steinbring, A. Watchman, P. Faulstich and P. Taçon (Editors) Time and Space: Dating and Spatial Considerations in Rock Art Research. Australian Rock Art Research Association, Melbourne: 57-63.

Clarkson, C. and David, B. 1995. The antiquity of blades and points revisited: investigating the emergence of systematic blade production south-west of Arnhem Land, northern Australia. *The Artefact*, 18: 22-44.

Cosgrove, R. 1995. Late Pleistocene behavioural variation and time trends: the case from Tasmania. *Archaeology in Oceania*, 30: 83-104.

Cosgrove, R., Allen, J. and Marshall, B. 1990. Palaeo-ecology and Pleistocene human occupation in south central Tasmania. *Antiquity*, 64: 59-78.

Coutts, P. 1978. Victorian Archaeological Surveys activities report 1977-1978. *Records of the Victorian Archaeological Survey*, 8: 1-23.

Coutts, P. and Witter, D. 1977. New radiocarbon dates for Victorian archaeological sites. *Records of the Victorian Archaeological Survey*, 4: 60-73.

Cundy, B. 1990. *An Analysis of the Ingaladdi Assemblage: A Critique of The Understanding of Lithic Technology*. Unpublished PhD thesis, Australian National University, Canberra.

David, B. 1991(a). Fern Cave, rock art and social formations: rock art regionalisation and demographic changes in southeastern Cape York Peninsula. *Archaeology in Oceania*, 26: 41-57.

David, B. 1991(b). Mitchell River Cave: a late Pleistocene-Holocene sequence from southeastern Cape York Peninsula. *Australian Aboriginal Studies*, 1991(1): 67-72.

David, B. 1993. Nurrabullgin Cave: preliminary results from a pre-37,000 year old rockshelter, north Queensland. *Archaeology in Oceania*, 28: 50-54.

David, B. 1994. *A Space-time Odyssey: Rock Art and Regionalisation in North Queensland Prehistory*. Unpublished Ph.D. thesis, Department of Anthropology and Sociology, University of Queensland, St Lucia.

David, B. and Cole, N. 1990. Rock art and inter-regional interaction in northeast Australian prehistory, *Antiquity*, 64: 788-806.

David, B. and Chant, D. 1995. Rock Art and Regionalisation in North Queensland Prehistory. *Memoirs of the Queensland Museum*, 37(2): 357-528.

David, B., Collins, J., Barker, B., Flood, J. and Gunn, R. 1995(b). Archaeological research in Wardaman country, Northern Territory: the Lightning Brothers Project 1990-91 field seasons. *Australian Archaeology*, 41: 1-8.

David, B. and Dagg, L. 1993. Two Caves. *Memoirs of the Queensland Museum*, 33: 143-162.

David, B. and Lourandos, H. 1997. 37,000 years and more in tropical Australia: investigating long-term archaeological trends in Cape York Peninsula. *Proceedings of the Prehistoric Society,* 63: 1-23.

David, B. and Lourandos, H. in press(a). Landscape as mind: land use, cultural space and change in north Queensland prehistory. *Quaternary International.*

David, B. and Lourandos, H. in press(b). Rock art and socio-demography in northeastern Australian prehistory. *World Archaeology.*

David, B., McNiven, I. and Flood, J. 1991. Archaeological excavations at Yiwarlarlay 1: site report. *Memoirs of the Queensland Museum,* 30: 373-80.

David, B., McNiven, I., Bekessy, L., Bultitude, R., Clarkson, C., Lawson, E., Murray, C. and Tuniz, C. in press. More than 37,000 years of human occupation. In: B. David (Editor) Ngarrabullgan: *Geographical Investigations in Djungan Country, Cape York Peninsula.* Monash Publications in Geography and Environmental Science No. 51. Monash University, Melbourne.

David, B., Chant, D. and Flood, J. 1992. Jalijbang 2 and the distribution of pecked faces in Australia. *Memoirs of the Queensland Museum,* 32: 61-77.

David, B., Tuniz, C., Lawson, E., Hua, Q., Jacobsen, G.E., Head, J. and Rowe, M. 1995(a). *New AMS Determinations for Chillagoe Rock Art, Australia, and their Implications for Northern Australian Prehistory.* Paper presented at the NEWS 95 International Rock Art Congress, Turin, Italy.

David, B., Roberts, R., Tuniz, C., Jones, R. and Head, J. 1997. New optical and radio-carbon dates from Ngarrabullgan cave, a Pleistocene archaeological site in Australia: implications for the comparability of time clocks and for the human colonisation of Australia. *Antiquity,* 71: 183-188.

Davidson, I., Sutton, S.A. and Gale, S.J. 1993. *The human occupation of Cuckadoo 1 rockshelter, northwest central Queensland.* In: M.A. Smith, M. Spriggs and B. Fank-hauser (Editors) Sahul in Review. Occasional Papers in Prehistory No. 24. Department of Prehistory, Australian National University, Canberra: 164-72.

Dodson, J. R. 1974. Vegetation and climatic history near Lake Keilambete, Western Victoria. *Australian Journal of Botany,* 22: 709-717.

Dodson, J.R. (Editor) 1992. *The Naive Lands: Prehistory and Environmental Change in Australia and the South-West Pacific.* Longman Cheshire, Melbourne: 258 pp.

Dortch, C. 1977. *Early and late stone industrial phases in Western Australia.* In R.V.S. Wright (Editor) Stone Tools as Cultural Markers. AIAS, Canberra: 104-32.

Ellen, R. 1982. *Environment, Subsistence and System.* Cambridge University Press. Cambridge: 324 pp.

Ellen, R. 1988. *Foraging, starch extraction and the sedentary lifestyle in the lowland rainforest of central Seram.* In: T. Ingold, D. Riches and J. Woodburn (Editors) Hunters and Gatherers: History, Evolution and Social Change. Berg, Oxford: 117-134.

Flood, J. 1980. *The Moth Hunters: Aboriginal prehistory of the Australian Alps.* AIAS, Canberra: 388 pp.

Flood, J., David, B., Magee, J. and English, B. 1987. Birrigai: a Pleistocene site in the southeastern highlands. *Archaeology in Oceania,* 22: 9-26.

Flood, J. and Horsfall, N. 1986. Excavations at Green Ant and Echidna Shelter, Cape York Peninsula. *Queensland Archaeological Research,* 3: 4-64.

Flood, J., David, B., Magee, J. and English, B. 1987. Birrigai: a 21,000 year old site from the ACT. *Archaeology in Oceania,* 22: 9-26.

Frankel, D. 1986. Excavations in the lower southeast of South Australia: November 1985. *Australian Archaeology,* 22: 75-87.

Frankel, D. 1988. Characterising change in prehistoric sequences: a view from Australia. *Archaeology in Oceania,* 23: 41-8.

Gamble, C. 1986. *The Palaeolithic Settlement of Europe*. Cambridge University Press, Cambridge: 471 pp.

Glover, E. 1974. *Report on the excavation of a second rock shelter at Curracurrang Cove, New South Wales*. In: J.V.S. Megaw (Editor) The Recent Archaeology of the Sydney District: Excavations 1964-1967. AIAS, Canberra: 13-18.

Gould, R.A. 1968. Preliminary report on excavations at Puntutjarpa rockshelter, near the Warburton Ranges, Western Australia. *Archaeology and Physical Anthropology in Oceania*, 3: 161-85.

Gould, R.A. 1977. Puntutjarpa rockshelter and the Australian desert culture. *Anthropological Papers of the American Museum of Natural History*, 54: 1-188.

Gould, R.A. 1980. James Range East Rockshelter, Northern Territory, Australia: a summary of the 1973 and 1974 investigations. *Asian Perspectives*, 21: 86-126.

Hall, J. 1986. Exploratory excavation at Bushrangers Cave (Site LA:A11), a 6000-year-old campsite in Southeast Queensland: preliminary results. *Australian Archaeology*, 22: 88-103.

Hall, J. and Hiscock, P. 1988. The Moreton Regional Archaeological Project (MRAP) - stage II: an outline of objectives and methods. *Queensland Archaeological Research*, 5: 4-24.

Harrison, S.P. and Dodson, J. 1993. *Climates of Australia and New Guinea since 18,000 yr BP*. In: H.E. Wright Jr., J.E. Kutzbach, T. Webb III, W.F. Ruddiman, F.A. Street-Perrott, and P.J. Bartlein (Editors) Global Climates since the Last Glacial Maximum. University of Minnesota Press, Minneapolis: 265-293.

Head, L. 1985. Pollen analysis of sediments from Bridgewater Caves archaeological site, southwestern Victoria. *Australian Archaeology*, 20: 1-15.

Henry, D.O. 1985. *Preagricultural sedentism: the Natufian example*. In: T. D. Price and J. A. Brown (Editors) Prehistoric Hunter-Gatherers: The Emergence of Cultural Complexity. Academic Press, Orlando: 365-384.

Hiscock, P. 1984. Preliminary report on the stone artefacts from Colless Creek Cave. Northwest Queensland. *Queensland Archaeological Research*, 1: 120-151.

Hiscock, P. 1986. Technological change in the Hunter River valley and the interpretation of late Holocene change in Australia. Archaeology in Oceania, 21(1): 40-50.

Hiscock, P. 1994. Technological responses to risk in Holocene Australia. *Journal of World Prehistory*, 8: 267- 292.

Hiscock, P. and Hall, J. 1988. Technological change at Bushranger's Cave (LA:A11), Southeast Queensland. *Queensland Archaeological Research*, 5: 90-112.

Hiscock, P. and Kershaw, A.P. 1992. *Palaeoenvironments and prehistory of Australia's tropical Top End*. In: J. Dodson (Editor) The Naive Lands: Prehistory and Environmental Change in Australia and the Southwest Pacific. Longman Cheshire, Melbourne: 43-75.

Hiscock, P. and Veth, P. 1991. Change in the Australian desert culture: a reanalysis of tulas from Puntutjarpa rockshelter, *World Archaeology*, 22: 332-345.

Holdaway, S. and Porch, N. 1995. Cyclical patterns in the Pleistocene human occupation of Southwest Tasmania. *Archaeology in Oceania*, 30: 74-82.

Horsfall, N. 1987. *Living in Rainforest: the Prehistoric Occupation of North Queensland's Humid Tropics*. Unpublished Ph.D thesis, Department of Behavioural Science, James Cook University, Townsville.

Hughes, P. and Djohadze, V. 1980. *Radiocarbon Dates from Archaeological Sites on the South Coast of New South Wales and the Use of Depth/Age Curves*. Occasional Papers in Prehistory No. 1, Department of Prehistory, Australian National University, Canberra:

Hughes, P. and Lampert, R. 1982. *Prehistoric population change in southern coastal New South Wales.* In: S. Bowdler (Editor), Coastal Archaeology of Eastern Australia. Australian National University, Canberra: 16-28.

Ingold, T. 1980. *Hunters, Pastoralists and Ranchers.* Cambridge University Press, Cambridge: 326 pp.

Ingold, T. 1988. *Notes on the foraging mode of production.* In: T. Ingold, D. Riches and J. Woodburn (Editors) Hunters and Gatherers: History, Evolution and Social Change. Berg, Oxford: 269- 285.

Jochim, M.A. 1983. *Palaeolithic cave art in ecological perspective.* In: G.N. Bailey (Editor) Hunter-Gatherer Economy and Prehistory. Cambridge University Press, Cambridge: 212-219.

Johnson, I. 1979. *The Getting of Data.* Unpublished Ph.D. thesis, Australian National University, Canberra.

Jones, R. 1990. *East of Wallace's line: issues and problems in the colonisation of the Australian continent.* In: P. Mellars and C. Stringer (Editors) The Human Revolution: Behavioural and Biological Perspectives on the Origins of Modern Humans, Volume 2. University of Edingburgh Press, Edinburgh: 743-782.

Jones, R. and Negerevich, T. 1985. *A review of previous archaeological work.* In: R. Jones (Editor) Archaeological Research in Kakadu National Park. ANPWS, Canberra:1-16.

Jones, R. and Johnson, I. 1985. *Rockshelter excavations: Nourlangie and Mt. Brockman massifs.* In: R. Jones (Editor), Archaeological Research in Kakadu National Park. ANPWS, Canberra: 39-77.

Kelly, M. 1982. *A Practical Reference Source to Radiocarbon Dates Obtained from Archaeological Sites in Queensland.* Cultural Resource Management Monograph No. 4. Archaeology Branch, Brisbane.

Kershaw, A.P. 1995. Environmental change in Greater Australia through the Pleistocene/ Holocene transition. *Antiquity,* 69: 656-675.

Kershaw, A.P. and Nanson, G.C. 1993. The last full glacial cycle in the Australian region. *Global and Planetary Change,* 7: 1-9.

Kershaw, A.P. and Nix, H.A. 1989. *The use of bioclimatic envelopes for estimation of quantitative palaeoclimate values.* In: T.H. Donnelly and R.J. Wasson (Editors) CLIMANZ 3: Proceedings of Symposium, Melbourne 1987. CSIRO, Institute of Natural Resources and Environment, Division of Water Resources, Canberra: 78-85.

Kohen, J.L., Stockton, E.D. and Williams, M.A.J. 1984. Shaw's Creek II rockshelter: a prehistoric occupation site in the Blue Mountains piedmont, eastern New South Wales. *Archaeology in Oceania,* 19: 57-73.

Lamb, L. 1993. *Fern Cave: a Technological Investigation of Increased Stone Artefact Deposition Rates.* Unpublished BA (Honours) thesis. The University of Queensland, Brisbane.

Lampert, R. 1966. An excavation at Durras North, New South Wales. *Archaeology and Physical Anthropology in Oceania* 1: 83-118.

Lampert, R. 1971. *Burrill Lake and Currarong. Terra Australis No. 1.* Department of Prehistory, Australian National University, Canberra: 86 pp.

Lampert, R. and Steele, D. 1993. *Archaeological studies at Bomaderry Creek, New South Wales.* Records of the Australian Museum Supplement, 17: 55-75.

Lewis, D. 1988. *The Rock Paintings of Arnhem Land: Social, Ecological and Material Culture Change in the Post-glacial Period.* British Archaeological Reports. International Series No. 415, Oxford: 425 pp.

Lloyd, P.J. and Kershaw, A.P. 1997. Late Quaternary vegetation and early Holocene quantitative climatic estimates from Morwell Swamp, Latrobe Valley, southeastern Australia. *Australian Journal of Botany,* 45: 549-563.

Lourandos, H. 1980. *Forces of change: Aboriginal Technology and Population in South-western Victoria.* Unpublished PhD thesis, University of Sydney, Sydney.

Lourandos, H. 1983. Intensification: a late Pleistocene-Holocene archaeological sequence from southwestern Victoria. *Archaeology in Oceania*, 18: 81-94.

Lourandos, H. 1985. *Intensification and Australian prehistory.* In: T.D. Price and J.A. Brown (Editors) Prehistoric Hunter-gatherers: the Emergence of Cultural Complexity. Academic Press, Orlando: 385-423.

Lourandos, H. 1987. *Pleistocene Australia: peopling a continent.* In: O. Soffer (Editor) The Pleistocene Old World: Regional Perspectives. Plenum, New York: 147-165.

Lourandos, H. 1988. *Palaeopolitics: resource intensification in Aboriginal Australia and Papua New Guinea.* In: T. Ingold, D. Riches and J. Woodburn (Editors) Hunters and Gatherers: History, Evolution and Social Change. Berg, Oxford: 148-160.

Lourandos, H. 1993. Hunter-gatherer cultural dynamics: long- and short-term trends in Australian prehistory. *Journal of Archaeological Research*, 1(1): 67-88.

Lourandos, H. 1996. *Change in Australian prehistory: scale, trends and frameworks of interpretation.* In: I. Lilley, A. Ross and S. Ulm (Editors) Proceedings of the 1995 Australian Archaeological Association Annual Conference. University of Queensland, Brisbane: 15-21.

Lourandos, H. 1997. *Continent of Hunter-Gatherers: New Perspectives in Australian Prehistory.* Cambridge University Press, Cambridge: 390 pp.

Lourandos, H. and Ross, A. (1994). The great 'intensification debate': Its history and place in Australian archaeology. *Australian Archaeology*, 39: 54-63.

Mardaga-Campbell, M. 1986. Prehistoric living-floors and evidence for them in north Queensland rockshelters. *Australian Archaeology*, 23: 42-61.

Marquardt W. H. 1985. *Complexity and scale in the study of fisher-gatherer-hunters: an example from the eastern United States.* In: T.D. Price and J.A. Brown (Editors) Prehistoric Hunter-gatherers: the Emergence of Cultural Complexity. Academic Press, Orlando: 59- 98.

Maynard, L. 1980. A Pleistocene date from an occupation deposit in the Pilbara region, Western Australia. *Australian Archaeology*, 10: 3-8.

McBryde, I. 1984. Kulin greenstone quarries: the social contexts of production and distribution for the Mt. William site. *World Archaeology*, 16: 267-285.

McCarthy, F.D. 1978. New light on the Lapstone Creek excavation. *Australian Archaeology*, 8: 49-60.

McDonald, J. 1991. *Archaeology and art in the Sydney region: context and theory in the analysis of a dual-medium style.* In: P. Bahn and A. Rosenfeld (Editors) Rock Art and Prehistory: Papers Presented to Symposium G of the AURA Congress, Darwin 1988. Oxbow Monograph No. 10, Oxford: 78-85.

McDonald, J. 1992. *The Archaeology of the Angophora Reserve Rock Shelter.* Environmental Heritage Monograph Series No. 1. NSW National Parks and Wildlife Service, Hurstville: 184 pp.

McDonald, J.J. 1994. *Sydney Basin Rock Art: The Analysis of a Regional Art Style in the Context of its Prehistory.* Unpublished PhD thesis, Department of Prehistory and Anthropology, Australian National University, Canberra.

McDonald, J. 1996. *AMS Dating Charcoal Drawings in the Sydney Region: Results and Issues.* Unpublished paper presented at the First Workshop on Australian Rock Picture Dating (1996), ANSTO, Lucas Heights.

McDonald, J., Officer, K., Jull, T., Donahue, D., Head, J. and Ford, B. 1990. Investigating [14]AMS: dating prehistoric rock art in the Sydney sandstone basin, Australia. *Rock Art Research*, 7(2): 83-92.

McDonald, J. and Ross, A. 1990. Helping the police with their enquiries: archaeology and politics at Angophora Reserve rockshelter. *Archaeology in Oceania*, 25(2): 114-21.

McNiven, I. 1988. Brooyar rockshelter: a late Holocene seasonal hunting camp from southeast Queensland. *Queensland Archaeological Research*, 5: 133-60.

McNiven, I. 1994. Technological organization and settlement in southwestern Tasmania after the glacial maximum. *Antiquity*, 68: 75-82.

McNiven, I., David, B. and Flood, J. 1992. Delamere 3: further excavations at Yiwarlarlay (Lightning Brothers site), Northern Territory. *Australian Aboriginal Studies*, 1992(1): 67-73.

Megaw, J.V.S. 1966. The excavation of an Aboriginal rock-shelter on Gymea Bay, Port Hacking, NSW. *Archaeology and Physical Anthropology in Oceania*, 1: 23-50.

Megaw, J.V.S. 1968. A dated cultural sequence for the south Sydney region of NSW. *Current Anthropology*, 9(4): 325-29.

Megaw, J.V.S. and Roberts, A. 1974. *The 1967 excavations at Wattamolla Cove, Royal National Park, NSW.* In: J.V.S. Megaw (Editor) The Recent Archaeology of the Sydney District: Excavations 1964-1967. AIAS, Canberra: 1-12.

Mellars, P.A. 1985. *The ecological basis of social complexity in the Upper Palaeolithic of Southwestern France.* In: T.D. Price and J.A. Brown (Editors) Prehistoric Hunter-gatherers: the Emergence of Cultural Complexity. Academic Press, Orlando: 271-297.

Moore, D.R. 1970. Results of an archaeological survey of the Hunter River Valley, New South Wales, Australia. Part I: the Bondaian industry of the upper Hunter and Goulburn River valleys. *Records of the Australian Museum*, 28: 25-64.

Moore, D.R. 1981. Results of an archaeological survey of the Hunter River Valley, New South Wales, Australia. Part II: problems of the lower Hunter and contacts with the Hawksbury Valley. *Records of the Australian Museum*, 33: 388-442.

Morse, K. 1993. *New radiocarbon dates from Northwest Cape, Western Australia: a preliminary report.* In: M. Smith, M. Spriggs and B. Fankhauser (Editors) Sahul in Review: Archaeology of Australia, New Guinea and Island Melanesia, Department of Prehistory, RSPaS, Australian National University, Canberra:155-63.

Morwood, M. 1980. Time, space and prehistoric art: a principal components analysis. *Archaeology and Physical Anthropology in Oceania*, 15: 98-109.

Morwood, M. 1981. Archaeology of the central Queensland highlands: the stone component. *Archaeology in Oceania*, 16:1-52.

Morwood, M. 1984. The prehistory of the central Queensland highlands. *Advances in World Archaeology*, 3: 325-80.

Morwood, M. 1986. The archaeology of art: excavations at Gatton and Maindenwell rockshelters, SE Queensland. *Queensland Archaeological Research*, 3: 88-132.

Morwood, M. 1987. The archaeology of social complexity in south-east Queensland. *Proceedings of the Prehistoric Society*, 53: 337-50.

Morwood, M. 1995(a). *Excavations at Giant Horse.* In: M. Morwood and D. Hobbs (Editors) Quinkan Prehistory: the Archaeology of Aboriginal Art in SE Cape York Peninsula, Australia. University of Queensland, Brisbane: 101-106.

Morwood, M. 1995(b). *Excavations at Red Bluff 1.* In: M. Morwood and D. Hobbs (Editors) Quinkan Prehistory: the Archaeology of Aboriginal Art in SE Cape York Peninsula, Australia. University of Queensland, Brisbane: 127-132.

Morwood, M. and Dagg, L. 1995. *Excavations at Yam Camp.* In: M. Morwood and D. Hobbs (Editors) Quinkan Prehistory: the Archaeology of Aboriginal Art in SE Cape York Peninsula, Australia, Tempus 3, University of Queensland, Brisbane: 107-115.

Morwood M. and Hobbs, D. (Editors) 1995. *Quinkan Prehistory: the Archaeology of Aboriginal Art in SE Cape York Peninsula Australia.* University of Queensland, Brisbane: 208 pp.

Morwood, M., Hobbs, D. and Price, D. 1995(a). *Excavations at Sandy Creek 1 and 2.* In: M. Morwood and D. Hobbs (Editors) Quinkan Prehistory: the Archaeology of Aboriginal Art in SE Cape York Peninsula, Australia. University of Queensland, Brisbane: 71-92.

Morwood, M., Loste-Brown, S. and Price, D. 1995(b). *Excavations at Mushroom Rock.* In: M. Morwood and D. Hobbs (Editors) Quinkan Prehistory: the Archaeology of Aboriginal Art in SE Cape York Peninsula, Australia. University of Queensland, Brisbane: 133-146.

Morwood, M. and Jung, S. 1995. *Excavations at Magnificent Gallery.* In: M. Morwood and D. Hobbs (Editors) Quinkan Prehistory: the Archaeology of Aboriginal Art in SE Cape York Peninsula, Australia. University of Queensland, Brisbane: 93-100.

Morwood, M. and L'Oste-Brown, S. 1995(a). *Excavations at Red Horse.* In: M. Morwood and D. Hobbs (Editors) Quinkan Prehistory: the Archaeology of Aboriginal Art in SE Cape York Peninsula, Australia. University of Queensland, Brisbane: 116-125.

Morwood, M. and L'Oste-Brown, S. 1995(b). Excavations at Hann River 1, central Cape York Peninsula. *Australian Archaeology,* 40: 21-28.

Mulvaney, D.J. 1962. Archaeological excavations on the Aire River, Otway Peninsula, Victoria. *Proceedings of the Royal Society of Victoria,* 75: 1-15.

Mulvaney, D. J. 1975. *The Prehistory of Australia.* Penguin Books, Harmondsworth: 327 pp.

Mulvaney, D.J. and Joyce, E.B. 1965. Archaeological and geomorphological investigations on Mt. Moffat Station, Queensland. *Proceedings of the Prehistoric Society,* 31: 147-212.

Napton, L.K. and Greathouse, E.A. 1985. Archaeological investigations at Pine Gap (Kuyunba), Northern Territory. *Australian Archaeology,* 20: 90-108.

O'Connell, J.F. and Allen, J. 1995. Human reactions to the Pleistocene-Holocene transition in Greater Australia: a summary. *Antiquity,* 69: 855-862.

O'Connor, S. 1987. The stone house structures of High Cliffy Island, north west Kimberley, WA. *Australian Archaeology,* 25: 30-39.

O'Connor, S. 1989. New radiocarbon dates from Koolan Island, West Kimberley, WA. *Australian Archaeology,* 28: 92-104.

O'Connor, S. 1992. The timing and nature of prehistoric island use in northern Australia. *Archaeology in Oceania,* 27: 49-60.

O'Connor, S. 1995. Carpenter's Gap rockshelter 1: 40,000 years of Aboriginal occupation in the Napier Ranges, Kimberley, WA. *Australian Archaeology,* 40: 58-9.

O'Connor, S., Veth, P. and Hubbard, N. 1993. *Changing interpretations of postglacial human subsistence and demography in Sahul.* In: M.A. Smith, M. Spriggs and B. Frankhauser (Editors) Sahul in Review: Pleistocene Archaeology in Australia, New Guinea and Island Melanesia. Department of Prehistory, Research School of Pacific Studies, Australian National University, Canberra: 95-105.

Ossa, P., Marshall, B. and Webb, C. 1995. New Guinea II Cave: a Pleistocene site on the Snowy River, Victoria. *Archaeology in Oceania,* 30: 22-35.

Pardoe, C. 1988. The cemetery as symbol. The distribution of prehistoric Aboriginal burial grounds in southeastern Australia. *Archaeology in Oceania,* 23: 1-16.

Pardoe, C. 1990. *The demographic basis of human evolution in southeastern Australia.* In: B. Meehan and N. White (Editors) Hunter-Gatherer Demography: Past and Present. Oceania Monograph No. 39, University of Sydney, Sydney: 59-70.

Poiner, G. 1974. *The trial excavation of an estuarine rock shelter at Yowie Bay.* In: J.V.S. Megaw (Editor) The Recent Archaeology of the Sydney District: Excavations 1964-1967. AIAS, Canberra: 28-34.

Polach, H., Stipp, J., Golson, J. and Lovering, J. 1967. ANU Radiocarbon date list 1. *Radiocarbon,* 9: 15-27.

Rick, J.W. 1987. Dates as data: an examination of the Peruvian Preceramic radiocarbon record. *American Antiquity,* 52: 55-73.

Roberts, R.G., Jones, R. and Smith, M.A. 1990. Thermoluminescence dating of a 50,000 year-old human occupation site in northern Australia. *Nature,* 345: 153-6.

Roberts, R.G., Jones, R., Spooner, N.A., Head, J., Murray, A.S. and Smith, M.A. 1994. The human colonisation of Australia: optical dates of 53,000 and 60,000 years bracket human arrival at Deaf Adder Gorge, Northern Territory. *Quaternary Science Reviews,* 13: 575-83.

Robins, R. 1995. The results of test excavations in three rockshelters in southwest Queensland. *Memoirs of the Queensland Museum,* 38 (2): 643-66.

Rosenfeld, A., Horton, D. and Winter, J. 1981. *Early Man in North Queensland.* Terra Australis No. 6. Australian National University, Canberra: 95 pp.

Rosenfeld, A., Winston-Gregson, J. and Maskell, K. 1983. Excavation at Nursery Swamp 2, Gudgenby Nature Reserve, ACT. *Australian Archaeology,* 17: 48-58.

Ross, A. 1981. Holocene environments and prehistoric site patterning in the Victorian Mallee. *Archaeology in Oceania,* 16: 145-55.

Ross, A. 1985. Archaeological evidence for population change in the middle to late Holocene in southeastern Australia. *Archaeology in Oceania,* 20: 81-89.

Ross, A., Donnelly, T. and Wasson, R. 1992. *The peopling of the arid zone: human-environment interactions.* In: J. Dodson (Editor) The Naive Lands: Prehistory and Environmental Change in Australia and the Southwest Pacific. Longman Cheshire, Melbourne: 76-114.

Ross, A. and Specht, J. 1976. An archaeological survey on Port Jackson, Sydney. *Australian Archaeology,* 5: 14-7.

Rowland, M.J. 1981. Radiocarbon dates for a shell fishook and disc from Mazie Bay, North Keppel Island. *Australian Archaeology,* 12: 63-9.

Rowland, M.J. 1982. Further radiocarbon dates from the Keppel Islands. *Australian Archaeology,* 15: 43-8.

Schrire, C. 1972. *Ethno-archaeological models and subsistence behaviour in Arnhem Land.* In: D.L. Clarke (Editor) Models in Archaeology. Methuen, London: 653-70.

Schrire, C. 1982. *The Alligator Rivers: Prehistory and Ecology in Western Arnhem Land.* Terra Australis No. 7, Australian National University, Canberra: 277 pp.

Smith, M.A. 1986. The antiquity of seedgrinding in central Australia. *Archaeology in Oceania,* 21: 29-39.

Smith, M.A. 1987. Pleistocene occupation in arid Central Australia. *Nature,* 328: 710-1.

Smith, M.A. 1988. *The Pattern and Timing of Prehistoric Settlement in Central Australia.* Unpublished PhD thesis, University of New England, Armidale.

Smith, M.A. 1989. The case for a resident human population in the Central Australian Ranges during full glacial aridity. *Archaeology in Oceania,* 24: 93-105.

Smith, M.A. and Sharp, N.D. 1993. *Pleistocene sites in Australia, New Guinea and Island Melanesia: geographic and temporal structure of the archaeological record.* In: M.A. Smith, M. Spriggs and B. Frankhauser (Editors) Sahul in Review: Pleistocene Archaeology in Australia, New Guinea and Island Melanesia. Department of Prehistory, Research School of Pacific Studies, Australian National University, Canberra: 37-59.

Soffer, O. 1985. *Patterns of intensification as seen from the Upper Paleolithic of the Central Russian Plain.* In: T.D. Price and J.A. Brown (Editors) Prehistoric Hunter-

Gatherers: the Emergence of Cultural Complexity. Academic Press, Orlando: 235-70.

Stockton, E.D. 1977. Review of early Bondaian dates. *Mankind*, 11: 48-51.

Stockton, E.D. and Holland, W. 1974. Cultural sites and their environment in the Blue Mountains. *Archaeology and Physical Anthropology in Oceania*, 9(1): 36-65.

Tracey, R. 1974. *Three minor sites near Curracurrang Cove with a preliminary note on a rock shelter at Newport, NSW.* In: J.V.S. Megaw (Editor) The Recent Archaeology of the Sydney District: Excavations 1964-1967. AIAS, Canberra: 17-9.

Taçon, P. 1993. Regionalism in the recent rock art of western Arnhem Land, Northern Territory. *Archaeology in Oceania*, 28: 112-20.

Taçon, P. and Chippindale, C. 1994. Australia's ancient warriors: changing depictions of fighting in the rock art of Arnhem Land, NT. *Cambridge Archaeological Journal*, 4(2): 211-48.

Ulm, S. 1995. *Fishers, Gatherers and Hunters on the Moreton Fringe: Reconsidering the Prehistoric Aboriginal Marine Fishery in Southeast Queensland, Australia.* Unpublished BA (Honours) thesis, University of Queensland, St. Lucia.

Veth, P. 1989. Islands of the interior: a model for the colonization of Australia's arid zone. *Archaeology in Oceania*, 24: 81-92.

Veth, P. 1993. *Islands in the Interior: the Dynamics of Prehistoric Adaptations within the Arid Zone of Australia.* International Monographs in Prehistory, Archaeological Series No. 3, 144 pp.

Veth, P. 1995. Aridity and settlement in northwest Australia. *Antiquity*, 69: 733-46.

Watchman, A. and Cole, N. 1993. Accelerator radiocarbon dating of plant-fibre binders in rock paintings from northeastern Australia. *Antiquity*, 67: 355-8.

Welch, D. 1997. Fight or dance? Ceremony and the spearthrower in northern Australian rock art. *Rock Art Research*, 14(2): 88-112.

Williams, E. 1987. Complex hunter-gatherers: a view from Australia. *Antiquity*, 61: 310-21.

Wobst, H.M. 1976. Locational relationships in Palaeolithic society. *Journal of Human Evolution*, 5: 49-58.

Woodroffe, C.D., Chappell, J., and Thom, B.G. 1988. Shell middens in the context of estuarine development, South Alligator River, Northern Territory. *Archaeology in Oceania*, 23: 95-103.

Wright, R.V.S. 1971. *Prehistory in the Cape York Peninsula.* In: D.J. Mulvaney and J. Golson (Editors). Aboriginal Man and Environment in Australia. Australian National University Press, Canberra: 133-40.

Zobel, D.E., Vanderwal, R.L. and Frankel, D. 1984. The Moonlight Head rockshelter. *Proceedings of the Royal Society of Victoria*, 96: 1-24.

Zvelebil, M. (Editor) 1986. *Hunters in Transition: Mesolithic Societies of Temperate Eurasia and Their Transition to Farming.* Cambridge University Press, Cambridge: 194 pp.

Forests and Grassland, Drought and Fire:
The Island of Borneo in the
Historical Environmental Record (post-1800)

Lesley M. Potter

Introduction

In a volume largely devoted to long-term environmental change as measured by the techniques of palaeoecology and archaeology, this paper takes a rather different focus and a shorter time-span. It is an attempt to examine aspects of the environmental history of the island of Borneo over the past two centuries, using evidence from Dutch and British historical records. Inevitably such a study must focus on change in the forests, both their species mix and the nature of their ecosystems, as well as their felling and conversion to other land uses. While it is common knowledge that the forests of Borneo have been subjected to accelerated changes over the past three decades, they are often believed to have been 'primary forests', largely untouched, before the advent of modern logging concessions, land settlements and plantations. When one examines the historical sources over the past two centuries, however, it is clear that the forests have been constantly evolving under the influence of both human and natural pressures. In some cases permanent vegetation change has taken place, with wide-scale transformation to grasslands or cash crops, including domestication of wild species. In other cases, clearing and conversion have been of a temporary nature and the intensity of human impact has slackened, so that regrowth of various kinds has blurred or obliterated the earlier patterns. Natural events have been occasionally cataclysmic, resulting in destructive episodes of drought and fire. In a short paper such as this, only selective aspects can be touched upon. While the sources are voluminous, there is much that is impressionistic and imprecise. The fact that the historian is forced to view the landscape through the lens of the colonial observer, rather than being able to rely on direct statements from the indigenous participants, inevitably distorts the picture. Nevertheless, the emphasis is heavily on the activities of the non-European population in changing local environments, sometimes in conjunction with extreme climatic events or with outside stimuli, such as high prices for forest products. The mines, plantations, timber concessions and ports which symbolised the modest, 'spotty' and uneven pace of enclave colonial development in the island have been written about elsewhere (e.g. Lindblad 1988; Potter 1988), as have the rapid transformations since the mid-1960s. It is hoped that an understanding of the present situation on the ground will assist in validating some of the earlier descriptions and comparisons with the current scene will be made where useful. The paper has

ISBN 3-923381-47-6

four sections: the impact of population and farming practices on the forest; collecting activities (especially gutta percha); grassland formation; drought and fire.

Population and farming practices

'Through the custom of the Dayak population after a succession of one or two years to continually exchange the land for their rice fields with new grounds, and to obtain these new grounds through the breaking up of the forests, we no longer find original forest in Borneo. The old forests which I have traversed are at the most 130 to 150 years old' (von Gaffron 1858: 224 (translated)). Writing from Sintang in the middle Kapuas basin of West Kalimantan, explorer von Gaffron presented his thoughts on the general position of the forests in the light of Dayak agricultural practices. His statement is interesting for its comparatively early date[1], its long time perspective and the inference that the 'original forest' had undergone major alteration by human hands. Unlike later observers, von Gaffron was not condemnatory of Dayak swidden systems. A similar kind of statement on the evolution of the forests, though with more botanical detail, was made by the Italian naturalist Beccari, who conducted scientific studies in Sarawak from 1865-8, then added an appendix on 'The Bornean Forest' to his 1904 account of his earlier exploits. 'The vegetation which covers an area where the primeval forest has been destroyed is utterly distinct from that of the latter, with its rich and specialised primitive forms. The species thus establishing themselves are quite different and are mostly those which have an extensive geographical distribution' (Beccari 1904: 381). While he acknowledged that it was likely that the 'primitive type' of forest would eventually be restored, Beccari also warned of numerous species extinctions if the destruction was extended and continued. He noted subsequently that secondary forest 'covers great areas in Sarawak, and appears quickly in all fertile localities where the primitive forest has been cut down, cultivated, and afterwards abandoned' (Beccari 1904: 410).These two quotations support the theory of Corlett, who argued that there is a vast difference in terms of biodiversity between a forest area which has been cleared and allowed to re-grow, and an area subjected to selected removal of species, as in logging or harvesting of forest products. Corlett suggested that unless the clearance was very small, the diversity of primary forest species was always less in secondary forest than in the original stand, while faunal habitats were fewer (Corlett 1994, 1995).

In nineteenth and twentieth century Borneo, and for perhaps 4000 years before that (Ipoi and Bellwood 1991; Bellwood *et al.* 1992), the predominant

---

[1] Movement into the interior of Borneo by Dutch officials, while beginning in the 1820s, became restricted after the outbreak of the Java War in 1825 forced a reduction of administration on the island. Travel inland was not permitted between 1833 and 1843. In 1824, Halewijn travelled in the SE between Banjarmasin and Kelua in the Hulu Sungai, and in Tanah Laut. It was 19 years before he was followed up the Barito by Schwaner (1843-7). Von Gaffron was the first European to cross from the south (Kotawaringin) to the west coast (Pontianak) in 1846-7 and was thus well qualified to comment on the state of the forest (Halewijn 1838; Schwaner 1853; Pijnappel 1860; Irwin 1955).

agricultural systems consisted of rice-based swidden cultivation in both its upland and lowland forms, including 'wet swiddens' in swampy areas[2]. All depended on forest clearance to create a space for farming and on burning to transfer nutrients from the cut vegetation to the soil. While there is a lack of historical ethnographic material to establish the details of many systems, it is nonetheless clear that there existed both the pioneering type, with frequent moves into new areas of primary forest (as among some of the Iban) and also the rotational type, in which longhouses moved infrequently, and land was rotated over a defined area (as among the Maloh) (Freeman 1955, 1970; King 1985). As Lowenhaupt-Tsing (1984) and Chin (1985) have noted, many swidden systems have been characterised by an amazing diversity of planted cultigens, including a great many varieties of rice, vegetables and fruits. What Lowenhaupt-Tsing has termed 'an aesthetic of diversity' (1984: 201) and a great willingness to experiment with new cultivars, appears as a common cultural pattern. The *tembawang* systems or complex agro-forests which have recently been subjected to scientific study are a common accompaniment to rotational swiddens These were commented upon by outside observers in the Kapuas basin of West Kalimantan in 1848 and described by Sather as being about 300 years old among groups of Iban in the Second Division of Sarawak (Kalamatan 1848; Sather 1990). In these systems there has been considerable domestication of forest species in fallows, while former longhouse sites became gardens specializing in *tengkawang* or illipe nuts, a wide range of fruits, especially durian, and many other edible or useful products (Padoch and Peters 1993). It has now been recognised that such systems are quite widespread, being found in the Sukadana and Melawi regions of West Kalimantan, for example, and in East Kalimantan. These sites exemplify the most permanent disruption of previous forest patterns, with the substitution of a new 'constructed forest', an entirely human artefact. While such agro-forests may not contain the 'specialised primitive forms' of ancient rain forest as described by Beccari, they are nevertheless very rich in species of interest to humans, including endemic and rare cultivars which sometimes survive only in such niches. Considerable biodiversity is thus maintained.

 The kind of warfare which existed among the different indigenous groups in the nineteenth century (and earlier) both encouraged and restricted migration, as more predatory or stronger groups moved into the territories of their weaker neighbours, causing them to flee to more secure locations. The establishment of peace and an end to headhunting under the Brooke regime in Sarawak and somewhat later in the Dutch territories, enabled groups which had existed under conditions of land shortage e.g. Kenyah in parts of the Apo Kayan, to break out from the uplands and move safely downriver, in this case to the Baram, where they could occupy land vacated by others and engage more easily in forest products trade (Whittier 1973; Chin 1985; Lian 1987). The frequent migrations and the more substantial populations in many districts before the advent of

---

[2] While the swamp-based systems were sometimes eventually turned into permanent fields in which water control (for rainfed systems) and even some forms of irrigation were practised, much of this type of farming was adventitious and based on the extent of the dry season, being able to expand further into the swamps in very dry years.

European diseases such as smallpox, had the effect of exposing more of the forests to occasional clearing than might have been expected[3]. While positioning of villages on or close to river banks would facilitate transportation, more remote sites above rapids might be preferred for reasons of security.

Areas of permanent agriculture, predominantly wet rice, were very limited until the twientieth century. Under such systems more lasting environmental changes, such as construction of 'landesque capital'[4] through canals or terraces may be expected to result, with wider areas being cleared from forest. Although larger settled populations certainly accompanied the establishment of wet rice systems, irrigation works tended to be small and local and most fields were basically rainfed. Three kinds of systems have been identified: the first type was established in coastal areas by immigrant populations of Chinese, Ilanun (from Mindanao) and Bugis (from Sulawesi); two other types existed in interior locations and might be considered 'indigenous'. Jackson believed that the carefully irrigated wet rice system on the north coast of West Kalimantan (Singkawang) had been brought directly from South China in the second half of the eighteenth century, as part of the establishment of extensive Chinese-operated gold-mining industry further inland. He suggested that the intricate systems of reservoirs, sluices and channels constructed to facilitate gold mining had also been borrowed from the ricefields of South China (Jackson 1970). Other wet rice systems perhaps introduced by Chinese along the north-west coast of Sabah may be considerably older. It is not clear whether the Ilanun agriculture located on the plain of Tampassak in northern Sabah should be included among these (St John 1862), but cattle appear to have been used for ploughing. The *sawahs* of the SE littoral in the Bugis sultanate of Pegatan equally stand out in their use of buffalo as plough animals and their communal nature (Netscher and von Dewall, 1853; van Hoeve 1936).

The indigenous wet rice agriculture practised by the peoples of the high plateaux of the Kerayan-Kelabit upland of East Kalimantan-Sarawak and the Keningau district of Sabah appear to be similar in type and may have evolved '*in situ*'. Rousseau (1990) suggests that the development of irrigated agriculture among the Kelabit may have been a way of coping with population pressure and invasion of swiddens by *Imperata* grass. The people themselves have a story which relates to the drainage of a lake formerly occupying the flat plateau area, making the land ideally suitable for wet rice (Schneeburger 1945). The Kelabit domesticated the local Sambur deer and later acquired buffalo. It is ironic that owing to recent outmigration their highlands are now underpopulated, the maintenance of the highly productive rice fields being almost entirely reliant on Indonesian labour, largely from Kerayan.

The final example of a permanent sawah system is in the Banjarese heartland of the Hulu Sungai in South Kalimantan. The Hulu Sungai is an inland alluvial

---

[3] St John (1862), travelling in Sarawak and present Sabah, noted both larger populations than now exist in some interior districts, and also areas where serious population decline had occurred (I, 10, 105). He often commented on the lack of 'old forest' and on the 'forests of fruit trees' surrounding houses.

[4] It was believed that Brookfield coined this term, but he has denied authorship (Brookfield 2001: 55).

basin occupying the middle reaches of the Negara river, lying between the foot-hills of the Meratus Mountains and the flood plain of the Barito. Since the eighteenth century boom in pepper exports from Banjarmasin, when a Banjarese[5] population began to replace pre-existing Dayak groups, the area has always supported a disproportionally large share of Borneo's population. When Halewijn visited Amuntai in November 1825 he found 1480 houses along the river banks, and extensive rice fields stretching away into the distance (Halewijn 1838: 195). These were in fact dry season fields, planted after the rainy season floods had subsided, and sometimes risky (Schuitemaker 1938). The slightly higher central districts of Alai and Amandit, named after the rivers which drained them, were later recognised as most important for producing a rice surplus from largely rain fed fields 'the most heavily populated and also the most blessed region ... the most fertile soil' (Happe 1867). It is interesting that there is no tradition of using plough animals in the Hulu Sungai. In 1880 the subdivision of Amuntai, which covered the region, was estimated to contain 250,000 people, making it by far the greatest population concentration in the island (Meijer 1880). Parts of the district were described as 'overpopulated' in the 1890s (Joekes 1893)[6], when the numbers had risen to over 360,000. Despite quite extensive wet rice production and a variety of other agricultural and small-scale industrial activities, lack of drainage and irrigation meant risks of crop failure were high in some sections and good land was limited. Large numbers of people from traditional slave classes or debt bondsmen worked as labourers only and had a high propensity to move once the opportunity arose. The migrations of Banjarese out of the area seeking new agricultural opportunities have been described in detail elsewhere (Potter 1993).

Such movements may be likened to those of the Iban in Sarawak, except that the Banjarese were mainly experts in swamp reclamation. Their techniques consisted of digging drains and clearing forests to open up previously empty areas of tidal swamp, not only in coastal South and Central Kalimantan but also in eastern Sumatra (Riau and Jambi) and on the Malay Peninsula. The swamps were planted to subsistence crops of rice and cash crops of coconuts or later rubber, on a constantly expanding moving frontier. In a similar manner, the Malays of southwest Sarawak are credited with constructing their present landscape of rice, coconut and rubber over the past 150 years (Harrison 1970). Further south in West Kalimantan, it was Bugis slaves who carried out the initial draining of the swamps near Pontianak and their substitution by a coconut and rice-based agriculture (Kater 1871).

Some Banjarese did spill over from the Hulu Sungai into the surrounding uplands, practising shifting cultivation of an equally pioneering kind to the Iban, but always with associated cash cropping. They utilised the hilly Meratus foothills along the eastern rim of the Hulu Sungai and further south in the twin valleys of the rivers Kiwa and Kanan, planting upland rice and pepper and actively creating *Imperata* grassland for deer hunting. Their invasions of the forested

---

[5] Banjarese are an Islamic group, a mixture of Javanese, Arabs, Malays from Sumatra or Malacca and local Dayaks.

[6] A similar comment was made by Perelaer in 1870.

Peter Kershaw, Bruno David, Nigel Tapper, Dan Penny and Jonathan Brown (Eds): Bridging Wallace´s Line

Dayak lands of the Meratus mountains were likened by Dutch authorities to invasions of *alang-alang* (Wentholt 1938). In the early twentieth century many of these lands began to be planted with a more permanent tree crop, *Hevea brasiliensis* or smallholder rubber. The Hulu Sungai foothills were transformed to forests of 'jungle rubber', in which the trees were planted very close together and much grassland was reclaimed. In the 1920s and 30s rubber also became the smallholder crop *par excellence* among the Dayak population of the Kapuas basin in West Kalimantan and throughout the island of Borneo, with major reductions in mobility among Banjarese and Dayaks alike (Ozinga 1940; Dove 1993, 1994).

Collecting activities: gutta percha and other forest products

In arguing that selective removal of forest products is not necessarily carried out without an impact on forest structure and ecosystem functioning, Peters has concluded that high intensity harvesting of a valuable product 'can be as devastating as logging in causing the disruption of local populations and species extinctions' (Peters 1996: 9). A good example of such harvesting was the collection of the latex gutta percha, the highest priced of all the non-timber forest products for the last thirty years of the nineteenth century and the first decade of the twentieth (Potter 1997b).

Gutta percha is a latex extracted from several species of the genera *Palaquium* and *Payena* from the family Sapotaceae, occurring primarily in the forests of Sumatra, Borneo and the Malaysian Peninsula. Being an excellent non-conductor of both heat and electricity, able to be moulded when heated, then harden and maintain its shape on cooling, gutta percha was found to be the most suitable natural substance for coating submarine cables. Although it was well known to local Malays, its qualities were first observed by Europeans in Singapore in the early 1840s and samples were sent to England. The market became interested almost immediately and steps were taken to organise the trade, which was controlled by Chinese interests in Singapore and their local agents, initially Malay but later increasingly Chinese, while the collection was in the hands of indigenous forest people. This particular latex occurs in intercellular cavities below the bark of the trees rather than in connected tubes, so it will not flow but oozes out slowly, with largest amounts being obtained from the upper limbs of the tall, straight, almost unclimbable trees. The universally preferred technique for harvesting was to fell the tree and make cuts in the bark all around the trunk. Because an economic yield of latex could be extracted only from the biggest trees, at least 35 years old, such harvesting methods quickly became unsustainable and the supply area shifted rapidly. From its original base in Singapore, it moved northwards to Johore, Pahang and Perak on the Malay Peninsula, south to the Riau and Lingga archipelagoes and the Sumatran coast, and eventually to the island of Borneo.

Gutta collecting had begun on the south-west coast of present West Kalimantan by 1852 and extended inland from Sambas and through the Kapuas basin within the following two or three years. By 1855, Dutch Borneo was already the largest source of supply, with the amounts imported into Singapore more than doubling from 1852-5 and outstripping the Malay Peninsula and Sumatra (Wong 1960). Prices were moderate, but high enough to persuade local collectors in the Sukadana-Matan region of the south-west that it was worth switching to this unknown commodity, away from traditional products such as rattan, damar (resin) and wax. The gutta seekers would leave for the bush in the dry season and stay there one or two months. By that time each group of five or six would have collected one pikul (68 kg) of latex, but one hundred trees would have been felled (De Vriese 1855-6). Not only was there thus potential for large-scale forest destruction in this trade, but as with other forest products in Borneo and throughout Malesia, the characteristic diversity of species was already causing confusion. From Sintang Von Gaffron described eight types of gutta percha latex, classifying them according to quality[7], while in Banjarmasin the ill-fated James Motley[8] identified nine varieties, all with different names from the Sintang list (Von Gaffron 1858; Motley, in de Vriese 1860). While local consensus might be reached as to which were the best species to trade, indigenous collectors soon became expert in adulterating the product: as much of it was obtained from deep in the 'uninhabited forest', no authority could exercise control over such activities.

By 1870, increased demand for cable construction led to price rises for gutta percha and a great expansion in collecting activity. The excitement was felt in Sarawak, with the newly established *Sarawak Gazette* complaining that the population was tempted away from their usual occupations of farming and gardening to rush into the forest (Smythies 1961). Gutta percha became the leading item by value in Sarawak's trade figures (Wong 1960). It was estimated that three million trees had been felled there between 1854 and 1875 (*Sarawak Gazette* May 1, 1875), with a tendency for younger and smaller trees to be cut as prices increased and supplies dwindled. While local collectors were able to barter their gutta for sought-after commodities, such as beads, brass wire, and ceramic jars, the bulk of the profits accrued to the traders and the Singapore merchants.

This period marked the beginning of the real boom in gutta percha collection, which lasted until about 1905 and was to result in a very thorough working over of the forests throughout most of Borneo, with the exception of remote mountain areas. Collecting activities were also intensified by the occurrence of a series of drought years and subsequent crop failure. The drought in 1877-78 was extremely widespread and severe, and followed by serious flooding. Dayaks in West Kalimantan bartered gutta percha for rice with the Chinese traders who controlled the industry, and exports of latex from Pontianak increased fivefold (Van Gorkum 1884). Further droughts at intervals throughout the 1880s kept pressure on the forests, as did the continuous price increases beyond the end of the century. As more accessible areas became worked out, Dayaks were having to penetrate much further into unfamiliar districts, far from their home territories. They usually assembled in large groups for protection, fearful of coming into

---

[7] Quality was determined by the proportion of resin in the gutta, which would determine how brittle it would eventually become. The best varieties had less than 10% resin.

[8] Motley was an English botanist and mining engineer who was killed with other Europeans at a Dutch-operated coal mine during the Banjarmasin War (1859-63).

conflict with warlike neighbours who resented such intrusions (van Basel 1880).

In Sarawak, young Ibans were encouraged by the Brooke regime to travel widely over the country, extracting the forest products on which the economy depended: 'These young fellows, for those who go are nearly all under eighteen years of age, are the working pioneers of jungle produce wherever it exists, from the mouth of Banjir Masin round northward to the Kina Batangan river ... about halfway round the whole island of Borneo' (*Sarawak Gazette,* 1 Dec, 1885). The fact that they often behaved in a truculent manner, taking heads as well as gutta, led to friction with resident groups, who also complained that their districts were being denuded. Pringle (1970) noted that the Ibans sometimes used these gutta expeditions to discover new areas for farming, to support the pioneering type of shifting cultivation for which they were famous. Dove agreed that this behaviour was also typical of the Ibanic Kantu' with whom he worked on the West Kalimantan side of the border (Dove 1994). Other indigenous groups, however, were more devoted to their own territories and not as interested in ranging far and wide. Witty, exploring the area south of Marudu Bay in the early years of British North Borneo (Sabah), discovered that the local Dusun carefully protected their enormous gutta trees, tapping them to extract the latex and imposing fines on anybody who felled a tree. Where he saw destruction of gutta trees near the coast, Witty blamed it on 'loafers' travelling to Labuan and Brunei (Witty 1881). These were probably also Iban, who had initially been invited to work gutta in Perak on the Malay Peninsula, until it was realized that they were rapidly eliminating the trees (Wray 1883; Foxworthy 1935). They were said to travel as far as Mindanao and West Sumatra (Baring Gould and Bampfylde 1909).

Is it possible to assess the ecological impact of these collection activities on the forests? As early as 1889 an article in the *Indian Forester* mentioned that a scarcity of large gutta percha trees in Borneo had led to new arrangements whereby each tree now had a permanent owner who tapped it twice a year. This system may possibly have operated in some areas of Sabah, but seems unlikely to have been adopted very widely, although the scarcity of large trees may well have been correct. More interesting is the comment by van Romburgh (1900) concerning the large amount of 'inferior guttas' included in the exports from both Banjarmasin and Pontianak in 1897 and 1898. This suggests that the trees producing the higher quality product were becoming harder to find. The attempts to discover a process for extracting the latex from the leaves was another answer to a real shortage of mature trees. It appeared that many areas of the forests were full of young trees, both re-sprouting from felled trunks and regenerating from seed, perhaps spread by bats (Combanaire 1910). There was optimism that the resource might after all be renewable. Once a successful process had been discovered, factories were established in a number of the producing areas, including Sarawak and West Kalimantan, which depended on a mechanical crushing of the leaves to extract the latex. Unfortunately the factories did not survive, as the necessary supplies of leaves were not regularly forthcoming. The only one to last for some years was a government-sponsored plant set up in Java with its own plantation attached, although this had encountered problems with acclimatization of the better species of *Palaquium,* as Java experienced a longer dry season than Borneo (Ham 1900; Combanaire 1910; van Gelder 1950).

Although attempts had been made by the Dutch authorities as early as the 1850s to establish plantations of the better kinds of gutta percha trees in parts of West Kalimantan, these had all failed. Successful domestication began to be undertaken by Dayak groups around the time of the highest prices and up until 1908, by which time prices had begun to fall. It was noted that Dayaks in the Kapuas were successfully growing *Palaquium* trees in their tembawangs (Ozinga 1940). This industry might be regarded as a nineteenth century equivalent of the recent 'timber boom' in the dipterocarp forests. Certainly, there was 'creaming' of the most desired species, while the surviving trees, once so tall, were reduced to mere poles. From the point of view of the excitement and heady adventure which it offered its young male participants, one can also find similarities with the modern collection of *gaharu* (*Aquilaria* sp.). Here the tree is also felled to obtain the product, and again the young men range far through the remaining forest. A study of gaharu in West Kalimantan in the 1930s also mentioned a wide range of different species which were tapped to produce the perfumed wood, while, as with gutta percha, adulteration of the product was common. Some types of gaharu were also planted out in the tembawangs along the Kapuas (Schuitemaker 1933).

From forest to grassland – and back?

A further important aspect of the environmental history of Borneo concerns the degradation of land cover. While gutta percha collecting has been likened to logging in its selective creaming of species, the opening up of the forest floor to penetration of light created suitable conditions for the establishment of invasive grassy weeds, of which *Imperata cylindrica* or *alang-alang* is the most infamous. Such grassland intrusion has sometimes led to a permanent vegetation change over wide areas, at other times has been a temporary phenomenon of limited scale and impact. Authorities have universally deplored the permanent transformations from forest to 'sheet lalang' and even the small-scale appearance of *Imperata* on individual farmers' plots, linking grassland formation with shifting cultivation as though the former had occurred through careless or ignorant behaviour. Studies in Borneo by anthropologists and geographers such as Seavoy (1975) and Dove (1981) have demonstrated that grassland plots were often deliberately created for purposes of deer hunting or pasturing cattle by Islamic populations, while sometimes they resulted from a more intensive sedentary agriculture. Particular kinds of cash cropping, especially pepper growing, are frequently reasons for the existence of widespread expanses of *Imperata*, which is then maintained through regular burning.

While many small patches of *alang-alang* occur through the forests of Borneo, very few areas are in fact large enough to show out on generalised maps of the island[9]. Undoubtedly the largest of these is in the Meratus foothills of

---

[9] When an earlier version of this paper, accompanied by slides, was given at a meeting in Brunei of the Borneo Research Council, many in the audience were amazed at the extent of grassland in Kalimantan: such grasslands are unheard of in Brunei or Sarawak.

South Kalimantan, extending from the north Hulu Sungai to the south coast through Tanah Laut. The South Kalimantan grasslands appear quite old and may date from the extensive pepper trade from Banjarmasin in the eighteenth century. It is argued that some of the grasslands were created through pepper growing, then maintained as appanage lands for deer-hunting nobles attached to the Banjarmasin sultanate (Idwar Saleh 1978). Halewijn discusses the extensive hunting activities of the sultan's court in the 1820s, and the export of dried deer meat to China (Halewijn 1838: 404). Mining activities could also promote grassland formation: Halewijn described the approaches to the area occupied by Chinese gold miners, near Pelaihari in Tanah Laut, as a 'bald and uncultivated region clothed in alang-alang' (Halewijin 1838: 410). A further wide expanse of grassland occurs in West Kalimantan, in the middle Kapuas and especially its southern tributary the Melawi. The Melawi and Sintang lands are still being studied, but evidence so far suggests that grasslands there are more recent. When a Dutch geographer visited the area towards the end of the last century, the vegetation was described as secondary forest, but by 1925 grass was widespread (Enthoven 1903; Scheuer 1932). The introduction of cattle from Madura and their pasturing by Islamicized Dayaks along the Melawi river seems to be one reason for the development of this extensive grassland, in a district of much heavier rainfall than in South Kalimantan (Brookfield *et al.* 1995). The former gold mining areas of Mandor and Montrado in West Kalimantan were described by Enthoven (1903) as a barren landscape covered in gravel, with a vegetation of scrub and coarse grass, totally uncultivable.

An interesting case of a grassland which was apparently created as a result of population pressure during the last century, is that of the Iwan river in the Apo Kayan highland of East Kalimantan. Lack of suitable land for farming and reduced crop yields because of *Imperata* invasion were cited by some Kenyah as a reason for their ancestors' decision to move westwards into the Baram basin of Sarawak, once it became safe for them to do so in the 1880s (Whittier 1973; Chin 1985).[10] The Iwan valley, formerly a great routeway used by many groups travelling through to the Baram, is now empty of population and part of the Kayan Mentarang Nature reserve. According to all accounts there is no longer any grassland there: it would be interesting to study the condition of the restored forest which now occupies most of the area. Above Apau Ping on the Upper Bahau river, again within the Kayan Mentarang Reserve, is another area of grassland which used to be settled by Sa'aban people. According to one source, the grassland was maintained to provide a source of feed for *banteng*, so that the wild cattle would not put pressure on the cultivated fields (Nazir Foead 1999). Deer were also encouraged by the existence of the grass. This area now too has been abandoned by its human occupants (who have moved to Sarawak), but the grassland is being seasonally burned and maintained as a hunting ground by people from villages downriver (author's fieldwork during 1995).

---

[10] A description in Verkuyl (1950) provides more recent evidence of a similar occurrence of 'overpopulation'. He stated that the 30,000 Kenyah in the neighbouring Kayan basin, east of the Iran range, had been completely isolated during World War II, which had forced them to over-farm their swiddens and had ruined the forest cover of the entire basin.

The current rate of attack on the forests of Borneo would appear to be creating the conditions, not only for widescale reversion to secondary formations, with concomitant loss of biodiversity and increased inflammability, but according to some observers, a likely transition to ever larger areas of grassland. One statement of impending doom is as follows: 'Currently, land use throughout Kalimantan is on an unstable trajectory of net conversion of rain forest to *alang alang*, without yet the development of sustainable systems of productive agriculture, of productive managed forestry and of protection of nature reserves' (Leighton and Peart 1988). I do not believe that such a scenario accurately represents the future, first because the grassland formations are a kind of land reserve, able fairly easily to be converted to more intensive agricultural systems when this is required. *Imperata* was once widespread on Java, but gave way under pressure of population and need to more intensive systems (Potter 1997a), which points to one possible direction of future development. In other areas the grasslands are under threat from the introduced composite weed *Chromolaena odorata*, which can shade out grass and is often the first colonizer of swidden fields. More important is the government pressure to build a pulp and paper industry, using as raw material the products of industrial forests, such as *Acacia mangium*, a fast-growing acacia which quickly develops a thick crown and shades out *alang-alang*. Many grassland areas, including both the Meratus foothills and Sintang-Sanggau are active reforestation sites, part of a determined commercial programme now taking shape on a large scale and backed by considerable international investment. Although there is some doubt about the sustainability of these plantations (and many social problems are being caused by their occupation of village lands), the pressures on the grasslands are nevertheless considerable. The days when *Imperata cylindrica* was regarded as a problem in Kalimantan will soon be past and it will come to be viewed as a scarce resource, especially for cattle feed[11]. Sabah also is reforesting with *Acacia* and reconverting some of its more limited grasslands back to trees. In Sarawak the areas of *Imperata* are small.

Drought and fire

As one studies the historical record in the light of increasing knowledge of the cyclic shift in climatic patterns known as the El Nino Southern Oscillation (ENSO), it becomes clear that many of the periodic droughts which appeared haphazardly to afflict parts of Borneo, are in fact related to this phenomenon. It has also been discovered that the natural rhythms of the dipterocarp trees, especially patterns of flowering and fruiting, are adjusted to the occurrence of periodic drought (Ashton *et al.* 1988). Migrations of wild pig and other important species such as orang-utan, follow the cyclic fruiting of such trees. Bruenig (1987) feels that some morphological features of the lowland rain forests, such as small, thick leaves in upper-storey trees, together with the adjustment of

---

[11] This statement was made before the onset of the Asian economic crisis led to a slowing of plantation development, for both pulpwood and oilpalm, in the areas in question. *Imperata* remains widespread in the Melawi basin of West Kalimantan.

flowering and fruiting, are a response to regular cyclic drought stress. Thus, despite the heavy rainfall which falls regularly on much of Borneo, especially Sarawak and West Kalimantan, and the occasional years when there is virtually no dry season, more significant are the regular episodes of drought. These may lead to actual tree death through dessication, preceded by the dropping of leaves of normally evergreen species, as happened in parts of Kutai in 1982-3 (Leighton 1984). Soil type can exacerbate drought stress, especially soils developed over limestone. The drought periods are often also periods of forest fire. Although it might have been expected that a rain forest would not burn readily, it will do so when subjected first to severe drought. In the analysis of the fire of 1982-3, the amount of trash lying on the ground as a result of logging has been partly blamed for the extent and severity of the conflagration, with the most severe burn damage occurring in the most intensively logged areas (Wirawan 1984; Schindele *et al.* 1989).[12] Nevertheless, we now have evidence that the forest also burned during earlier periods, which again coincided with ENSO events, and there is additional evidence of burning much further back into the past, as is revealed by the charcoal layers found in several locations in East Kalimantan (Goldammer and Siebert 1989). In this province, the existence of coal seams close to the surface has led to the persistence of burning for long periods beyond the immediate trigger of a drought. One can also draw a distinction between burning of forest over peat swamps, which dry out in times of drought and become very easily combustible, and burning in the main areas of lowland dipterocarp forest.

Human agency has sometimes been shown to be directly responsible for starting the fires accompanying the more severe drought events, while at other times such a connection, while suspected, remains unproven. One large fire with no apparent human origin was that reported by von Gaffron in 1846 during a drought in Kotawaringin, Central Kalimantan. Von Gaffron, exploring the district for coal and gold, reported a huge fire in the coastal swamp forest, which he estimated to cover 600,000 hectares and believed to have been burning for months. He suggested that natural friction from the dry scrub had started the blaze, which was in a remote and apparently uninhabited area (Pijnappel 1860). During the very severe ENSO of 1877-8 there is evidence of burning from several sources, only one of which was directly linked to human activity. From the Upper Kapuas basin of West Kalimantan, Gerlach noted that the lakes had dried out, that Dayaks had here and there set fire to the forest and that there were extensive areas of dead forest (Gerlach 1881). Bock, on the opposite side of the mountain divide in Kutai (East Kalimantan), did not mention fire, though he did say that one third of the trees in the lakes region of the Middle Mahakam had died (Bock 1881). Looking back on the effects of this drought in Kutai, Tromp argued that a sure sign of the poverty of the soil was the fact that the woods which had burned 'in the enormous drought of 1878' had still not recovered, while with good soil, a burned area would regenerate to thick bush within seven years (Tromp 1890: 732). Burbidge, travelling by boat along the north-west

---

[12] The forests burned again in the ENSO of 1997-8, mainly because of plantation establishment.

coast of Sabah in 1878, described 'jungle fires raging along the coast' after nearly five months without rain, this being recognized as a period of exceptional drought (Burbidge 1880: 183). The total impact of such events as 1878 on the forests awaits the piecing together of further evidence. The effect on the human population was severe, but the forest at least provided some subsistence when crops failed. There was no great famine, as took place in India at the time, although some starvation may well have occurred. Dayaks were said to have existed on wild forest yams and fruits, while in Kutai the Sultan distributed rice (Bock 1881). Conditions in the Hulu Sungai were described as 'critical' by Dutch authorities (Meijer 1880), and steps were taken to give out work digging drains so those whose crops had failed could buy food. It has been mentioned that forest products were also bartered for rice in West Kalimantan. Some groups of people living in the remote mountainous interior were outside the reach of such assistance. A great drought, with accompanying starvation, features in the mythology of the Kenyah Leppo' Ke in the Upper Bahau river of East Kalimantan, but it has not yet been accurately dated (author's fieldwork during 1995).

There is further evidence for Borneo of extensive burning in another drought which occurred in 1914. This 1914 ENSO event, while devastating in Indonesia, appears to have had only a moderate impact elsewhere in the region (Kiladis and Diaz 1986; Allen *et al.* 1989; Brookfield and Allen 1991). The Resident of South and East Borneo described the drought in 1914 as long and fierce, with great forest complexes being destroyed by fire (Rijckmans 1916). Again, the forest in Kutai seemed to take a long time to recover. Endert, travelling with the 'Middle-East Borneo Expedition' of 1925, described hill forest which was thin-stemmed and poor in species, showing signs of having been burnt 'after a fire in a very dry year, roughly ten years ago' (Endert 1927: 233). In Sabah a large conflagration destroyed the forests over a wide area south of Mount Kinabalu, estimated by the recently appointed Conservator of Forests at 100 square miles. In an interesting letter, he noted that the fire '... in places climbed the trunks of the trees to the crowns, a very rare occurrence in tropical forests' (Matthews 1917). This fire resulted in the formation of the famous Sook Plain, in which the forest did not grow back, but was replaced by permanent *Imperata* grass (Cockburn 1974). All along the Borneo coast, fire and smoke were reported. The harbour master at Samarinda wrote that 'the haziness, often becoming mists, was largely caused by bush fires which could be smelled and caused pain to one's eyes' (Braak 1915). The *Straits Times* of 31 October reported a large fire in 'West-Central Borneo', adding 'Huge columns of smoke have spread themselves for miles around, hanging low over the sea and inconveniencing the ship routes in the Muntok and Karimata Straits'. These descriptions have a familiar ring!

Enough evidence has been provided here to indicate the importance of drought and fire as relatively frequently occurring events in the recent past as well as the present. Such events constituted other forces shaping the forests of Borneo. The prehistoric studies of earlier charcoal deposits have pushed back the incidence of fire well into the Holocene (Goldammer and Siebert 1989; Kartawinata 1993). One might conclude 'It now seems evident that drought and fire have always been a part of the natural environment in Borneo, at least for the last several thousand years. The effect of this finding for the ecology of the tropical rain forest as a whole, especially in this wettest of regions, remains to be written' (Brookfield *et al.* 1995: 178).

Conclusion

To what extent may the surviving Borneo forest be considered to be a human artefact? It is only in interior Brunei and in remote mountain areas, some of which are now protected, that the forest has survived the worst recent ravages of both logging and conversion. Such refuges provide a unique laboratory for examining the contrast between stands which remain much more 'natural', in terms of lightness of impact, and the surviving forest elsewhere. There is no doubt, however, that the bulk of the forests have already been subjected to several centuries of human impact, and in some areas have been completely reconstructed by their human inhabitants. Comparative studies will help provide the answers to the way in which the forests have been forced to adapt and adjust, in terms of biodiversity and ecosystem maintenance. Perhaps it will be disc- overed that the Borneo forests are after all, quite resilient under human pressure, or perhaps the reverse finding will be made, as secondary species become more dominant over wide areas, burning becomes more frequent and many of the unique attributes of 'primary rainforest' disappear for ever.

References

Allen, B., Brookfield, H. and Byron, Y. 1989. Frost and drought through time and space. Part II: The written, oral and proxy records and their meaning. *Mountain Research and Development, (Special issue on frost and drought in the Highlands of Papua New Guinea)* 9: 279-305.

Anon. 1889. Scarcity of gutta percha. *The Indian Forester*, 15: 478.

Ashton, P., Givnish, T. and Appanah, S. 1988. Staggered flowering in the dipterocarpa- ceae. New insights into floral induction and the evolution of mast fruiting in the aseasonal tropics. *The American Naturalist*, 132(1): 44-66.

Baring-Gould, S. and Bampfylde, C.A. 1909. *A History of Sarawak under its White Rajahs, 1839-1908.* Henry Sotheran, London: 464 pp.

Beccari, O. 1904. *Wanderings in the great Forests of Borneo.* (English translation). Archibald Constable & Co., London: 423 pp.

Bellwood, P., Gillespie, R., Thompson, G.B., Vogel, J.S., Ardika, I.W. and Datan, I. 1992. New Dates for Prehistoric Asian Rice. *Asian Perspectives,* 31(2): 161-70.

Bock, C. 1881. *The Headhunters of Borneo: a Narrative of Travel up the Mahakam and Down the Barito, also Journeying to Sumatra.* Sampsom Low, Marston, Searle and Rivington, London. Reprinted: (1985) Oxford University Press, Singapore: 344 pp.

Braak, C. 1915. De sterke nevel in de oostmoesson van 1914 (The dense fog in the east monsoon of 1914). *Natuurkundig Tijdschrift voor Nederlandsch Indie,* 74: 131-40.

Brookfield, H.C. 2001. Exploring Agrodiversity. New York, Columbia University Press.

Brookfield, H.C. and Allen, B. 1991. Environmental and human responses to climatic variability in the west Pacific 'maritime continent' region. In: J.E. Hay (Editor) *South Pacific Environments: Interactions with Weather and Climate. Proceedings of a Conference Organized by Environmental Science, University of Auckland, 2-7 September 1991.* University of Auckland, Auckland: 71-82.

Brookfield, H.C., Potter, L.M. and Byron, Y. 1995. *In Place of the Forest: Environ-*

mental and Socio-Economic Transformation in Borneo and the Eastern Malay Peninsula. United Nations University Press, Tokyo: 310 pp.

Bruenig, E.F. 1987. The forest ecosystem: tropical and boreal. *Ambio,* 16(2-3): 68-79

Burbidge, F.W. 1880. *Gardens of the Sun: or A Naturalist's Journal on the Mountains and in the Forests and Swamps of Borneo and the Sulu Archipelago.* John Murray, London: 364 pp.

Chin, S.C. 1985. Agriculture and resource utilization in a lowland rainforest Kenyah community. *Sarawak Museum Journal,* 35(56): 322 pp.

Cockburn, P. 1974. The origin of Sook Plain, Sabah. *Malaysian Forester,* 37: 61-3.

Combanaire, A. 1910. *Au Pays des Coupeurs de Tetes: a travers Borneo.* Plon-Nourrit et Cie, Paris. Reprinted: (1993) Editions Pagodes, Singapore: 319 pp.

Corlett, R.T. 1994. What is secondary forest? *Journal of Tropical Ecology,* 10: 445-7.

Corlett, R.T. 1995. Tropical secondary forests. *Progress in Physical Geography,* 19(2): 159-72.

de Vriese, W.H. 1855-6. *Tuinbouw-flora van Nederland en zijne overzeesche bezittingen,* I-III, Leiden.

de Vriese, W.H. 1860. Aanteekeningen betreffende getah-pertja-boomen (sapoteen) en getah pertcha van Zuid-oostlijk Borneo, naar aanleiding van ontdekkingen van James Motley. *Natuurkundig Tijdschrift voor Nederlandsch Indie,* 21: 299-315.

Dove, M.R. 1981. Symbiotic relationships between human populations and *Imperata cylindrica.* The question of ecosystemic succession and preservation in South Kalimantan. In: M. Nordin (Editor) *Conservation inputs from life sciences.* Universiti Kebangsaan, Bangi, Malaysia: 187-200.

Dove, M.R. 1993. Smallholder rubber and swidden agriculture in Borneo: a sustainable adaptation to the ecology and economy of the tropical forest. *Economic Botany,* 47: 136-47.

Dove, M.R. 1994, Transition from native forest rubbers to *Hevea brasiliensis* (Euphorbiaceae) among tribal smallholders in Borneo. *Economic Botany,* 48(4): 382-96.

Endert, F. 1927. Floristisch Verslag (Report on Flora). *Midden-Oost Borneo Expeditie, 1925 (Expedition to Central-East Borneo, 1925)* D.Buijs, H. Witcamp, F. Endert and T. Siebers, Weltevreden. G. Kolff and Co: 200-83.

Enthoven, J.J.K. 1903. *Bijdragen tot de Geographie van Borneo's Westerafdeeling.* Brill, Leiden: 2 Volumes.

Foxworthy, F.W. 1935. Palaquium. In: I.H. Burkill (Editor) *A Dictionary of the Economic Products of the Malay Peninsula.* Crown Agents for the Colonies, London: 1623-43.

Freeman, J.D. 1955. *Iban Agriculture: a Report on the Shifting Cultivation of Hill Rice by the Iban of Sarawak.* Colonial Research Studies No. 18. Her Majesty's Stationery Office, London: 148 pp.

Freeman, J.D. 1970. *Report on the Iban.* Athlone Press, London: 317 pp.

Gerlach, L.W.C. 1881. Reis naar het Meergebied van den Kapoeas in Borneo's Westerafdeeling (Journey to the Lake District of the Kapoeas in Borneo's Western Division). *Bijdragen tot de Taal- Land- en Volkenkunde van Nederlandsch Indie,* 29: 285-322.

Goldammer, J. and Siebert, B. 1989. Natural rain forest fires in Eastern Borneo during the Pleistocene and Holocene. *Natuurwissenschaften,* 76: 518-20.

Halewijn, M. 1838. Borneo. Eenige reizen in de Binnenlanden van het Eiland, door eenen Ambtenaar van het Gouvernement, in het jaar 1824. *Tijdschrift voor Neerland's Indie,* 1(2: 1-25); 81-200; 401-13.

Ham, S.P. 1900. De Proefaanplantingen van Caoutchouc en Getah-pertja leverende houtsoorten te Tjipetir onder de leiding van het Boschwezen. *Tijdschrift voor Nijverheid en Landbouw in Nederlandsch Indie,* 61: 64-77.

Happe, E. 1867. Uittreksel uit een rapport van den Kolonal, resident der Zuider en Ooster Afdeeling van Borneo, van 30 September 1863. *Tijdschrift voor Indische Taal-, Land- en Volkenkunde*, 81-86.

Harrison, T. 1970. *The Malays of Southwest Sarawak Before Malaysia: a Socio-ecological Survey*. Michigan State University Press, East Lansing.

Idwar Salleh, M. 1978. Pepper trade and the ruling class of Banjarmasih in the seventeenth century. *Papers of the Dutch-Indonesian Historical Conference 1976*. Bureau of Indonesian Studies, Leiden and Jakarta: 203-21.

Ipoi, D. and Bellwood, P. 1991. Recent Research at Gua Sireh (Serian) and Lubang Angin (Gunung Mulu National Park), Sarawak. *Indo-PacificPrehistory Association Bulletin*, 10: 386-405.

Irwin, G. 1955. *Nineteenth Century Borneo: A Study in Diplomatic Rivalry*, Verhandelingen van het Kokinklijk Instituut voor Taal-Land-en Volkenkunde, Deel XV, the Hague, Nijhoff.

Jackson, J.C. 1970. *Chinese in the West Borneo Goldfields: a Study in Cultural Geography*. Occasional Papers in Geography No. 15. University of Hull Publications, Hull: 88 pp.

Joekes, A. 1893. *Koloniaal Verslag*, bijlage B: Zuider en Oosterafdeeling van Borneo.

'Kalamatan' 1848. Miscellaneous notices, contributions and correspondence: letters from the interior of Borneo (west coast). *Journal of the Indian Archipelago and Eastern Asia* 2(3): v-viii.

Kartawinata, K. 1993. A wider view of the fire hazard. In: H. Brookfield and Y. Byron (Editors) *South-east Asia's Environmental Future: The Search for Sustainability*. Oxford University Press and United Nations University Press, Singapore and Tokyo: 261-6.

Kater, C. 1871. Iets over het pandelingschap in de Westerafdeeling van Borneo, *Tijdschrift voor Nederlandsch Indie*, 3e serie 5, 2: 294-305.

Kiladis, G. and Diaz, H. 1986. An analysis of the 1877-78 ENSO episode and comparison with 1982-3. *Monthly Weather Review,* 114: 1035-47.

King, V.T. 1985. *The Maloh of West Kalimantan: an Ethnographic Study of Social Inequality and Social Change among an Indonesian Borneo People*. Verhandelingen van het Koninklijk Instituut voor Taal- Land- en Volkenkunde, No 108. Cinnaminson, USA, Foris.

Leighton, M. 1984. *The El Nino-Southern Oscillation event in Southeast Asia: Effects of Drought and Fire in Tropical Forest in Eastern Borneo*. WWF-US No. 293 Project report.

Leighton, M. and Peart, D. 1988. Ecological research for sustainable land management and yield of forest products on the poor soil of West Kalimantan, mimeo.

Lian, F.J. 1987. *Farmers' perceptions and economic change: the case of Kenyah swidden farmers in Sarawak*. Unpublished PhD thesis, Human Geography, Australian National University, Canberra.

Lindblad, J.T. 1988. *Between Dayak and Dutch. The Economic History of Southeast Kalimantan, 1880-1942. Verhandelingen van het Taal-Land-en Volkenkunde*, Foris, Dordrecht: 282 pp.

Lowenhaupt-Tsing, A. 1984. *Politics and Culture in the Meratus Mountains*. Unpublished PhD thesis, Stanford University.

Lumholtz, C. 1920. *Through Central Borneo*. Charles Scribners Sons, New York: 2 Volumes.

Matthews, D.M. 1917. Conservator of Forests, British North Borneo to Chairman North Borneo Company, 20 April 1917. CO 874/714. *British North Borneo Company Papers*, Singapore University Library.

Meijer, J. 1880. Memorie van Overgave, Zuider-en Ooster Afdeeling van Borneo

(Handing-over Memorial, Southern and Eastern Divisions of Borneo), Algemein Rijksarchief (ARA) The Hague: 267 pp.

Nazir Foead, 1999. The Upper Bahau grasslands in East Kalimantan: a traditional ecosystem management or pest control? In: K.W. Sorenson (Editor) *People and Plants in Kayan Mentarang.* UNESCO (MAB), Jakarta.

Netscher, E. and von Dewall, H. 1853 Historische, geographische en statistieke aanteekeningen betreffende Tanah Boemboe; aangetroffen onder de bij het gouvernement van Nederlandsch-Indie berustende papieren van C.M. Schwaner. *Tijdschrift voor Indische Taal-Land-en Volkenkunde,* 21(1): 335-71.

Ozinga, J. 1940. *De Economische Ontwikkeling der Westerafdeeling van Borneo en de Bevolkingsrubbercultuur* (The Economic Development of the Western Division of Borneo and Native Rubber Cultivation), Wageningen: N.V.Gebr.Zomer en Keunig's Uitgeversmaatschappij.

Padoch, C. and Peters, C. 1993. Managed forest gardens in West Kalimantan, Indonesia. In: C.S. Potter, J.I. Cohen and D. Janczewski (Editors) *Perspectives on Biodiversity: case studies of genetic resource conservation and development.* American Association for the Advancement of Science Press, Washington, D.C.: 167-76.

Perelaer, M.T.H. 1870. *Ethnographische Beschrijving der Dayaks.* Zalt-Bommel, Netherlands, Joh. Noman.

Peters, C.M. 1996. Observations on the sustainable exploitation of non-timber tropical forest products. An ecologist's perspective. In: *Current Issues in Non-Timber Forest Products Research.* CIFOR/ODA, Jakarta: 19-40.

Pijnappel, J. 1860. Beschrijving van het Westelijke Gedeelte van de Zuider en Ooster-Afdeeling van Borneo (De Afdeeling Sampit en de Zuidkust). *Bijdragen tot de Taal-Land-en Volkenkunde,* 3: 243-346.

Potter, L.M. 1988. Indigenes and colonisers: Dutch forest policy in South and East Borneo. In: John Dargavel (Editor) *Changing Tropical Forests,* ANU Press, Canberra: 127-53.

Potter, L.M. 1993. Banjarese in and beyond the Hulu Sungai, South Kalimantan. A study of cultural independence, economic opportunity and mobility. In: T. Lindblad (Editor) *New Challenges in the Modern Economic History of Indonesia: Proceedings of the 1ˢᵗ Conference on Indonesia's Modern Economic History.* Bureau of Indonesian Studies, University of Leiden, Leiden: 264-93.

Potter, L. M. 1997(a). The dynamics of *Imperata.* Historical overview and modern farmer perspectives, with an example from South Kalimantan, Indonesia. *Agroforestry Systems,* 36(1-3): 31-51.

Potter, L. 1997(b). 'A forest product out of control: gutta percha in Indonesia and the wider Malay world, 1845-1915'. Paper Landscapes: Explorations in the Environmental History of Indonesia In: P. Boomgaard, F. Colombijn and D. Henley (Editors) *Verhandelingen van het Taal ,-Land en Volkenkunde.* KITLV Press, Leiden: 281-308.

Pringle, R. 1970. *Rajahs and Rebels: The Ibans of Sarawak under Brooke Rule, 1841-1941.* Macmillan, London: 410 pp.

Rijckmans, L.F.J. 1916. Memorie van overgave van de Zuider-en-Ooster Afdeeling van Borneo, No. 271, Algemeen Rijksarchief (ARA), The Hague.

Rousseau, J. 1990. *Central Borneo: Ethnic Identity and Social Life in a Stratified Society.* Clarendon Press, Oxford: 380 pp.

Sather, C. 1990. Trees and tree tenure in Paku Iban society: the management of secondary forest resources in a long-established Iban community. *Borneo Review* 1(1):16-40.

Scheuer, W.H.E. 1932. Memorie van Overgave van de Afdeeling Sintang. Fiche 424-5, MvO KIT, microfiche series, Switzerland, ITC.

Schindele, W., Thoma, W. and Panzer, K. 1989. The forest fire 1982/3 in East Kalimantan Part 1: The fire, the effects, the damage and technical solutions. FR-Report No 5.

Investigation of the steps needed to rehabilitate the areas of East Kalimantan seriously affected by fire. Jakarta: German Forest Inventory Service Ltd for German Agency for Technical Co-operation (GTZ) and International Tropical Timber Organization (ITTO), typescript.

Schneeberger, W.F. 1945. The Kerayan-Kelabit Highland of central northeast Borneo. *Geographical Review*, 25(4): 544-62.

Schuitemaker, J.P. 1933. Het garoehout van West Borneo. Bijlage V van het Jaarverslag van den dienstkring West-Borneo over het jaar 1931, bijgewerkt tot en met medio 1933 *Tectona*, 26: 851-92.

Schuitemaker, J.P. 1938. Korte schets van de landbouw in het Oostmoesson-rijstgebied in de Zuider - en Oosterafdeling van Borneo. *Landbouw* 14, 714-778.

Schwaner, C.A.L.M. 1853. *Borneo: Beschrijving van het Stroomgebied van den Barito en Reizen Langs Eenige voor-name Rivieren van het Zuid-oostilijk Gedeelte van het Eiland: oplast van het Gouvernement Nederl. Indie Gedaan in de Jaaren 1843-1847.* van Kampen, Amsterdam.

Seavoy, R.E. 1975. The origin of tropical grasslands in Kalimantan, Indonesia. *Singapore Journal of Tropical Geography,* 40: 48-52.

Smythies, B.E. 1961. History of Forests in Sarawak. *Sarawak Gazette*, September 30: 167-74.

St John, S.B. 1862. *Life in the Forests of the Far East, or Travels in Northern Borneo.* Smith Elder, London: 2 Volumes.

Tromp, S.W. 1890. Mededeelingen uit Borneo (Information from Borneo). *Tijdschrift voor de Aardrijkskundig Genootschap van Nederlandsch Indie (Journal of the Geographical Society of the Netherlands Indies),* 7: 728-63.

van Basel, Senn W.H. 1880. De onbebouwde gronden op Borneo's Westkust. *Tijdschrift voor Nederlandsch Indie*, 1(2): 81-101.

van Gelder, A. 1950. Guttapercha. In: C.J.J. van Hall and C. van der Koppell (Editors) *De Landbouw in den Indischen Archipel, (Agriculture in the Indian Archipelago).*Vol 3, van Hoeve, The Hague: 476-519.

van Gorkum, K.W. 1884. *De Oost-Indische Cultures, in betrekking tot Handel en Nijverheid*, de Bussy, Amsterdam: 596-617.

van Hoeve, P. 1936. Memorie van Overgave Poeloe Laut en Tanah Boemboe, No. 470 (KIT series) the Hague, Algemeen Rijksarchief (ARA).

van Romburgh, P. 1900. Caoutchouc en getah pertja in Nederlandsch-Indie. *Mededeelingen uit's Lands Plantentuin*, 39.

Verkuyl, A.H. 1950. The natural timber reserves of Borneo. Annual Report, 1949, Head of the Forest Planning Brigade for Kalimantan, mimeo.

von Gaffron, W.G. 1858. Nota omtrent den getah-pertja-boom. *Natuurkundig Tijdschrift voor Nederlandsch Indie,* 16: 224-8.

Wentholt, J. 1938. Memorie van Overgave, afdeeling Martapura, Zuider en-Ooster Afdeeling van Borneo, No 482, the Hague, Algemeen Rijksarchief (ARA).

Whittier, H.L. 1973. *Social organization and symbols of social differentiation: an ethnographic study of the Kenyah Dayak of East Kalimantan (Borneo)*, Unpublished PhD. thesis, Michigan State University, East Lansing.

Wirawan, N. 1984. *Good forests within the burned forest area in East Kalimantan.* Report for the World Wildlife Fund, Field Report No 1637.

Witty F., 1880-1. Diary, Journey to Marudu Bay (8 and 9 May) CO874/74 1880-81 *British North Borneo Company Papers*, Singapore University Library.

Wong, L.K. 1960. The trade of Singapore, 1819-1869. *Journal of the Malaysian Branch of the Royal Asiatic Society,* 33(4): 1-315.

Wray, L. Jr 1883. Gutta-producing trees. *Journal of the Straits Branch of the Royal Asiatic Society*, 12: 207-21.

Contributors

Dr Ken P. Aplin

Western Australian Museum.
Francis St., Perth, WA 6000, Australia
ken.aplin@museum.wa.gov.au

Dr Lynda K. Ayliffe

Laboratoirc des Sciences du Climat et de l´Environement
(CNRS-CEA)
Domains du CNRS-Bt 12
91198 Gif-sur-Yvette Cedex, Frankreich

Timothy T. Barrows

Department of Geology
(The Australian Marine Quaternary Program)
The Australian National University
Canberra, ACT 0200, Australia
and
Research School of Physical Sciences
and Engineering
The Australian National University
Canberra, ACT 0200, Australia
tim.barrows@anu.edu.au

Dr Bruno David

School of Geography and Environmental Science
PO Box 11a
Monash University, Vic. 3800, Australia
bruno.david@arts.monash.edu.au

Dr Patrick De Deckker

Department of Geology
(The Australian Marine Quaternary Program)
The Australian National University
Canberra, ACT 0200, Australia
patrick.dedeckker@anu.edu.au

David Godley

School of Geography and Environmental Science
PO Box 11a
Monash University, Vic. 3800, Australia
present address:
Department of Geography
University of Wisconsin
Madison, WI 53706, USA
godley@stumpy.geography.wisc.edu

Dr. John Grindrod

School of Geography and Environmental Science,
PO Box 11a,
Monash University, Vic. 3800, Australia
john.grindrod@arts.monash.edu.au

Wahyoe Hantoro   Pusat Geoteknologi
                          Lembaga Ilmu Pengatahuan Indonesia
                          Bandung, Republic of Indonesia
                          hantoro@embon.lipi.go.id

Professor Geoffrey S. Hope Department of Archaeology and Natural History
                          Research School of Pacific and Asian Studies
                          The Australian National University
                          Canberra, ACT 0200, Australia
                          geoff.hope@coombs.anu.edu

Professor A. Peter Kershaw School of Geography and Environmental Science
                          PO Box 11a
                          Monash University, Vic. 3800, Australia
                          peter.kershaw@arts.monash.edu.au

Dr Kam-biu Liu   Department of Geography and Anthropology
                          Louisiana State University
                          Baton Rouge, Louisiana 70803, USA
                          Kliu1@unix1.sncc.lsu.edu

Dr Harry Lourandos   Department of Anthropology and Sociology
                          University of Queensland, Qld 4072, Australia
                          h.lourandos@mailbox.uq.edu.au

Dr J. Ignacio Martinez   Department of Geology
                          (The Australian Marine Quaternary Program)
                          The Australian National University
                          Canberra, ACT 0200, Australia
                          and
                          Department of Geology
                          Universidad EAFIT, A.A.
                          3300 Medellin, Colombia

Dr. Andrew L. Maxwell   Department of Geography and Anthropology
                          Louisiana State University
                          Baton Rouge, Louisiana 70803, USA
                          sralao@bellsouth.net

Assoc. Prof. Ian Metcalfe Asia Centre
                          University of New England
                          Armidale, NSW 2351, Australia
                          imetcalf@met3.une.edu.au

Dr Robert Morley   Palynova, Woodstock
                          1 Mow Fen Road
                          Littleport, Cambs CB6 1PY, UK
                          pollenpower@dial.pipex.com

Dr Patrick Moss   School of Geography and Environmental Science
                               PO Box 11a
                               Monash University, Vic. 3800, Australia
                               patrick.moss@arts.monash.edu.au
                               and
                               Department of Geography
                               University of Iowa
                               Iowa City, Iowa 52242-1316, USA
                               patrick-moss@uiowa.edu

Dr Sue O´Connor   Department of Archaeology and Natural History
                               Research School of Pacific and Asian Studies
                               The Australian National University
                               Canberra, ACT 0200, Australia
                               soconnor@coombs.anu.edu.au

Assoc. Prof. Jim Peterson   School of Geography and Environmental Science
                               PO Box 11a
                               Monash University, Vic. 3800, Australia
                               jim.peterson@arts.monash.edu.au

Dr Netty Polhaupessy   Geological Research and Development Centre,
                               Jl. Diponegoro no. 57
                               40122 Bandung, Indonesia

Dr Lesley Potter   Department of Geography and Environmental Studies
                               University of Adelaide, S.A. 5005, Australia
                               lesley.potter@adelaide.edu.au

Dr Mike Prentice   Climate Change Research Center
                               Institute for the Study of Earth, Oceans and Space
                               University of New Hampshire
                               Durham, NH 03824, USA
                               mike.prentice@unh.edu

Prof. Matthew Spriggs   School of Archaeology and Anthropology
                               A.D. Hope Building
                               The Australian National University
                               Canberra, ACT 0200, Australia
                               matthew.spriggs@anu.edu.au

Prof. Nigel J. Tapper   School of Geography and Environmental Science
                               PO Box 11a
                               Monash University, Vic. 3800, Australia
                               nigel.tapper@arts.monash.edu.au

Dr Alan Thorne   Department of Archaeology and Natural History
                               Research School of Pacific and Asian Studies
                               The Australian National University
                               Canberra, ACT 0200, Australia
                               alan.thorne@coombs.anu.edu.au

Dr Sander van der Kaars   School of Geography and Environmental Science
                              PO Box 11a
                              Monash University, Vic. 3800, Australia
                              sander.vanderkaars@arts.monash.edu.au

Dr Nicola van Dijk   Department of Archaeology and Natural History
                              Research School of Pacific and Asian Studies
                              The Australian National University
                              Canberra, ACT 0200, Australia
                              andnik@ozemail.com.au

Dr Peter Veth   School of Anthropology, Archaeology and Sociology
                              James Cook University
                              Townsville, Qld 4811, Australia
                              peter.veth@jcu.edu.au

Dr Ian Walters   Director, Centre for Southeast Asian Studies
                              Northern Territory University, NT 0909, Australia
                              Ian.Walters@ntu.edu.au

Sue Wang   School of Geography and Environmental Science
                              PO Box 11a
                              Monash University, Vic. 3800, Australia
                              present address:
                              School of Geosciences
                              University of Wollongong
                              Wollongong, NSW 2522, Australia
                              swang@uow.edu.au

Dr Trevor Whiffin   School of Botany
                              La Trobe University
                              Bundoora, Vic. 3083, Australia
                              t.whiffin@latrobe.edu.au

John Wolseley   4/25 Dalgety St
                              St Kilda, Vic. 3182, Australia

CATENA SUPPLEMENT 25

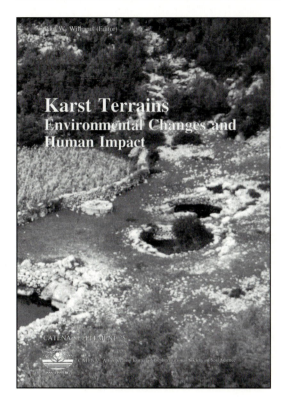

Paul W. Williams (Editor)

Karst Terrains
Environmental
Changes
and
Human Impact

CATENA SUPPLEMENT 25

268 pages, numerous
figures, photos and tables
list price EUR 95,50 /
US$ 96,50

1993
ISBN 3-923381-34-4

The collection of essays in this volume sets out explicitly to achieve an international perspective, as a step towards gaining a more global and generalized understanding of human impact on karst.

CATENA VERLAG GMBH   *GeoScience Publisher*
Ärmelgasse 11   D-35447 Reiskirchen   Germany
Phone + fax (+49)6408-64978   E-mail: catenaverl@aol.com
Visit us on the Internet:
http://members.aol.com/catenaverl

Advances in GeoEcology 30

(follow-up series of CATENA SUPPLEMENTS)

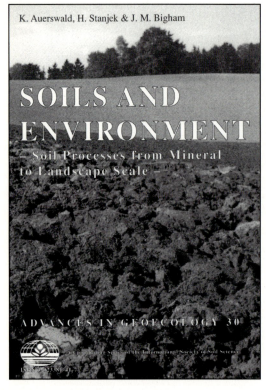

K. Auerswald, H. Stanjek &
J.M. Bigham (Editors)

SOILS AND ENVIRONMENT

Soil Processes from
Mineral to Landscape
Scale

Advances in GeoEcology 30

422 pages, numerous figures
and tables
list price EUR 96,50 / US$ 96,50

1997

ISBN 3-923381-41-7

Soils play an important role in the function of both natural and managed ecosystems. Understanding the impact of environmental factors on these ecosystems requires a knowledge of soil processes ranging from molecular to landscape in scale. The properties and spatial distributions of soils and their mineral constituents are especially useful as indicators of both present and past environmental conditions. Soils and soil minerals form or transform in direct response to biogeochemical factors that are driven by local, regional or global processes. Through a series of case studies, this text demonstrates that linkages can be made between measurable soil and mineral properties and the processes that shape the weathering environment. Examples include layer silicates, carbonates, sulfides, and both iron and aluminium oxides. Several papers also emphasize the role of biotic factors in soil systems, the sensitivity of these factors to management practices, and the frequently overlooked interactions that exist between organic and inorganic compartments of soils and surface waters. Finally, this text recognizes that spatial and temporal variability are inherent attributes of soils and landscapes that must be quantified and properly managed if ecological and environmental harmony are to be maintained. The buffering function of soils and the importance of this function to the release and flux of nutrients in terrestrial ecosystems are given special consideration.

CATENA VERLAG GMBH   *GeoScience Publisher*
Ärmelgasse 11   D-35447 Reiskirchen   Germany
Phone + fax (+49)6408-64978   E-mail: catenaverl@aol.com
Visit us on the Internet:
http://members.aol.com/catenaverl

Advances in GeoEcology 33

(follow-up series of CATENA SUPPLEMENTS

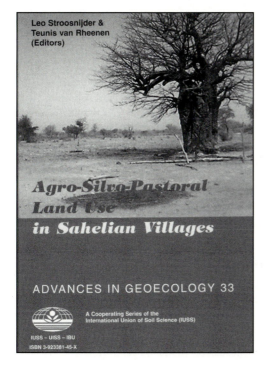

Leo Strosnijder & Teunis van Rheenen
(Editors)

Agro-Silvo-Pastoral Land Use in Sahelian Villages

Advances in GeoEcology 33
408 pages, numerous photos,
figures and tables
list price EUR 101,00 / US$ 101.-
2001
ISBN 3-923381-45-X

The book **Agro-Silvo-Pastoral Land Use in Sahelian Villages** synthesises results of 10 years research by the Wageningen University in a partnership with the University of Ouagadougou. The aim was to ascertain the conditions under which Sahelian villages are able and willing to improve the productivity of their natural resources and subsequently use and manage these resources in a sustainable way.

Major findings of disciplinary nature were brought into a framework that allowed the holistic analysis of land use and this is now used as a tool for guiding rural development.

This book is of interest for a wide public, Africanists and others. Social scientists will learn about the perception of land use in Sahelian villages. Soil and water experts and ecologists will be interested in the dynamics of Sahelian Natural resources and economists and policy makers will hope to find viable development paths to improve food security. Finally, a lot of people will be interested in the well-documented struggle from disciplinary knowledge to a multi-disciplinary understanding. (*from Editorial*)

CATENA VERLAG GMBH   *GeoScience Publisher*
Ärmelgasse 11   D-35447 Reiskirchen   Germany
Phone + fax (+49)6408-64978   E-mail: catenaverl@aol.com

http://members.aol.com/catenaverl

GeoEcology textbook

D. Nielsen & O. Wendroth

SPATIAL AND TEMPORAL STATISTICS
Sampling Field Soils and their Vegetation

About 400 pages, numerous figures and tables
2002   ISBN 3-923381-46-8
list price EUR 50,00 / US $ 50.-

Many methods of statistical analysis are available for examining experimental data observed at different points in time and space relative to describing and understanding soil-plant-atmospheric processes within the landscape (e.g. Cahill et al., 1999). For observations that are temporally or spatially independent, parametric and nonparametric statistical methods are available. For those that manifest temporal or spatial dependence, methods derived from regionalized variable analysis and applied time series may be selected.

Hence, the question arises, "How can these regionalized variable and applied time series theories frequently being used successfully in other scientific disciplines be applied to agricultural research?" This book is intended to introduce such concepts and theories to scientists already familiar with classical statistics and one or more disciplines of agricultural science. Each chapter introduces one concept and its application to several sets of field-measured data.

Examples of data from various field studies observed by colleagues and ourselves are used as a frame for explaining basic concepts of spatial statistics and how to apply them within and between fields. The original data, the analysis and the interpretation are followed by a discussion of issues and concerns associated with the underlying assumptions of the analysis.

CATENA VERLAG GMBH   *GeoScience Publisher*
Ärmelgasse 11   D-35447 Reiskirchen   Germany
phone + fax (+49)6408-64978   E-mail:catenaverl@aol.com
Visit us on the Internet:
http://members.aol.com/catenaverl